Rarefied Gas Flows and Dynamic Plasma Phenomena in Electric Propulsion Systems

Rarefied Gas Flows and Dynamic Plasma Phenomena in Electric Propulsion Systems

Vom Fachbereich Produktionstechnik
der
UNIVERSITÄT BREMEN

zur Erlangung des Grades
Doktor-Ingenieur
genehmigte

Dissertation

von

M. Sc. Juan Esteban Gomez Herrera

Gutachter:
Prof. Dr. rer. nat. Claus Lämmerzahl
Prof. Dr. Claus Braxmaier

Tag der mündlichen Prüfung:
7. Februar 2020

Bibliografische Information der Deutschen Nationalbibliothek
Die Deutsche Nationalbibliothek verzeichnet diese Publikation in der Deutschen Nationalbibliographie; detaillierte bibliographische Daten sind im Internet über http://dnb.d-nb.de abrufbar.
1. Aufl. - Göttingen: Cuvillier, 2021
Zugl.: Bremen, Univ., Diss., 2020

Nonnenstieg 8, 37075 Göttingen
Telefon: 0551-54724-0
Telefax: 0551-54724-21
www.cuvillier.de

1. Auflage, 2021
Gedruckt auf umweltfreundlichem, säurefreiem Papier aus nachhaltiger Forstwirtschaft.

ISBN 978-3-7369-7324-4
eISBN 978-3-7369-6324-5

Zusammenfassung

Zu den aktuellen Entwicklungen in der Raumfahrtindustrie zählen das stetig wachsende Interesse an miniaturisierten Satelliten sowie der immer häufigere Einsatz elektrischer Antriebssysteme zu allgemeinen Lage- und Bahnregelungszwecken. Die Entwicklung miniaturisierter Satelliten, welche koordinierte Formationsflüge durchführen können, erfordert ihrerseits den Einsatz von Antriebssystemen, die sehr kleine und präzise zu steuernde Schubkräfte erzeugen. Vor diesem Hintergrund stellen elektrische Triebwerke eine attraktive Option dar, die Antriebsanforderungen von Satelliten sowohl in herkömmlichen als auch in miniaturisierten Größen langfristig zu erfüllen. Bei miniaturisierten Satelliten sind die Schubanforderungen oft mit niedrigen Treibstoff-Massenstromwerten und verhältnismäßig kleinen geometrischen charakteristischen Längen verbunden. Dies kann zu verdünnten Gaszuständen innerhalb der Triebwerksdüsen führen. Wegen der hohen Komplexität der Plasmaphänomene innerhalb elektrischer Triebwerke sowie der typischerweise hohen Rechenanforderungen, die mit der Plasmamodellierung einhergehen, werden elektrische Antriebssysteme oft auf Basis empirischer Modelle und experimenteller Daten entwickelt. In Anbetracht der stetig wachsenden Rechenleistung und des verfügbaren technischen und wirtschaftlichen Optimierungspotenzials gewinnt die Untersuchung neuartiger Modellierungsansätze immer mehr an Bedeutung. Der Fokus der vorliegenden Arbeit liegt auf den oben beschriebenen Herausforderungen und den dazugehörigen Forschungsfeldern: der Untersuchung verdünnter Gaszustände in transsonischen Strömungen sowie der Entwicklung numerischer Modellierungsansätze zur Beschreibung des Plasmaverhaltens innerhalb elektrischer Antriebssysteme.

Verdünnte Gaszustände in transsonischen Mikrodüsenströmungen werden in der vorliegenden Arbeit anhand des experimentellen Versuchsaufbaus INGA III am Zentrum für angewandte Raumfahrttechnologie und Mikrogravitation (ZARM) untersucht. Hierzu wird die Kaltgas-Expansion von Edelgasen durch eine Lavaldüse bei Knudsen-Zahlen (Kn) im Bereich $0.02 < Kn < 0.33$ sowohl experimentell als auch numerisch analysiert. Die an der Brennkammer der Lavaldüse experimentell ermittelten Druckwerte werden mit numerischen Ergebnissen aus Finite-Volumen-Simulationen sowie mit DSMC-Rechnungen verglichen.

Im Bereich $Kn > 0.1$ wird eine sehr gute Übereinstimmung zwischen den DSMC-Ergebnissen und den experimentellen Druckwerten festgestellt. Darüber hinaus liefern die Navier-Stokes-Simulationen gute Ergebnisse im Bereich $Kn < 0.05$. Mit relativen Abweichungen von über 50% liegen die Ergebnisse der kontinuumsbasierten Simulationen bei höheren Knudsen-Zahlen jedoch erheblich über den experimentell ermittelten Druckwerten. Diese Beobachtung wird durch die Diskrepanz zwischen der angewandten Haftbedingung und der hohen Knudsen-Zahlen, welche auf verdünnte Gaszustände innerhalb der Gleitströmungs- und Transitionsgebiete hinweisen, erklärt. Anhand des Vergleiches zwischen den experimentellen und den numerischen Ergebnissen wurde ein mathematischer Ausdruck für den Zusammenhang zwischen Druckverlustabweichung innerhalb der Lavaldüse in Abhängigkeit von der Knudsen-Zahl entwickelt. Die ermittelte Knudsen-Funktion kann sowohl mit gasabhängigen als auch mit gasunabhängigen Koeffizienten gebildet werden. Eine genaue Analyse der Knudsen-Funktion zeigt, dass Gase mit kleineren Wirkungsquerschnitten stärker von der Haftbedingung-Annahme abweichen. Der dazugehörige Zusammenhang wird durch eine Polynomfunktion zweiten Grades dargestellt. Bei Gasen mit relativ großen Wirkungsquerschnitten wird die Abweichung von der Haftbedingung-Annahme durch eine von Kn abhängige lineare Funktion abgebildet. Die entwickelte Knudsen-Funktion ermöglicht die Korrektur der Druckergebnisse aus Navier-Stokes-Simulationen mit Haftbedingung. Hierzu müssen keine Fallunterscheidungen zwischen den Kontinuums-, Gleitströmungs- und Transitionsgebieten durchgeführt werden. Die aus der dimensionslosen Betrachtung errechneten Knudsen-Koeffizienten ermöglichen die Vorhersage des Druckverlustverhaltens bei vergleichbaren, skalierten Triebwerkssystemen höherer Leistungsklassen.

Um die numerische Modellierung der Plasmavorgänge innerhalb elektrischer Antriebssysteme wie des INGA III Lichtbogentriebwerks zu ermöglichen, wird im zweiten Hauptteil der vorliegenden Arbeit ein übergeordnetes Modellierungskonzept entwickelt. Der hybride numerische Ansatz beruht auf der Kopplung eines kinetischen Submodells für plasmadynamische Prozesse und eines Fluid-Submodells für die Behandlung schwerer Neutralteilchen sowie die Gaserhitzung durch die Stromwärme. Im Rahmen der vorliegenden Arbeit wird das kinetische Plasma-Submodell vollständig entwickelt und validiert. Der entwickelte PIC-MCC Solver, der als *dsmcPlasmaFoam* bezeichnet wird, stellt einen ersten Schritt im Hinblick auf die Bewältigung der zahlreichen Herausforderungen im Bereich der Plasmamodellierung in elektrischen Antriebssystemen dar. Zu den Haupteigenschaften des Modells gehören klassische PIC-Elemente, die es ermöglichen, Wechselwirkungen langer Reichweite zu berücksichtigen. Darüber hinaus sind Coulomb-Kollisionen zur Modellierung von Wechselwirkungen kurzer Reichweite sowie Ionisierungs- und Rekombinationsmodelle Bestandteil des Solvers. Um die Rechenanforderungen in akzeptablen Bereichen zu halten, wurde eine dynamische Anpassung der numerischen Partikelgewichte ebenfalls konzipiert und implementiert. Die Validierungsergebnisse für Testfälle wie das Verhalten von Plasmawellen, die Thermalisierung von Elektronenstrahlen und die Ionisierung von Neutralgasen durch Kollisionen mit Elektronen belegen die Qualität und Flexibilität des Solvers *dsmcPlasmaFoam*, welcher zur Untersuchung eines breiten Spektrums an plasmadynamischen Vorgängen eingesetzt werden kann. Neben dem offensichtlichen Nutzen zur Auslegung und Optimierung von elektrischen Antriebssystemen ist es möglich, durch die Entwicklung eines präzisen, hybriden Modells auf Basis von *dsmcPlasmaFoam* ein besseres Verständnis von plasmadynamischen Phänomenen, Energieaustausch- und Transportmechanismen nicht nur innerhalb elektrischer Triebwerke, sondern auch in weiteren Anwendungen wie dem Plasmaschweißen und der Elektronenstrahlbehandlung zu erlangen.

Abstract

New trends regarding fundamental design approaches of orbital spacecraft have been developing in the space industry in recent years. They include an increased interest in miniaturized satellites with the ability to perform spacecraft formation flying as well as a general rise in the use of electric propulsion systems for orbit and attitude control. The successful implementation of miniaturized satellites requires the use of propulsion devices able to provide small and precise thrust and impulse levels. One technical solution able to meet the requirements of both standard-sized as well as miniaturized spacecraft involves the use of highly efficient and precise electric propulsion systems. In the particular case of miniaturized satellites, the propulsion requirements are often associated with low propellant mass flow rates and small characteristic geometrical lengths, potentially leading to the appearance of rarefied conditions inside the nozzles of the propulsion devices. Because of the high complexity of the plasma phenomena taking place inside such systems and the usually very high computational requirements associated with their numerical modelling, electric propulsion systems for space applications are usually designed based on empirical models and experimental data. However, considering the always increasing computational power as well as the potential in terms of technical optimisation and development costs reduction, the exploration of new approaches for the modelling of such systems becomes more attractive every day. The present work focuses on two key aspects outlined above: rarefied gas conditions in transonic micronozzle flows as well as the numerical modelling of plasma phenomena inside electric propulsion systems.

Rarefied gas conditions in transonic micronozzle flows are examined in the present work taking advantage of the INGA III experimental setup at the Center of Applied Space Technology and Microgravity (ZARM). To this end, the cold-gas expansion of several noble gases through a millimetre-scale Laval nozzle with Knudsen numbers (Kn) in the range $0.02 < Kn < 0.33$ has been studied experimentally and numerically. The experimentally measured pressure in the ionisation chamber is compared with the numerical results from Navier-Stokes based simulations with a no-slip boundary condition as well as with DSMC results.

The DSMC and the experimental results are in good agreement in the range $Kn > 0.1$, while the Navier-Stokes results describe the experimental data more accurately than the DSMC method for $Kn < 0.05$. For high Kn numbers, the pressure values in the ionisation chamber estimated through continuum-based simulations are considerably higher than the experimental results and relative deviations of over 50% are established. The high deviations are determined to be the result of the discrepancy between the no-slip boundary condition used for the Navier-Stokes simulations and the high experimental Knudsen numbers corresponding to the slip-flow and transition regimes. Based on the comparison between the experimental and numerical results, a relation describing the deviation of the pressure drop through the nozzle as a function of Kn with gas-dependent and gas-independent accommodation coefficients is obtained. The Knudsen-function shows that for gases with small collision cross sections, the experimental pressure results strongly deviate from the no-slip assumption and the relation is mathematically best described through a second degree polynomial. For gases with large collision cross sections, the deviation as a function of Kn is closer to a linear function. Through the proposed Knudsen functions it is possible to correct the pressure results obtained with a continuum-based model while at the same time, avoiding a numerical case distinction between continuum, slip-flow and transition regimes. Since the accommodation coefficients are determined from dimensionless numbers, the results are expected to be qualitatively valid for other comparable Laval nozzle geometries.

In the next step of this work, a global modelling concept is designed in order to enable the numerical description of plasma phenomena inside electric propulsion systems like the INGA III arcjet thruster. The hybrid modelling concept is based on the coupling of a kinetic submodel for plasma phenomena with a fluid submodel for ohmic gas heating and neutral gas handling. In the frame of the present thesis, the kinetic submodel is fully developed. The obtained PIC-MCC solver, referred to as *dsmcPlasmaFoam*, constitutes a first step aiming to tackle the numerous challenges concerning plasma modelling for electric spacecraft propulsion systems. The described model includes, among others, classic PIC elements for long-range interactions between charge carriers, short-range Coulomb charge collisions as well as ionisation and recombination interactions. Additional features like dynamic weighting have been conceived and implemented in the solver in order to keep the computational requirements at acceptable levels. The numerical validation results presented in this thesis for cases like plasma wave behaviour, electron beam thermalization and neutral gas collisional ionisation confirm the solver *dsmcPlasmaFoam* as a powerful, flexible and accurate modelling tool covering a wide range of typical plasma phenomena. Besides the obvious benefits in terms of electric thruster design and optimization, the development of an accurate hybrid numerical model based on *dsmcPlasmaFoam* has the potential to provide new insights into plasma phenomena, energy exchange and transport mechanisms not only in electric propulsion devices but also in further plasma-related applications like plasma arc welding and electron beam processing.

Preface

First I would like to express my sincere gratitude to my project supervisor PD Dr.-Ing. habil. Rodion Groll for giving me the opportunity to work in his team. This PhD thesis would not have been possible without his valuable guidance and the numerous technical discussions I enjoyed during my time at ZARM.

I would like to thank Professor Dr. rer. nat. Claus Lämmerzahl and Prof. Dr. Claus Braxmaier for their role as supervisors and reviewers of this PhD thesis as well as Prof. Dr. Marc Avila for kindly supporting my work in the last months of my stay at ZARM.

During my time at the institute, I had the opportunity to share ideas and learn from many fellow colleagues. I owe special thanks to my fellow team members Till Frieler, Charles Chelem and Christoph Kühn, to our students Denis Zimmer, André Schmerglatt and Björn Klose as well as to André Pingel, Holger Faust, Ronald Mairose and Thorsten Coordes for their support during my time at ZARM. I would like to specially thank my office mate Želimir Marojević for his continuous help during my time at ZARM. Both the coffee maker and the door opening angle incidents will never be forgotten.

I owe special thanks to Ramy Shoeib for his incredible support in the late stages of this PhD thesis.

I would like to also thank Susanne, Markus and Wiebke for their year-long support.

This work would not have been possible without support from the North-German Supercomputing Alliance (HLRN) and I would like to thank HLRN for providing free access to their super-computing facilities.

My special thanks are extended to my close friends Fiedje, Lena, Mikelis, Jonathan, Daniel, Alejandro and Stefanie (Stf) for their time, support and patience as well as for continuously motivating me during my work on this thesis.

I owe my eternal gratitude to Mareike for her continuous kindness, support and patience.

Finally, I would like to specially thank my family Martha, Conrado and Claudia for their love and support.

Contents

Nomenclature

Roman Symbols

Symbol	Description	Unit
a	Speed of sound	$m\ s^{-1}$
A	Surface area	m^2
A	Nanbu's collision parameter	
A_1	Dimensionless Knudsen coefficient	
A_2	Dimensionless Knudsen coefficient	
A_s	Sutherland coefficient	$Pa\ s\ K^{-1/2}$
b	Channel height	m
b	Collisional impact parameter	m
B	Magnetic field vector	T
c	Particle's velocity vector	$m\ s^{-1}$
$\bar{c}$	Most probable velocity magnitude	$m\ s^{-1}$
c_0	Mean kinetic velocity magnitude	$m\ s^{-1}$
C_1	Dimensionless Knudsen coefficient	
C_2	Dimensionless Knudsen coefficient	
c'_m	Most probable velocity magnitude	$m\ s^{-1}$
c_p	Heat capacity at constant pressure	$J\ kg^{-1}\ K^{-1}$
c_r	Relative velocity between two particles	$m\ s^{-1}$
c_r	Relative velocity magnitude	$m\ s^{-1}$
c_v	Heat capacity at constant volume	$J\ kg^{-1}\ K^{-1}$
d	Particle's diameter	m
D_L	Diffuser length	m
D_t	Throat diameter	m
e	Internal energy per unit mass	$J\ kg^{-1}$
e	Elementary charge	A s
E	Energy	J
E	Electron field vector	$N\ A^{-1}\ s^{-1}$
e_{tr}	Internal translational energy per unit mass	$J\ kg^{-1}$
f	Volumetric force	$N\ m^{-3}$
f	Molecular velocity distribution function	
F	Force	N
$F(\|c\|)$	Maxwell distribution	
f^c	Convective flux of arbitrary property	

f^d	Diffusive flux of arbitrary property	
f_{pe}	Plasma frequency	Hz
F_t	Thrust	N
g	Plasma parameter	
g	Degeneracy factor	
g	Relative pre-collision velocity	m s^{-1}
g'	Relative post-collision velocity	m s^{-1}
G	Random scalar between 0 and 1	
$g(c_i)$	Probability of particle with velocity c_i	
$g(\chi)$	Probability distribution function for scattering angle χ	
h	Enthalpy per unit mass	J kg^{-1}
h	Planck constant	m^2 kg s^{-1}
$\dot{I}$	Momentum flux	N
I_{sp}	Specific impulse	s
J	Current density	A m^{-2}
k_B	Boltzmann constant	J K^{-1}
k	Thermal conductivity	W m^{-1} K^{-1}
K	Three-body recombination coefficient	m^6 s^{-1}
l^*	Geometrical length	m
l	Characteristic length	m
L	Characteristic length	m
m	Mass	kg
M	Molar mass	kg mol^{-1}
M	Mass of neutral particle	kg
$\dot{m}$	Mass flow rate	kg s^{-1}
M_1^*	Gas-independent Knudsen coefficient	kg kmol^{-1}
M_2^*	Gas-independent Knudsen coefficient	kg kmol^{-1}
n	Amount of matter	mol
n	Number density	m^{-3}
$\dot{n}$	Particle flux	s^{-1}
n	Surface normal vector	m
n_e	Electron number density	m^{-3}
n_i	Ion number density	m^{-3}
N	Total number of particles	
N	Total number of collisions	
N	Number density	m^{-3}
$\dot{N}$	Particle flux	s^{-1} m^{-2}

N_A	Avogadro's number	mol^{-1}
N'_α	Number of particles α colliding with particles β	
N''_α	Number of particles α colliding with particles α	
p	Pressure	Pa
$\bar{p}$	Mechanical pressure	Pa
p	Momentum vector	$kg\ m\ s^{-1}$
P	Probability	
P	Heating power	W
q	Charge	A s
q	Vector rate of heat flux per unit area	$W\ m^{-2}$
$\dot{q}$	Joule heating rate	$W\ m^{-3}$
Q	Heat	J
Q_v	Volumetric flow rate	$m^3\ s^{-1}$
r	Distance	m
r	Radius	m
r	Spatial position vector	m
R	Specific gas constant	$J\ kg^{-1}\ K^{-1}$
R_u	Universal gas constant	$J\ mol^{-1}\ K^{-1}$
s	Entropy	$J \cdot K^{-1}$
s	Molecular speed ratio	
s	Nanbu's collision parameter	
S	Surface	m^2
s_{ij}	Symmetric tensor	
t	Time	s
t_{ij}	Arbitrary tensor	
T	Temperature	K
T_{pe}	Plasma period	s
T_s	Sutherland temperature	K
T_{tr}	Translational temperature	K
T_W	Wall temperature	K
u	Velocity vector	$m\ s^{-1}$
U	Velocity vector magnitude	$m\ s^{-1}$
U	Energy level	J
U	Random number	
U_W	Wall velocity vector	$m\ s^{-1}$
v	Volume	m^{-3}
v	Macroscopic velocity	$m\ s^{-1}$
V	Volume	m^{-3}

V	Neutral particle velocity vector	m s^{-1}
v_d	Drift velocity	m s^{-1}
w	Channel width	m
W	Work	J
W	Numerical particle weight	
w_{ij}	Antisymmetric tensor	
x	Spatial position vector	m
X	Normalized length in x Cartesian coordinate	
X	Spatial coordinate	m
x,y,z	Cartesian coordinates	m
Y	Normalized length in y Cartesian coordinate	
Y	Lagrangian coordinate	m
Z	Random number	

Greek Symbols

Symbol	Description	Unit
α	Ionisation degree	
α	Recombination coefficient	m^3 s^{-1}
α	Electron impact ionisation coefficient	m^{-1}
α_E	Accommodation coefficient of energy	
α_p	Accommodation coefficient of momentum	
β_1	Gas-independent Knudsen exponent	
β_2	Gas-independent Knudsen exponent	
γ	Heat capacity ratio	
Γ	Diffusion coefficient	
Γ	Gamma function	
ε	Strain rate	s^{-1}
ε	Parent electron energy	J, eV
ε_0	Vacuum permittivity	F m^{-1}
ε_p	Progeny electron energy	J, eV
ε	Electron energy	J, eV
η	Efficiency	
$\eta_{\alpha\beta}$	Energy ratio	
θ	Polar deflection angle	rad
λ	Mean free path	m
λ	Coefficient of bulk viscosity	Pa s
λ_e	Linear interpolation factor	

λ_D	Debye length	m
λ_w	Linear interpolation factor	
μ	Dynamic viscosity	Pa s
μ	Particle mass	kg
$\mu_{\alpha\beta}$	Reduced mass	kg
ν	Kinematic viscosity	$m^2\ s^{-1}$
ν_c	General collision frequency	s^{-1}
ν_{en}	Collision frequency between electrons and neutrals	s^{-1}
ν_{pe}	Plasma frequency	s^{-1}
ξ	Kinetic particle's velocity vector	$m\ s^{-1}$
ρ	Mass density	$kg\ m^{-3}$
ρ	Charge density	$A\ s\ m^{-3}$
σ	Particle's diameter	m
σ	Electrical conductivity	$S\ m^{-1}$
σ_T	Total collision cross section	m
τ	Mean time between electron-neutral collisions	s
τ	Arbitrary time value	s
τ_0	Reference time	s
τ_{ij}	Viscous stress tensor	$N\ m^{-2}$
ϕ	Electric potential	V
ϕ_C	Coulomb potential	V
Φ	Dissipation function	$W\ m^{-3}$
$\Phi(\xi)$	Arbitrary function of particle's velocity	
φ	Arbitrary property	
φ	Azimuthal deflection angle	rad
χ	Deflection/scattering angle	rad
χ_N	Cumulative deflection angle	rad
ψ	Arbitrary intensive material property	
ψ	Azimuthal scattering angle for electron-neutral collision	rad
Ψ	Arbitrary extensive material property	
ω_{pe}	Plasma frequency	s^{-1}
ω	Viscosity index	

Dimensionless Quantities

Symbol	Description	Definition
Co	Courant number	$Co = \frac{u\Delta t}{\Delta x}$
Kn	Knudsen number	$Kn = \frac{\lambda}{l}$
Ma	Mach number	$MA = \frac{u}{a}$
Pe	Peclet number	$Pe = \frac{\rho u \Delta x}{\Gamma}$
Re	Reynolds number	$Re = \frac{\rho u L}{\mu}$
S	Dimensionless pressure drop	$S = \frac{\Delta p w b^3}{Q_v \mu L}$

Subscripts

Symbol	Description
0	Initial value
0	Total or stagnation property
A	Species A
α	Species α
$\alpha\beta$	Reduced property of Species $\alpha\beta$
B	Species B
β	Species β
$Cath$	Cathode
ch	Property in ionisation chamber
$corr$	Corrected result
cr	Collisional-radiative
$diff$	DSMC result with diffusive wall interaction model
e	Exit
e	Electron
exp	Experimental result
f	Final value
g	Gas
i	Ion
i	i^{th} cell
j	j^{th} cell
k	k^{th} collisional event
kin	Kinetic
max	Maximum
n	Neutral

$neut$	Neutral
ns	Neutral stabilized
p	Species P
pot	Potential
q	Species Q
r	Reduced property
ref	Property at reference temperature
sim	Continuum-based numerical result
$spec$	DSMC result with specular wall model
$therm$	Thermal
tot	Total
vac	Property in vacuum chamber
$\perp$	Perpendicular component

Superscripts

Symbol	Description
exp	Experimental
sim	Continuum-based numerical result
$spec$	Specific
$+$	Lagrangian property
$*$	Property at nozzle throat
$*$	Post-collision property
$*$	Property obtained from Knudsen function with gas-independent coefficients
$'$	Post-collision property
$\bar{}$	Normalized property
$\hat{}$	Normalized property

Abbreviations

Acronym	Description
1D	One-dimensional
2D	Two-dimensional
3D	Three-dimensional
BC	Boundary condition
BDS	Backward Difference Scheme
CDS	Central Difference Scheme

CFD	Computational Fluid Dynamics
CIC	Cloud In Cell
cp.	Compare
CV	Control volume
DF	Degrees of Freedom
DSMC	Direct Simulation Monte Carlo Method
e.g.	*exempli gratia* (for example)
Eq	Equation
etc.	*Et cetera* (and so forth)
FDM	Finite Difference Method
FDS	Forward Difference Scheme
Fig	Figure
FVM	Finite Volume Method
i.e.	*id est* (that is)
INGA III	Ionised Noble Gas Accelerator III
LHS	Left Hand Side
LUDS	Linear Upwind Interpolation Scheme
MCC	Monte Carlo Collision
MEMS	Micro-Electro-Mechanical-Systems
MPD	Magneto Plasma Dynamic
n/a	Not available/undefined
NGP	Nearest Grid Point
NTC	No Time Counter method
ODE	Ordinary Differential Equation
PIC	Particle-In-Cell
PISO	Pressure Implicit with Splitting of Operator
PPT	Pulsed Plasma Thruster
QUICK	Quadratic Upwind Interpolation Scheme for Convective Mathematics
RHS	Right Hand Side
s.	See
s.P.	Symmetry plane boundary condition
Sec	Section
SIMPLE	Semi-Implicit Method for Pressure-Linked Equations
UDS	Upwind Interpolation Scheme
VHS	Variable Hard Sphere model
VSS	Variable Soft Sphere model
z.G.	Zero-gradient boundary condition

List of Figures

List of Tables

Chapter 1

Introduction

For over a decade, the group Thermofluid Dynamics at the Center of Applied Space Technology and Microgravity (ZARM) has performed both experimental and numerical work in a variety of research fields including turbulent and convective flows, electrodynamics and magneto-fluid dynamics. One of the technical applications offering the greatest potential in terms of research scope, relevance and applicability is the field of propulsion systems for spacecraft. Besides the undeniable importance from an engineering standpoint, the study of thermodynamic phenomena in cold-gas and electric propulsion devices is closely related to complex scientific research topics like rarefied gas conditions, transonic and supersonic flows, magnetohydrodynamics and plasma dynamics. In these particular fields, several experimental studies using diverse thermo-electric arcjet thruster setups as well as numerical investigations employing custom models and numerical approaches have been performed at ZARM. The present work takes advantage of an existing experimental setup for a miniaturised arcjet thruster referred to as *Ionised Noble Gas Accelerator* INGA III in an attempt to gain a deeper understanding of transonic micronozzle flows in rarefied gas conditions, as well as to develop and validate a numerical model able to describe plasma-dynamic phenomena inside electric propulsion systems for spacecraft applications.

1.1 Motivation

In recent years, several new trends regarding fundamental characteristics of orbital spacecraft have originated inside the space industry. One of them is the increased interest in miniaturized satellites (e.g., CubeSats) with the ability to perform spacecraft formation flying. As a result, the development of small propulsion systems based on Micro-Electro-Mechanical-Systems (MEMS) and able to provide small and precise thrust and impulse levels has become an important research topic. The required low propellant mass flow rates and the small characteristic lengths of such devices lead to gas rarefaction effects in the micronozzles of the thrusters becoming dominant factors affecting both general behaviour and performance of the propulsion systems. For this reason, the study of the effects of the associated high Knudsen numbers in micronozzle transonic flows as well as their numerical modelling constitute research fields of growing importance and applicability.

In addition to the increased interest in miniaturized spacecraft as a viable option to meet orbital satellite requirements, recent years have also seen a considerable rise in the use of electric propulsion systems for orbit and attitude control of standard-sized satellites, complementing classic chemical and cold-gas devices. Electric propulsion devices enable the generation of very low and precise thrust values, a feature which at the same time is key for the attitude control of miniaturized satellites. In addition, the much higher specific impulse values I_{sp} achievable with electric devices highlight their inherent superior efficiency compared to cold-gas and chemical systems. Because of the high complexity of the plasma dynamic phenomena taking place inside electric propulsion systems and the usually very high computational requirements associated with their numerical modelling, electric propulsion systems for space applications are usually designed based on empirical models and experimental data. However, considering the always increasing computational power as well as the potential in terms of technical optimisation and development costs reduction, the exploration of new approaches for the modelling of such systems becomes more attractive every day. From a scientific standpoint, research work in this field offers plenty of opportunities to improve the understanding of complex dynamic plasma phenomena, energy exchange and transport mechanisms, not only inside electric propulsion devices, but also in other plasma systems.

1.2 Basic setup

In the present work, both experimental and numerical tools are employed extensively. The INGA III thruster setup at ZARM shown in Fig. 1.1 is used for the experimental study of transonic cold-gas flows under rarefied gas conditions. The main component of the setup is the Laval nozzle shown schematically in Fig. 1.2 which in turn, is installed inside a spherical vacuum chamber with a total volume of approximately 10 litre. In cold-gas operation mode, the thruster is capable of accelerating the noble gases xenon, krypton, argon and neon to supersonic conditions, transforming this way the thermal energy of the compressed propellant gas into kinetic energy. Vacuum conditions in the spherical chamber are achieved using a commercial rotary vane pump. The setup includes, among others, mass flow rate and pressure measurement devices allowing the determination of the gas pressure drop through the Laval nozzle as a function of the propellant type and its mass flow rate. In the frame of this work, the obtained results are compared with numerically obtained data in order to examine in detail the effects of rarefied gas conditions on the system as well as to identify potential efficiency improvement strategies.

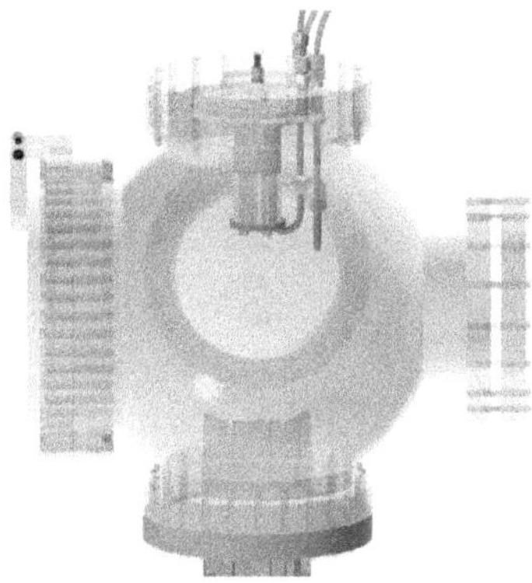

Figure 1.1: INGA III experimental setup with Laval nozzle fixed on the top lid of the spherical vacuum chamber.

Cold-gas numerical simulations are performed using standard continuum-based and kinetic Direct Simulation Monte Carlo solvers included in the software package *OpenFOAM®*. Furthermore, a Particle-in-Cell with Monte-Carlo collisions numerical model is developed and validated in the present work. The model, which includes key elements for the accurate modelling of plasma-dynamic behaviour inside electric propulsion systems, is based on the solver *dsmcFoam* of *OpenFOAM®*. In this thesis, the developed solver is shown to reproduce basic plasma dynamic phenomena as expected inside electric propulsion devices like the INGA III thruster. The numerical simulations with high computational requirements for both the cold-gas operation as well as for the validation of the developed custom plasma solver were performed in the ZARM computational cluster as well as in the super computer of the North-German Supercomputing Alliance (*Norddeutscher Verbund zur Förderung des Hoch- und Höchstleistungsrechnens - HLRN*).

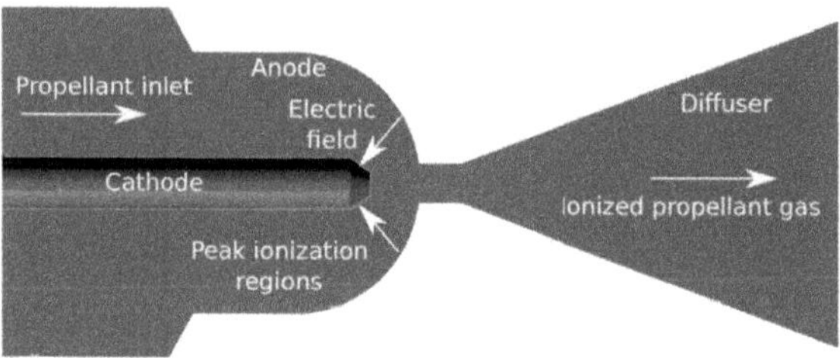

Figure 1.2: Schematic representation of the Laval nozzle in the INGA III setup and basic elements of the hot-gas arcjet operation mode.

1.3 Goals and thesis outline

The main goals of the present thesis are the investigation of gas rarefaction effects in transonic micronozzles flows and their impact on system behaviour and performance as well as the development of a numerical kinetic modelling concept able to accurately describe dynamic plasma phenomena inside electric propulsion devices with acceptable computational requirements. The detailed structure of this work is described in the following.

In Chapter 2, theoretical principles linked to the performed studies are outlined. The focus of the chapter lies on the definition of the different flow regimes as a function of the Knudsen number. Based on this, conservation equations for the description of flows in the continuum regime are developed and discussed. Furthermore, the working principle of the Laval nozzle, a common element of chemical, cold-gas and thermoelectric arcjet thrusters, is described. In order to establish a theoretical basis for the development of a kinetic plasma model, fundamental concepts of the plasma state and the kinetic description of gases are introduced in the final sections of this chapter.

Chapter 3 is dedicated to the computational methods employed for the numerical studies performed in the present work. The chapter introduces typical spatial and temporal discretization schemes for conservation equations valid in the continuum regime, as well as commonly used solution algorithms. In addition, the Direct Simulation Monte Carlo method (DSMC), widely employed for flows under rarefied conditions, is outlined. The chapter ends with a description of the Particle-In-Cell (PIC) algorithm, which combined with the DSMC method, constitutes the basis for the numerical plasma model developed in the frame of this thesis.

The experimental and numerical study of cold-gas transonic flows across multiple flow regimes is presented in Chapter 4. Here, gas rarefaction conditions in transonic micro-nozzle flows and their impact on system behaviour and performance are examined taking advantage of the INGA III experimental setup. Furthermore, comparisons between different numerical approaches in the slip-flow and transition regimes are performed. Finally, the chapter introduces a new gas-independent correcting relation as a function of the Knudsen number and applicable to continuum-based Navier-Stokes results under rarefied gas conditions.

Chapter 5 is dedicated to the development of a global numerical modelling concept for the simulation of plasma phenomena inside electric propulsion systems like the INGA III arcjet thruster in hot-gas operation mode. The hybrid modelling concept is based on the coupling of a kinetic submodel for plasma phenomena with a fluid submodel for ohmic gas heating and neutral gas handling. The kinetic plasma submodel is fully developed in the frame of this thesis based on the Particle-In-Cell (PIC) algorithm and Monte Carlo collisional schemes. The obtained kinetic solver, referred to as *dsmcPlasmaFoam*, includes short-range Coulomb collisions between charge carriers, electron-neutral interactions linked to gas ionisation as well as dynamic test particle weighting for computational optimisation.

The main components of the developed solver *dsmcPlasmaFoam* are analysed and validated in Chapter 6. They include the Maxwell solver for the determination of the electric field from a kinetic particle distribution, the Lorentz solver describing the charge carriers motion in an electric field as well as the weighting steps for the computation of the particle density from the particle distribution and the electric force acting on the charge carriers. Furthermore, validation cases for the short-range Coulomb interactions model as well as for the implemented electron-neutral collisional approach are presented. The results discussed in this chapter confirm the solver *dsmcPlasmaFoam* as a powerful, flexible and accurate modelling tool covering a wide range of typical plasma phenomena. The developed solver is expected to serve as one of two main submodels in the envisioned hybrid PIC-MCC-FVM plasma model currently in development at the Center of Applied Space Technology and Microgravity (ZARM).

Chapter 2

Theoretical Principles

In this chapter, the theoretical principles and main governing equations required for the modelling of the cold-gas and hot-gas operating modes of an electric propulsion system are presented. The chapter begins with a brief introduction of the Lagrangian and Eulerian specification of the flow field followed by the mass and momentum conservation equations as well as the ideal gas law. The final sections of the chapter are dedicated to the Laval nozzle, fundamentals of plasma phenomena and the kinetic theory of gases.

2.1 Knudsen number and flow regimes

Let λ represent the mean free path in a fluid, i.e., the mean distance covered by a particle or molecule before a collision with a second particle takes place. In this case, the dimensionless Knudsen number Kn is defined as the ratio between the mean free path λ and a characteristic geometrical length l as follows:

$$Kn = \frac{\lambda}{l} \tag{2.1}$$

For low values of Kn, the mean distance covered by a particle before a collision with a second particle takes place is small in comparison with the characteristic length of the system. This is the case for gases with relatively high densities and a regular molecular distribution or for problems with large geometrical scales. On the other hand, high values of Kn are present when the particle density in the system is low or when the problem involves small geometrical scales. In this case, the probability of a particle interacting with a domain boundary before it collides with a second particle is relatively high. Based on the above, the Knudsen number Kn enables the definition of the flow regimes shown in Table 2.1.

Continuum regime	0	$< Kn <$	10^{-2}
Slip flow	10^{-2}	$< Kn <$	10^{-1}
Transition regime	10^{-1}	$< Kn <$	10
Free molecular flow	10	$< Kn <$	∞

Table 2.1: Flow regimes [Hän04].

In the **continuum regime** (low Knudsen numbers), the mean free path of the gas particles is much smaller than the characteristic length of the problem and the fluid can be mathematically handled as a continuum. Macroscopic quantities are obtained through averaging of the corresponding molecular quantities in a volume of the order l^{*3}. Here, l^* represents a geometrical length between the mean free path of the fluid λ and the characteristic length l of the studied system ($\lambda < l^* < l$) and can be imagined as the transition scale between the molecular and the continuum length scales (cp. [Pop00]). The control volume can therefore be divided into mathematical infinitesimal volumes, allowing the description of macroscopic quantities and their spatial and temporal changes through sets of partial differential equations. The most important set of partial differential equations for the flow description in the continuum regime are the Navier-Stokes-Equations [Hän04]. They are introduced in Section 2.4 of this work.

Relatively high Knudsen numbers in the range $10^{-2} < Kn < 10$ correspond to the **slip flow** and **transition regimes**. Here, the results from the Navier-Stokes-Equations with classic no-slip boundary conditions deviate from experimentally obtained data and a local thermodynamic non-equilibrium region known as the Knudsen layer becomes important. In this layer, the diffusive nature of the collisions between gas atoms and walls leads to a loss of momentum in the flow direction. This effect is, however, weaker than for gases in the continuum regime. Classic approaches for the description of flows with these characteristics are based on the formulation of modified boundary conditions for the near-wall velocity profile. These include first-order and second or higher-order slip boundary conditions.

The extreme case of high Knudsen numbers corresponds to the **free molecular flow**. Because of the very high values of λ and small geometrical dimensions, the particles rarely collide with each other before interacting with the domain boundaries. As a consequence, a thermodynamic equilibrium cannot be reached in the gas and the macroscopic state of the medium depends strongly on the state of individual particles. In this case, differential equations become inadequate for the flow description. Instead, flows in this regime can be accurately described through the kinetic theory of gases and the Boltzmann equation, which is, in theory, valid for all flow regimes. Since mathematical solutions of the Boltzmann equation are complex, numerical methods such as the Direct Simulation Monte Carlo method (s. [Bir94]) are usually employed [Hän04]. In Sections 2.2 to 2.6 of this work, the theoretical principles and main equations for the description of flows in the continuum regime at low Knudsen numbers are introduced. Furthermore, Section 2.9 is dedicated to the kinetic theory of gases, commonly employed for the description of flows at high Knudsen numbers.

2.2 Lagrangian and Eulerian specification of the flow field

Let us consider a flow in the continuum regime. The *Eulerian* specification of the flow field corresponds to the case in which specific locations in space are observed as the flow passes by. The *Eulerian* density and velocity fields can be defined here as $\rho(x,t)$ and $u(x,t)$ respectively, where x denotes an specific location in space. In contrast, in the *Lagrangian* specification of the flow field, the observer moves with the local velocity of the fluid. Let Y be the position of a specific fluid particle at a reference time t_0. As shown in Fig. 2.1, the position of the particle at any different time can therefore be defined as $X^+(t,Y)$ (cp. [Pop00]).

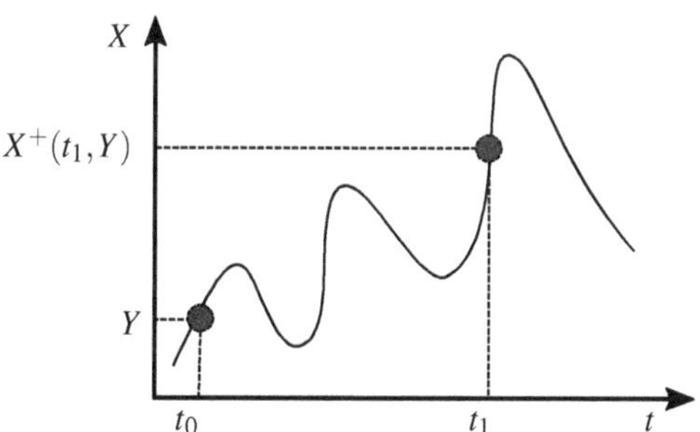

Figure 2.1: Trajectory $X^+(t,Y)$ of a fluid particle. Y represents the particle's position at the reference time t_0 (cp. [Pop00]).

Thus, any given fluid particle can be defined by an equation for the particle's position Y at the reference time t_0,

$$X^+(t_0,Y) = Y \tag{2.2}$$

and an equation for its velocity,

$$\frac{\partial}{\partial t}X^+(t,Y) = u(X^+(t,Y),t) \tag{2.3}$$

or alternatively,

$$u^+(t,Y) = u(X^+(t,Y),t) \tag{2.4}$$

Equation 2.4 establishes the relation between the *Lagrangian* velocity u^+ and the *Eulerian* velocity u and indicates that the particle's velocity u^+ is the same as the local flow velocity u [Pop00]. Furthermore, a similar expression can be formulated for the density:

$$\rho^+(t,Y) = \rho(X^+(t,Y),t) \tag{2.5}$$

Since the Lagrangian fields are defined by the particle's position Y at the reference time t_0, Y is known as the *Lagrangian* or material coordinate [Pop00]. The partial derivative of the *Lagrangian* density represents the time-dependent change of the density for a constant value of Y, i.e., following a specific fluid particle:

$$\begin{aligned}
\frac{\partial}{\partial t}\rho^+(t,Y) &= \frac{\partial}{\partial t}\rho(X^+(t,Y),t) \\
&= \left(\frac{\partial}{\partial t}\rho(x,t)\right)_{x=X^+(t,Y)} + \frac{\partial}{\partial t}X_i^+(t,Y)\left(\frac{\partial}{\partial x_i}\rho(x,t)\right)_{x=X^+(t,Y)} \\
&= \left(\frac{\partial}{\partial t}\rho(x,t) + u_i(x,t)\frac{\partial}{\partial x_i}\rho(x,t)\right)_{x=X^+(t,Y)} \\
\frac{\partial}{\partial t}\rho^+(t,Y) &= \left(\frac{D}{Dt}\rho(x,t)\right)_{x=X^+(t,Y)}
\end{aligned} \tag{2.6}$$

where,

$$\frac{D}{Dt} = \frac{\partial}{\partial t} + u_i\frac{\partial}{\partial x_i} = \frac{\partial}{\partial t} + u\cdot\nabla \tag{2.7}$$

represents the *material* or *substantial* derivative. The rate of change of velocity following a fluid particle (i.e., its acceleration) can be formulated in a similar way:

$$\frac{\partial}{\partial t}u^+(t,Y) = \left(\frac{D}{Dt}u(x,t)\right)_{x=X^+(t,Y)} \tag{2.8}$$

Therefore, the rate of change of flow fields following a fluid particle is described by both the partial derivative of the *Lagrangian* fields and the material derivative of the *Eulerian* fields (cp. [Pop00]).

2.3 Conservation of mass

The transport equation for mass conservation, also known as the continuity equation, can be obtained through consideration of the total mass fluxes in a control volume (s. [VM95], Chapter 2), i.e., by taking advantage of the Reynolds transport theorem[1]. Let Ψ denote an arbitrary extensive material property and ψ, the corresponding intensive property. Hence, Ψ can be calculated as:

$$\Psi = \int_V \rho \psi \mathrm{d}V \tag{2.9}$$

Furthermore, the rate of change of the material property Ψ can be expressed as:

$$\begin{aligned}\frac{\mathrm{d}\Psi}{\mathrm{d}t} &= \frac{\mathrm{d}}{\mathrm{d}t}\int_V \rho \psi \mathrm{d}V \\ \frac{\mathrm{d}\Psi}{\mathrm{d}t} &= \int_V \frac{\partial \rho \psi}{\partial t}\mathrm{d}V + \int_{\delta V} \rho \psi (un)\,\mathrm{d}A\end{aligned} \tag{2.10}$$

The surface integral at the RHS of Eq. 2.10 describes the flux of the vector field $\rho \psi n$ through the boundary surface δV of the control volume. This term can be transformed into a volume integral using the divergence theorem[2] leading to:

$$\frac{\mathrm{d}\Psi}{\mathrm{d}t} = \int_V \frac{\partial \rho \psi}{\partial t}\mathrm{d}V + \int_V \nabla\cdot(\rho \psi u)\,\mathrm{d}V \tag{2.11}$$

By setting $\psi = 1$, Ψ becomes the total mass m in the control volume. Assuming that mass is neither created nor destroyed in the control volume, the rate of change $\mathrm{d}\Psi/\mathrm{d}t$ must equal zero. Eq. 2.11 then becomes:

$$\begin{aligned}\frac{\mathrm{d}m}{\mathrm{d}t} = \int_V \frac{\partial \rho}{\partial t}\mathrm{d}V + \int_V \nabla\cdot(\rho u)\,\mathrm{d}V = 0 \\ \int_V \left(\frac{\partial \rho}{\partial t} + \nabla\cdot(\rho u)\right)\mathrm{d}V = 0\end{aligned} \tag{2.12}$$

Taking the derivative of Eq. 2.12 produces, in vector notation:

$$\frac{\partial \rho}{\partial t} + \nabla\cdot(\rho u) = 0 \tag{2.13}$$

[1] See Appendix A
[2] See Appendix A

or alternatively, in indicial notation:

$$\frac{\partial \rho}{\partial t} + \frac{\partial \rho u_j}{\partial x_j} = 0 \tag{2.14}$$

Equations 2.13 and 2.14 are known as the **mass continuity equation**. For an incompressible flow with ρ = constant, Eq. 2.13 can be rewritten to produce:

$$\nabla \cdot u = 0 \tag{2.15}$$

2.4 Conservation of momentum

Let us define ψ as the flow velocity u. Substitution of $\psi = u$ into Eq. 2.9 yields the momentum p:

$$\Psi = p = \int_V \rho u \, \mathrm{d}V \tag{2.16}$$

Newton's Second Law of Motion states that the rate of change of momentum of a body equals the net force applied to it. By taking advantage of Eq. 2.11, this can be written as:

$$\frac{\mathrm{d}p}{\mathrm{d}t} = \int_V \frac{\partial \rho u}{\partial t} \mathrm{d}V + \int_V \nabla \cdot (\rho u \cdot u) \, \mathrm{d}V = \sum_k F_k \tag{2.17}$$

or alternatively, in indicial notation:

$$\frac{\mathrm{d}p_i}{\mathrm{d}t} = \int_V \frac{\partial \rho u_i}{\partial t} \mathrm{d}V + \int_V (\rho u_i u_j) \, \mathrm{d}V = \sum_k F_{k,i} \tag{2.18}$$

Let f_i^{spec} represent the net intensive force [N/kg], so that the expression,

$$\sum_k F_{k,i} = \int_V \rho f_i^{\mathrm{spec}} \, \mathrm{d}V \tag{2.19}$$

holds. Substitution of Eq. 2.19 into Eq. 2.18 and subsequent differentiation yields:

$$\frac{\partial \rho u_i}{\partial t} + \frac{\partial \rho u_i u_j}{\partial x_j} = \rho f_i^{\mathrm{spec}} = f_i \tag{2.20}$$

where f_i represents the force on a unit volume basis [N/m^3]. Now, let us examine the force term on the RHS of Eq. 2.20. In general, the forces acting on a control volume may be classified as surface forces and body forces. While surface forces are exerted to the surfaces of a control volume, body forces, such as gravity and electromagnetic forces, act throughout the full volume element. As a consequence, the RHS of Eq. 2.20 can be written as:

$$f_i = \rho g_i + f_{\text{surface},i} \tag{2.21}$$

where ρg_i corresponds to the volumetric, gravitational body force. For the definition of the surface forces, the *viscous stress tensor* τ_{ij}, is introduced:

$$\tau_{ij} = \begin{pmatrix} \tau_{xx} & \tau_{xy} & \tau_{xz} \\ \tau_{yx} & \tau_{yy} & \tau_{yz} \\ \tau_{zx} & \tau_{zy} & \tau_{zz} \end{pmatrix}. \tag{2.22}$$

The components of the stress tensor are shown in Fig. 2.2. As can be seen, the first index of the component indicates the direction of the normal vector of the surface on which the stress acts, while the second index specifies the direction of the applied stress. Furthermore, in order for equilibrium of moments to be satisfied, the stress tensor must be symmetric, i.e., $\tau_{ij} = \tau_{ji}$. Therefore, the total force acting on the front faces of the control volume $\mathrm{d}V = \mathrm{d}x\mathrm{d}y\mathrm{d}z$ in Fig. 2.2 in each direction can be written as follows:

$$\begin{aligned}
\mathrm{d}F_x &= \tau_{xx}\mathrm{d}y\mathrm{d}z + \tau_{yx}\mathrm{d}x\mathrm{d}z + \tau_{zx}\mathrm{d}x\mathrm{d}y \\
\mathrm{d}F_y &= \tau_{xy}\mathrm{d}y\mathrm{d}z + \tau_{yy}\mathrm{d}x\mathrm{d}z + \tau_{zy}\mathrm{d}x\mathrm{d}y \\
\mathrm{d}F_z &= \tau_{xz}\mathrm{d}y\mathrm{d}z + \tau_{yz}\mathrm{d}x\mathrm{d}z + \tau_{zz}\mathrm{d}x\mathrm{d}y
\end{aligned}$$

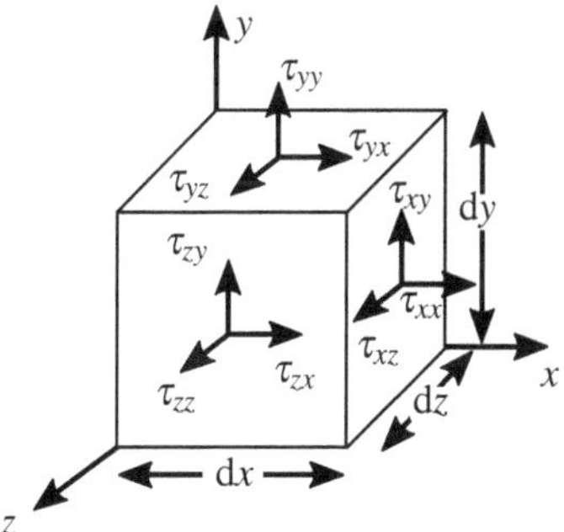

Figure 2.2: Components of the stress tensor τ_{ij} acting on the surfaces of a control volume $\mathrm{d}V = \mathrm{d}x\mathrm{d}y\mathrm{d}z$.

If the control element is in equilibrium, these forces are balanced by equal and opposite forces on the back surfaces of the control volume. However, if the element is accelerating, the stresses on the back faces would deviate by differential amounts. For instance, for τ_{xx} we obtain:

$$\tau_{xx,\text{front}} = \tau_{xx,\text{back}} + \frac{\partial \tau_{xx}}{\partial x}\mathrm{d}x \tag{2.23}$$

The net force in the x direction can therefore be formulated as:

$$\begin{aligned} \mathrm{d}F_{x,\text{net}} &= \mathrm{d}F_{x,\text{front}} - \mathrm{d}F_{x,\text{back}} \\ &= \left(\frac{\partial \tau_{xx}}{\partial x}\mathrm{d}x\right)\mathrm{d}y\mathrm{d}z + \left(\frac{\partial \tau_{yx}}{\partial y}\mathrm{d}y\right)\mathrm{d}x\mathrm{d}z + \left(\frac{\partial \tau_{zx}}{\partial z}\mathrm{d}z\right)\mathrm{d}x\mathrm{d}y \end{aligned} \tag{2.24}$$

which divided by the differential volume $\mathrm{d}x\mathrm{d}y\mathrm{d}z$ and because of the symmetry of the stress tensor, may be rewritten as:

$$f_x = \frac{\partial \tau_{xx}}{\partial x} + \frac{\partial \tau_{xy}}{\partial y} + \frac{\partial \tau_{xz}}{\partial z} \tag{2.25}$$

According to Eq. 2.25, the net surface force in the x direction f_x corresponds to the divergence of the vector $\{\tau_{xx}, \tau_{xy}, \tau_{xz}\}$ (first row of the stress tensor). Similarly, the forces f_y and f_z equal the divergences of the second and third row of the stress tensor respectively. The total vector surface force f can therefore be defined as:

$$f_{\text{surface}} = \nabla \cdot \tau_{ij} = \frac{\partial \tau_{ij}}{\partial x_j} \tag{2.26}$$

Substituting Eqs. 2.26 and 2.21 into Eq. 2.20 yields:

$$\frac{\partial \rho u_i}{\partial t} + \frac{\partial \rho u_i u_j}{\partial x_j} = \rho g_i + \frac{\partial \tau_{ij}}{\partial x_j} \tag{2.27}$$

In order to further develop the equation of motion, let us consider a fluid at rest. In this case, the velocities and shear stresses disappear and surface forces are linked only to hydrostatic pressure. Fig. 2.3 shows the corresponding balance of forces for a differential volume. The net force in the x direction as a result of the hydrostatic pressure is:

$$F_x^{hyd} = p\,\mathrm{d}y\mathrm{d}z - \left(p + \frac{\partial p}{\partial x}\right)\mathrm{d}y\mathrm{d}z$$

which divided by the volume $\mathrm{d}V = \mathrm{d}x\mathrm{d}y\mathrm{d}z$ yields:

$$f_x^{hyd} = -\frac{\partial p}{\partial x} \tag{2.28}$$

The surface forces in the y and z directions can be formulated in a similar way. As a consequence, the stress tensor for a fluid at rest takes the form:

$$\tau_{ij} = -p\delta_{ij} = \begin{pmatrix} -p & 0 & 0 \\ 0 & -p & 0 \\ 0 & 0 & -p \end{pmatrix} \tag{2.29}$$

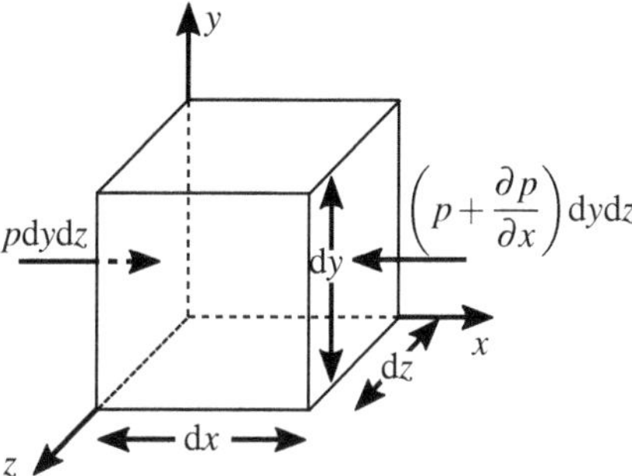

Figure 2.3: Balance of forces resulting from the hydrostatic pressure for a differential volume $\mathrm{d}V = \mathrm{d}x\mathrm{d}y\mathrm{d}z$ at rest.

In order to fully develop the stress tensor, we must now consider a fluid element under shear stress. Stokes (1845) derived a deformation law based on the following postulates:

1. The fluid is continuous and its stress tensor τ_{ij} is a linear function of the strain rates.
2. The fluid is isotropic, i.e., its properties are independent of direction.
3. If the strain rates are zero, the deformation law must reduce to the hydrostatic pressure condition $\tau_{ij} = -p\delta_{ij}$ (s. Eq. 2.29).

Figure 2.4 shows a fluid element located between two large parallel plates with surface area A. A tangential force F_s is applied on one of the plates, so that it moves relative to the other. The applied shear stress $\tau = F_s/A$ sets the fluid in motion and leads to the appearance of the strain rate $\varepsilon = U/\Delta h$. The simplest assumption for the relation between the viscous stress and the strain rate is a linear function. This corresponds to the first assumption postulated by Stokes. Fluids which obey this law are called *Newtonian fluids*. The proportionality factor is the dynamic viscosity μ. Mathematically:

$$\tau = \frac{F_s}{A} = \mu \frac{U}{\Delta h} \tag{2.30}$$

or in differential form,

$$\tau = \mu \frac{\partial u}{\partial y} \tag{2.31}$$

which in a general three-dimensional form, can be introduced into the stress tensor to produce:

$$\tau_{ij} = -p\delta_{ij} + \mu \frac{\partial u_i}{\partial x_j} = \begin{pmatrix} -p & 0 & 0 \\ 0 & -p & 0 \\ 0 & 0 & -p \end{pmatrix} + \begin{pmatrix} \mu \frac{\partial u_1}{\partial x_1} & \mu \frac{\partial u_1}{\partial x_2} & \mu \frac{\partial u_1}{\partial x_3} \\ \mu \frac{\partial u_2}{\partial x_1} & \mu \frac{\partial u_2}{\partial x_2} & \mu \frac{\partial u_2}{\partial x_3} \\ \mu \frac{\partial u_3}{\partial x_1} & \mu \frac{\partial u_3}{\partial x_2} & \mu \frac{\partial u_3}{\partial x_3} \end{pmatrix} \tag{2.32}$$

with,

$$\frac{\partial u_i}{\partial x_j} = \nabla u = L \tag{2.33}$$

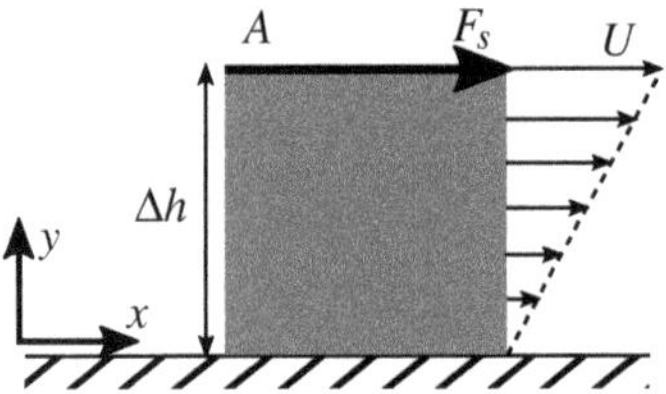

Figure 2.4: Differential volume under shear stress.

As already stated, the stress tensor must be symmetric, i.e., $\tau_{ij} = \tau_{ji}$. This condition is, however, not satisfied if the velocity gradients are antisymmetric, i.e.,

$$\frac{\partial u_i}{\partial x_j} \neq \frac{\partial u_j}{\partial x_i} \tag{2.34}$$

As a consequence, only the symmetric part of the velocity gradient must be taken into account. The tensor $L = \nabla \cdot u$ may be expressed as the sum of a symmetric tensor S and an antisymmetric tensor W. Hence:

$$l_{ij} = s_{ij} + w_{ij} \tag{2.35}$$

$$s_{ij} = s_{ji} \rightarrow s_{ij} = \frac{1}{2}(l_{ij} + l_{ji}) \tag{2.36}$$

$$w_{ij} = -w_{ji} \rightarrow w_{ij} = \frac{1}{2}(l_{ij} - l_{ji}) \tag{2.37}$$

Note that the definitions in Eqs. 2.36 and 2.37 satisfy the condition $l_{ij} = s_{ij} + w_{ij}$. Hence, a provisional stress tensor can be formulated as follows:

$$t_{ij} = -p\delta_{ij} + \mu s_{ij} + \Delta t_{ij} \tag{2.38}$$

where the tensor Δt_{ij} contains additional terms yet to be defined. Based on the three postulates stated above, Stokes derived the following deformation law for Newtonian (linear) viscous flow:

$$t_{ij} = -p\delta_{ij} + 2\mu s_{ij} + \lambda\left(\delta_{ij}\frac{\partial u_k}{\partial x_k}\right)$$

$$t_{ij} = -p\delta_{ij} + \mu\left(\frac{\partial u_i}{\partial x_j} + \frac{\partial u_j}{\partial x_i}\right) + \lambda\left(\delta_{ij}\frac{\partial u_k}{\partial x_k}\right) \tag{2.39}$$

in which λ, commonly known as the *coefficient of bulk viscosity*, is linked to volume expansion. There is an interesting consequence of Eq. 2.39. Let us define the mechanical pressure $\bar{p}$ as the average compression stress experienced by a fluid $p = -1/3(\tau_{xx} + \tau_{yy} + \tau_{zz})$. This equals the product of the first invariant of the stress tensor in Eq. 2.39 and the scalar -1/3:

$$\bar{p} = -\frac{1}{3}(\tau_{xx} + \tau_{yy} + \tau_{zz}) = -\frac{1}{3}(-3p + 2\mu\nabla \cdot u + 3\lambda\nabla \cdot u)$$

$$= p - (\lambda + \frac{2}{3}\mu)\nabla \cdot u \tag{2.40}$$

Equation 2.39 therefore suggests the existence of a deviation between the mechanical pressure $\bar{p}$ and the thermodynamic pressure p. Since the divergence of the velocity field $\nabla \cdot u$ is usually small, this distinction does not play an important role in common flow problems. Furthermore, for truly incompressible flows, $\nabla \cdot u = 0$ and the discrepancy between $\bar{p}$ and p disappears. In a generally valid approach, Stokes resolved this inconsistency by assuming:

$$\lambda + \frac{2}{3}\mu = 0 \rightarrow \lambda = -\frac{2}{3}\mu \tag{2.41}$$

The expressions in Eq. 2.41 are known as the *Stokes's hypothesis*. The stress tensor can now be formulated by substituting the definition of the bulk viscosity λ given in Eq. 2.41 into Eq. 2.39. This yields:

$$\tau_{ij} = -p\delta_{ij} + \mu\left(\frac{\partial u_i}{\partial x_j} + \frac{\partial u_j}{\partial x_i}\right) - \frac{2}{3}\mu\left(\delta_{ij}\frac{\partial u_k}{\partial x_k}\right) \tag{2.42}$$

which in turn can be substituted into Eq. 2.27 to obtain the equation of motion:

$$\frac{\partial \rho u_i}{\partial t} + \frac{\partial \rho u_i u_j}{\partial x_j} = \rho g_i - \frac{\partial p}{\partial x_i} + \frac{\partial}{\partial x_j}\left[\mu\left(\frac{\partial u_i}{\partial x_j} + \frac{\partial u_j}{\partial x_i} - \frac{2}{3}\delta_{ij}\frac{\partial u_k}{\partial x_k}\right)\right] \tag{2.43}$$

The set of three expressions (one per spatial dimension) contained in Eq. 2.43 are known as the **Navier-Stokes equations** and are of central importance for the study of viscous flows (cp. [Whi06]). The Navier-Stokes equations are used for the description of a great variety of fluid related phenomena and are applicable in a wide range of technical and engineering problems. Despite this and the fact that they were formulated in the 19th century, the Navier-Stokes equations are, from a mathematical standpoint, still poorly understood. There is (at the time of writing) no proof that solutions always exist in three dimensions and respectively, that these solutions are unique. As a consequence, the Clay Mathematics Institute formulated the existence and smoothness of the Navier-Stokes equations as one of its seven so-called Millennium Prize Problems [Ins, Fef06].

A well-known form of the Navier-Stokes equations can be obtained by applying the product rule of the derivative to the LHS of Eq. 2.43 which, after rearranging, leads to:

$$\rho\frac{\partial u_i}{\partial t} + \underbrace{u_i\left(\frac{\partial \rho}{\partial t} + \frac{\partial \rho u_j}{\partial x_j}\right)}_{\substack{=0 \\ \text{mass continuity} \\ \text{equation}}} + \rho u_j\frac{\partial u_i}{\partial x_j} = \rho g_i - \frac{\partial p}{\partial x_i} + \frac{\partial}{\partial x_j}\left[\mu\left(\frac{\partial u_i}{\partial x_j} + \frac{\partial u_j}{\partial x_i} - \frac{2}{3}\delta_{ij}\frac{\partial u_k}{\partial x_k}\right)\right]$$

The Navier-Stokes equations can therefore be written as follows:

$$\rho\frac{\partial u_i}{\partial t} + \rho u_j\frac{\partial u_i}{\partial x_j} = \rho g_i - \frac{\partial p}{\partial x_i} + \frac{\partial}{\partial x_j}\left[\mu\left(\frac{\partial u_i}{\partial x_j} + \frac{\partial u_j}{\partial x_i} - \frac{2}{3}\delta_{ij}\frac{\partial u_k}{\partial x_k}\right)\right] \tag{2.44}$$

Note that the LHS of Eq. 2.44 equals the product of density and the material derivative of the velocity. Thus, Eq. 2.44 can also be formulated in the following way:

$$\rho\frac{Du_i}{Dt} = \rho g_i - \frac{\partial p}{\partial x_i} + \frac{\partial}{\partial x_j}\left[\mu\left(\frac{\partial u_i}{\partial x_j} + \frac{\partial u_j}{\partial x_i} - \frac{2}{3}\delta_{ij}\frac{\partial u_k}{\partial x_k}\right)\right] \tag{2.45}$$

2.5 Conservation of energy

According to the first law of thermodynamics, the net energy increase of a system equals the sum of the work and heat added to it. This can be mathematically expressed as (cp. [Whi06]):

$$\mathrm{d}E_{tot} = \delta Q + \delta W \tag{2.46}$$

where E_{tot} denotes the total energy of a system, Q the added heat and W, the work applied to the system. For a system in motion, the total energy per unit mass consists of a mass specific internal energy e, a mass specific kinetic energy $(1/2)U^2$ and a mass specific potential energy $g \cdot r$. Hence, the total energy on a unit volume basis can be expressed as:

$$E_{tot} = \rho \left(e + \frac{1}{2}U^2 - g \cdot r \right) \tag{2.47}$$

Note that the scalar U represents the magnitude of the velocity vector u. The change of the total energy can be written as the rate of change of E_{tot} following a fluid particle by taking advantage of the material derivative in Eq. 2.7. Therefore,

$$\frac{DE_{tot}}{Dt} = \frac{DQ}{Dt} + \frac{DW}{Dt} \tag{2.48}$$

Furthermore, differentiation of Eq. 2.47 yields:

$$\frac{DE_{tot}}{Dt} = \rho \left(\frac{De}{Dt} + U \frac{DU}{Dt} - g \cdot u \right) \tag{2.49}$$

In order to fully develop the RHS of Eq. 2.48, the quantities Q and W must be expressed in terms of fluid properties. The heat rate Q can be defined by taking advantage of Fourier's law, which states that the heat flux is proportional to a temperature gradient. Hence,

$$q = -\kappa \nabla T \tag{2.50}$$

where the vector q represents the vector rate of heat flow per unit area. Moreover, the proportionality constant κ is the thermal conductivity [W/(m·K)]. For fluids, κ is an isotropic thermodynamic property. The heat flow into the fluid element $\mathrm{d}V$ in the x direction can be defined based on Fig. 2.5 as:

$$q_x \mathrm{d}y \mathrm{d}z \tag{2.51}$$

while the heat flow out of the element can be expressed as:

$$\left(q_x + \frac{\partial q_x}{\partial x} \mathrm{d}x \right) \mathrm{d}y \mathrm{d}z \tag{2.52}$$

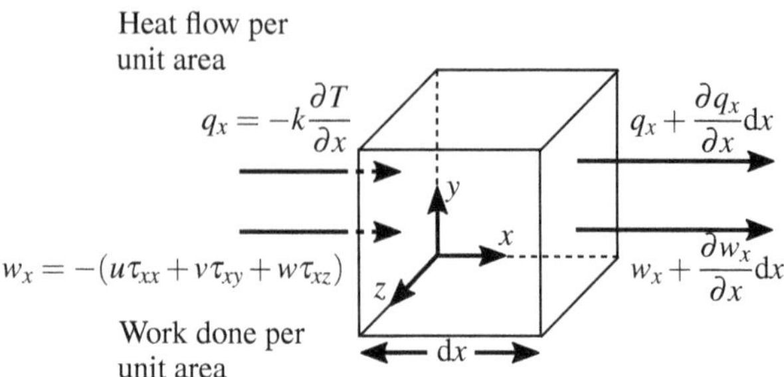

Figure 2.5: Heat and work balance in the x direction of the volume element dV [Whi06].

The heat flows in the y and z directions can be defined in a similar way. Furthermore, common sign convention assumes the heat flows to be positive when going into the control volume and negative when exiting it. Hence, the net heat transfer into the element can be expressed as:

$$-\left(\frac{\partial q_x}{\partial x} + \frac{\partial q_y}{\partial y} + \frac{\partial q_z}{\partial x}\right)\mathrm{d}x\mathrm{d}y\mathrm{d}z \tag{2.53}$$

Division of the expression in Eq. 2.53 by the volume of the fluid element and subsequent combination with Fourier's law yields the following equation for the heat transfer:

$$\frac{DQ}{Dt} = -\left(\frac{\partial q_x}{\partial x} + \frac{\partial q_y}{\partial y} + \frac{\partial q_z}{\partial x}\right) = -\nabla \cdot q = \nabla \cdot (\kappa \nabla T) \tag{2.54}$$

Note that potential source terms associated with internal heat generation have been neglected. Based on Fig. 2.5, the rate of work per unit area applied to the element by the stresses on the left face of the fluid element is given by,

$$w_x = -(u\tau_{xx} + v\tau_{xy} + w\tau_{xz}) \tag{2.55}$$

where u, v and w represent the individual components of the velocity vector u in the x, y and z directions respectively. Furthermore, the rate of work resulting from stresses on the right face is:

$$-\left(w_x + \frac{\partial w_x}{\partial x}\mathrm{d}x\right) \tag{2.56}$$

The rates of work done on the element in the y and z direction can be formulated in a similar way. The net rate of work applied to the element equals the sum of the rates of work on each of the element faces multiplied with the respective face area. Hence:

$$-\left(\frac{\partial w_x}{\partial x}+\frac{\partial w_y}{\partial y}+\frac{\partial w_z}{\partial z}\right)\mathrm{d}x\mathrm{d}y\mathrm{d}z \tag{2.57}$$

where the same sign convention as for the heat flow has been followed. The net rate of work applied to the element on a unit volume basis is therefore:

$$\frac{DW}{Dt}=-\left(\frac{\partial w_x}{\partial x}+\frac{\partial w_y}{\partial y}+\frac{\partial w_z}{\partial z}\right)=-\nabla\cdot w \tag{2.58}$$

$$=\nabla\cdot(u\cdot\tau_{ij}) \tag{2.59}$$

The expression in Eq. 2.59 can be decomposed as follows:

$$\nabla\cdot(u\cdot\tau_{ij})=u\cdot(\nabla\cdot\tau_{ij})+\tau_{ij}\frac{\partial u_i}{\partial x_j} \tag{2.60}$$

Furthermore, the equation of motion (Eq. 2.45) can be expressed in terms of the stress tensor in vector notation as follows:

$$\rho\frac{Du}{Dt}=\rho g+\nabla\cdot\tau_{ij} \tag{2.61}$$

The first term on the RHS of Eq. 2.60 can be rewritten through combination with Eq. 2.61 to yield:

$$u\cdot(\nabla\cdot\tau_{ij})=u\cdot\left(\rho(\frac{Du}{Dt}-g)\right)$$

$$=\rho\left(U\frac{DU}{Dt}-g\cdot u\right) \tag{2.62}$$

which corresponds to the terms for kinetic and potential energy in Eq. 2.49. Substitution of Eqs. 2.49, 2.54, 2.60 and 2.62 into Eq. 2.48 leads to:

$$\rho\left(\frac{De}{Dt}+U\frac{DU}{Dt}-g\cdot u\right)=\nabla\cdot(\kappa\nabla T)+\rho\left(U\frac{DU}{Dt}-g\cdot u\right)+\tau_{ij}\frac{\partial u_i}{\partial x_j}$$

$$\rho\left(\frac{De}{Dt}\right)=\nabla\cdot(\kappa\nabla T)+\tau_{ij}\frac{\partial u_i}{\partial x_j} \tag{2.63}$$

Equation 2.63 represents the first law of thermodynamics for fluid motion in terms of the internal energy e. Alternatively, it is possible to formulate the conservation of energy in terms of enthalpy. To this end, let us split the stress tensor τ_{ij}, defined in Eq. 2.42, into a pressure and a viscous term as follows:

$$\tau_{ij}\frac{\partial u_i}{\partial x_j}=-p\frac{\partial u_i}{\partial x_i}+\tau'_{ij}\frac{\partial u_i}{\partial x_j}$$

$$=-p\nabla\cdot u+\tau'_{ij}\frac{\partial u_i}{\partial x_j} \tag{2.64}$$

where $\tau'_{ij} = \tau_{ij} + p\delta_{ij}$. Furthermore, the mass continuity equation in Eq. 2.14 can be rearranged to give:

$$\frac{\partial \rho}{\partial t} + u_j \frac{\partial \rho}{\partial x_j} + \rho \frac{\partial u_j}{\partial x_j} = 0$$
$$\frac{D\rho}{Dt} + \rho \nabla \cdot u = 0 \qquad (2.65)$$

Multiplication of Eq. 2.65 with p and division by ρ produces the following expression:

$$p\nabla \cdot u = -\frac{p}{\rho}\frac{D\rho}{Dt} = \rho \frac{D}{Dt}\left(\frac{p}{\rho}\right) - \frac{Dp}{Dt} \qquad (2.66)$$

which combined with Eqs. 2.63 and 2.64 yields:

$$\rho \frac{D}{Dt}\left(e + \frac{p}{\rho}\right) = \frac{Dp}{Dt} + \nabla \cdot (\kappa \nabla T) + \tau'_{ij}\frac{\partial u_i}{\partial x_j} \qquad (2.67)$$

Furthermore, substitution of the definition of enthalpy,

$$h = e + \frac{p}{v} \qquad (2.68)$$

and the dissipation function Φ,

$$\Phi = \tau'_{ij}\frac{\partial u_i}{\partial x_j} \qquad (2.69)$$

into Eq. 2.67 yields the energy equation in terms of enthalpy (cp. [Whi06]).

$$\rho \frac{Dh}{Dt} = \frac{Dp}{Dt} + \nabla \cdot (\kappa \nabla T) + \Phi \qquad (2.70)$$

2.6 Ideal gas

The term *ideal gas* is used to describe a gas with the following characteristics:

- The gas atoms or molecules can be represented by hard spheres.
- The collisions between the gas particles are perfectly elastic. Collisions do not lead to energy losses.
- The gas particles move in random directions. The velocity of the gas particles can be described by a velocity distribution function.
- Intermolecular attractive forces can be neglected.

- The mean distance between the gas particles is much larger than the characteristic size of the particles.

The classic form of the *ideal gas law* describes the relation between the thermodynamic properties pressure, temperature, volume and the amount of substance of an ideal gas and is obtained from the combination of several empirical laws and observations. The first of these laws, known as *Boyle's law*, describes the pressure increase of a gas as a result of a volume reduction and is named after the scientific Robert Boyle, who published it in 1662. Edme Mariotte described, independently of Boyd, the same law in 1679. Hence, the law is also known as the *Boyle-Mariotte law*. This relation can be expressed as:

$$p \propto \frac{1}{V} \tag{2.71}$$

where p stands for the gas pressure and V, for its absolute volume. A second empirical law, known as *Charles' law*, describes the tendency of gases to expand when heated. It was postulated by Jacques Charles in the 1780s and can be written as:

$$V \propto T \tag{2.72}$$

where T represents the gas temperature. In 1811 Amedeo Avogadro postulated the hypothesis that two gas samples with the same values of temperature, pressure and volume contain the same amount of matter. The resulting *Avogadro's law* can be formulated as follows:

$$V \propto n \tag{2.73}$$

with n representing the amount of matter (moles) of gas. These empirical observations can be combined to obtain the *ideal gas law*:

$$pV = nR_uT \tag{2.74}$$

The proportionality constant $R_u = 8.314$ J$\cdot$K$^{-1}\cdot$ mol^{-1} is known as the universal gas constant and its value has been estimated from experimental data. Based on the definition of the amount of substance $n = m/M$, where m stands for the mass of gas in the system and M represents its molar mass, the ideal gas law can also be written in the following form:

$$pV = mRT \tag{2.75}$$

where $R = R_u/M$ stands for the specific gas constant. Substitution of the definition of density $\rho = m/V$ into Eq. 2.75 yields another useful form of the ideal gas law:

$$\frac{p}{\rho} = RT \tag{2.76}$$

Alternatively, the ideal gas law can be formulated in terms of the Boltzmann constant $k_B = 1.3806 \times 10^{-23}$J·K^{-1} by taking advantage of the definition $R_u = k_B N_A$, where $N_A = 6.0221 \times 10^{23}$ mol^{-1} stands for Avogadro's number:

$$\begin{aligned} pV &= nk_B N_A T \\ &= \frac{N}{N_A} k_B N_A T \\ &= Nk_B T \\ \rightarrow p &= n_n k_B T \end{aligned} \tag{2.77}$$

Here, $N = nN_A$ represents the total number of gas particles (atoms or molecules) in the system and $n_n = N/V$, the number density. The ideal gas law in Eq. 2.77 is particularly useful in the field of gas kinetics. Careful analysis of the conditions necessary for a gas to be described as *ideal* leads to some interesting conclusions regarding the applicability of this theory. At high temperatures, the intermolecular forces between gas particles are small in comparison to their kinetic energy. Furthermore, low pressure (or density) values result in a large distance between the individual gas particles when compared to their physical dimensions. Thus, the assumption of an ideal gas works well for high temperatures and low pressures. For low temperatures and high pressure values, deviations from the ideal gas behaviour are observed. These deviations become stronger as the gas approaches the liquid state (cp. [Win08]). In these cases, the ideal gas law is no longer accurate and alternative models which take into account the effects of molecular interaction forces and molecular size must be used. Examples of these models include the *van der Waals equation of state* and the *Virial equation of state*.

2.7 The Laval nozzle

A *Laval nozzle* is a device consisting of a convergent and a divergent tube pipe or cylinder with the ability to accelerate a gas to supersonic speeds and commonly used in rocket and supersonic jet engines. Through the convergent part of the nozzle, the gas accelerates as a result of a cross section reduction so that the speed of sound is reached at the throat of the nozzle. The cross section increase in the divergent part of the nozzle leads to a further increase of the gas velocity. This way, supersonic velocities at the nozzle exit can be achieved. The energy of the propellant gas, closely related to its pressure and temperature, is transformed into kinetic energy along the nozzle. A Laval nozzle is shown schematically in Fig. 2.6, where Ma stands for the Mach number, A^* for the nozzle throat area and the subscripts $_0$ and $_e$ denote inlet and exit properties respectively. The working principle as well as several key relations for the description of

Laval nozzle flows are presented in this section.

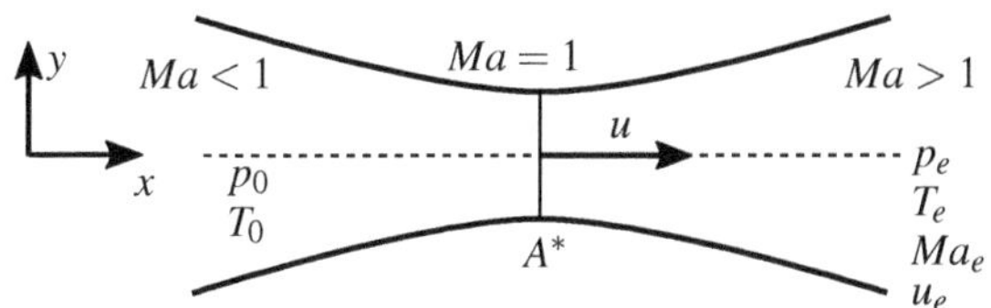

Figure 2.6: Schematic representation of a Laval Nozzle.

Analytic expressions for nozzle flows can be obtained from the Navier-Stokes and continuity equations using the following assumptions:

- The flow is stationary: $\frac{\partial}{\partial t} = 0$.
- The flow can be assumed to be nearly one-dimensional: $u_x >> u_y$.
- The expansion process through the nozzle is isentropic: $\mathrm{d}s = 0$.
- The flow velocity in the main flow direction (x-axis in Fig. 2.6) is constant along the nozzle's radial axis: $\frac{\partial u_x}{\partial y} = 0$.
- The fluid viscosity is negligible: $\mu \approx 0$.

Neglecting the gravitation term, the Navier-Stokes equations (Eq. 2.44) can be reformulated for nozzle flows as follows:

$$\rho u \frac{\partial u}{\partial x} = -\frac{\partial p}{\partial x} \tag{2.78}$$

where u_x has been replaced by u for simplicity. Because of the stationarity and one-dimensionality assumptions, Eq. 2.78 can be written as:

$$\rho u \mathrm{d}u = -\mathrm{d}p \tag{2.79}$$

Furthermore, the speed of sound a, in general given by the following equation:

$$a^2 = \frac{\partial p}{\partial \rho} \tag{2.80}$$

can be calculated for an ideal gas using the much simpler expression:

$$a^2 = \gamma \frac{p}{\rho} \tag{2.81}$$

$$= \gamma \frac{R_u T}{M} \tag{2.82}$$

where γ stands for the heat capacity ratio. Multiplication of the pressure term in Eq. 2.79 with $\mathrm{d}\rho/\mathrm{d}\rho$, subsequent combination with the definition of the speed of sound for an ideal gas and division of the resulting expression by ρu^2 yields:

$$\frac{1}{u}\mathrm{d}u + \left(\frac{a^2}{u^2}\right)\left(\frac{1}{\rho}\right)\mathrm{d}\rho = 0$$
$$\frac{1}{u}\mathrm{d}u + \left(\frac{1}{Ma^2}\right)\left(\frac{1}{\rho}\right)\mathrm{d}\rho = 0 \tag{2.83}$$

which can be rewritten to give:

$$\frac{Ma^2}{u}\mathrm{d}u = -\frac{1}{\rho}\mathrm{d}\rho \tag{2.84}$$

where $Ma = u/a$ represents the *Mach number*. Now let us consider mass conservation for nozzle flows. Under the assumptions above, mass continuity can be described by the simple expression $\rho u A = \text{const.}$, with A representing the cross section of the nozzle. This expression can be further developed as follows:

$$\begin{aligned} \text{const} &= \ln(\rho u A) \\ &= \ln\frac{\rho}{\rho_0} + \ln\frac{u}{u_0} + \ln\frac{A}{A_0} \end{aligned} \tag{2.85}$$

Taking the derivative of Eq. 2.85 results in:

$$\begin{aligned} 0 &= \mathrm{d}\left(\ln\frac{\rho}{\rho_0}\right) + \mathrm{d}\left(\ln\frac{u}{u_0}\right) + \mathrm{d}\left(\ln\frac{A}{A_0}\right) \\ &= \frac{1}{\rho}\mathrm{d}\rho + \frac{1}{u}\mathrm{d}u + \frac{1}{A}\mathrm{d}A \\ \rightarrow -\frac{1}{\rho}\mathrm{d}\rho &= \frac{1}{u}\mathrm{d}u + \frac{1}{A}\mathrm{d}A \end{aligned} \tag{2.86}$$

Substitution of Eq. 2.86 into Eq. 2.84 yields:

$$\begin{aligned} 0 &= (1 - Ma^2)\frac{1}{u}\mathrm{d}u + \frac{1}{A}\mathrm{d}A \\ &= (1 - Ma^2)\mathrm{d}\left(\ln\frac{u}{u_0}\right) + \mathrm{d}\left(\ln\frac{A}{A_0}\right) \\ &= (Ma^2 - 1)\mathrm{d}\left(\ln\frac{u}{u_0}\right) - \mathrm{d}\left(\ln\frac{A}{A_0}\right) \end{aligned} \tag{2.87}$$

Thus, Eq. 2.87 is satisfied by the expression:

$$\begin{aligned} \frac{(u/u_0)^{Ma^2-1}}{A/A_0} &= \text{const.} \\ \rightarrow u^{Ma^2-1} &\propto A \end{aligned} \tag{2.88}$$

The relation in Eq. 2.88 is useful for the description of the working principle of the Laval nozzle. In the convergent part of the device, the Mach number lies always below one. As evident from Eq. 2.88, this leads to an inverse proportionality relation between the flow velocity u and the cross-sectional area A. Hence, the reduction of A towards the nozzle throat results in an increase of velocity and a drop in pressure. This phenomenon is known as the *Venturi effect*. When the flow velocity in the nozzle throat equals the local speed of sound, the flow reaches a limiting state known as *chocked flow*. In this case, a pressure decrease downstream of the nozzle throat will not result in higher upstream flow velocities. As can be inferred from Eq. 2.88, Mach numbers higher than one result in a positive proportionality between the flow velocity and cross-sectional area. As a consequence, an increase of A in the region downstream of the nozzle throat, a section also know as the nozzle's *diffuser*, leads to further acceleration of the working gas to supersonic speeds, provided that a chocked flow state has already been reached in the nozzle throat and the diffuser is free of shock waves.

In an ideal Laval nozzle, flow properties such as temperature, pressure and density obey the isentropic functions. These functions are based on the energy equation for a **steady**, **adiabatic**, **inviscid** flow involving a **calorically perfect** gas. In the following, the relevant isentropic functions are presented. Their detailed derivation can be found in [AJ91]. In order to examine the isentropic functions, let us introduce the concept of *total enthalpy* mathematically defined as follows:

$$h_0 = h + \frac{u^2}{2} \tag{2.89}$$

In an isentropic process involving a flow, the total enthalpy h_0 is constant along any streamline. If a flow is isentropically decelerated until its velocity reaches zero (known as the flow's *stagnation point*), its kinetic energy is completely converted into internal energy. This is, in turn, associated with an increase of the local static temperature T along the streamline. The temperature value reached at the stagnation point is called *total temperature* T_0 and its value is, as for h_0, constant along any streamline inside the flow field. The relation between temperature and Mach number in an ideal Laval nozzle flow is given by the following isentropic function:

$$\frac{T_0}{T} = 1 + \frac{\gamma - 1}{2} Ma^2 \tag{2.90}$$

Here, $\gamma = c_p/c_v$ stands for the heat capacity ratio and c_p and c_v represent the heat capacity at constant pressure and constant volume respectively. Eq. 2.90 can be used to calculate the temperature T along any streamline as a function of its total temperature T_0 and the local Mach number. A similar isentropic relation can be formulated between the total pressure p_0, the local pressure p and the local Mach number:

$$\frac{p_0}{p} = \left(\frac{T_0}{T}\right)^{\gamma/(\gamma-1)} = \left(1 + \frac{\gamma-1}{2}Ma^2\right)^{\gamma/(\gamma-1)} \tag{2.91}$$

Moreover, the ratio of total density ρ_0 to local density ρ as a function of Ma can be calculated as follows:

$$\frac{\rho_0}{\rho} = \left(\frac{T_0}{T}\right)^{1/(\gamma-1)} = \left(1 + \frac{\gamma-1}{2}Ma^2\right)^{1/(\gamma-1)} \tag{2.92}$$

An additional useful relation for the description of the ideal nozzle flow links the local Mach number to the nozzle geometry (s. [AJ91]):

$$\left(\frac{A}{A^*}\right)^2 = \frac{1}{Ma^2}\left[\frac{2}{\gamma+1}\left(1 + \frac{\gamma-1}{2}Ma^2\right)\right]^{(\gamma+1)/(\gamma-1)} \tag{2.93}$$

In Eq. 2.93, A represents the local cross-sectional area of the nozzle and A^*, the narrowest cross-sectional area of the nozzle, i.e., the area of the nozzle throat. One interesting consequence of Eq. 2.93 is that the Mach number for an **ideal nozzle flow** depends only on the ratio of local cross-sectional area to throat area.

Recall that the total properties T_0, p_0, ρ_0 are *stagnation properties*. As a consequence, in the case of a nozzle flow, they can be considered to represent the fluid properties in a large reservoir where the gas can be assumed to be nearly at rest (e.g. *propellant tank*). Furthermore, due to the relatively low flow velocities, the values of the stagnation properties can also be interpreted as the flow properties at the nozzle inlet. Another important aspect of Eqs. 2.90 to 2.93 is that, because of the quadratic nature of Ma, they are satisfied by two different values of the Mach number. The first one, $Ma < 1$, corresponds to the subsonic solution in the convergent section of the nozzle. The second value, $Ma > 1$, describes the supersonic solution in the diffuser. Fig. 2.7 shows, schematically, the Mach number, pressure and temperature profiles along the longitudinal axis (x-axis) of an ideal Laval nozzle. Here, the Ma profile is obtained from Eq. 2.93 as a function of the ratio of cross-sectional area $A(x)/A^*$, i.e., of the nozzle geometry, with the solution $Ma < 1$ applying for the convergent portion of the nozzle and $Ma > 1$, for the divergent one. The pressure ratio p/p_0 and the temperature ratio T/T_0 are subsequently computed from the corresponding isentropic functions based on the calculated local Mach numbers. As evident from Fig. 2.7, the energy (pressure and temperature) of the gas at the inlet of the nozzle is converted into kinetic energy along the nozzle, with the flow reaching supersonic velocities at the nozzle's exit.

According to Eqs. 2.90 to 2.93, the profiles of Ma, T, p and ρ depend only on the nozzle geometry. However, an ideal nozzle flow can only exist under specific pressure conditions. Clearly, the propellant gas will only flow trough the nozzle if the inlet pressure p_0

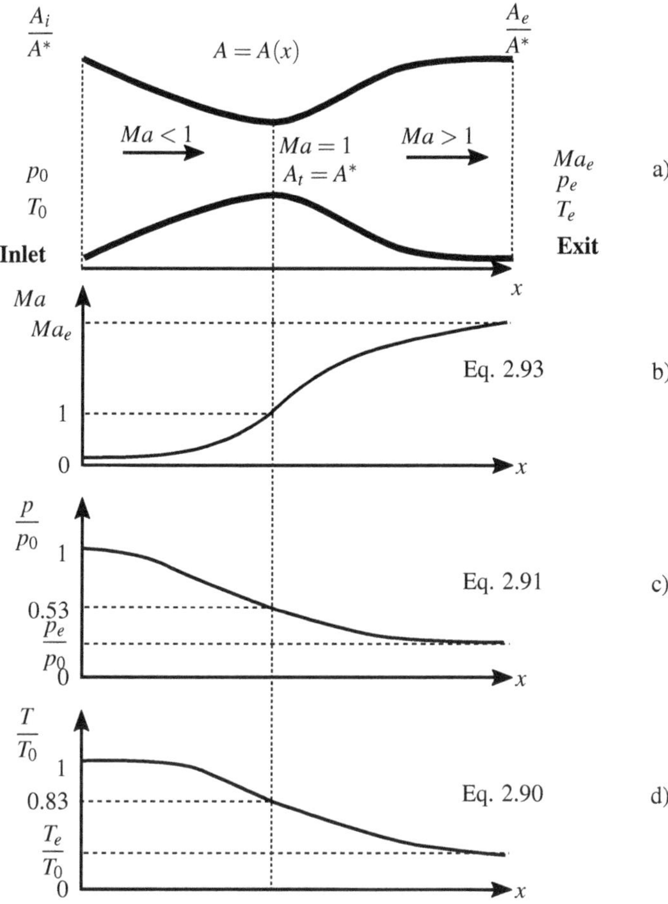

Figure 2.7: Isentropic nozzle flow [AJ91].

is higher than the pressure at the nozzle exit p_e. In addition, if the pressure difference is not high enough, then $Ma = 1$ cannot be reached at the throat and supersonic conditions will not be achieved at the nozzle exit. In fact, for the nozzle flow to behave as depicted in Fig. 2.7 the pressure ratio p_0/p_e must exactly match the value predicted by Eq. 2.91 with $Ma = Ma_e$ obtained from Eq. 2.93. Otherwise, the ideal and actual flow profiles deviate from one another (cp. [AJ91]).

Let us briefly discuss nozzle flows for which the pressure ratio p_e/p_0 does not exactly

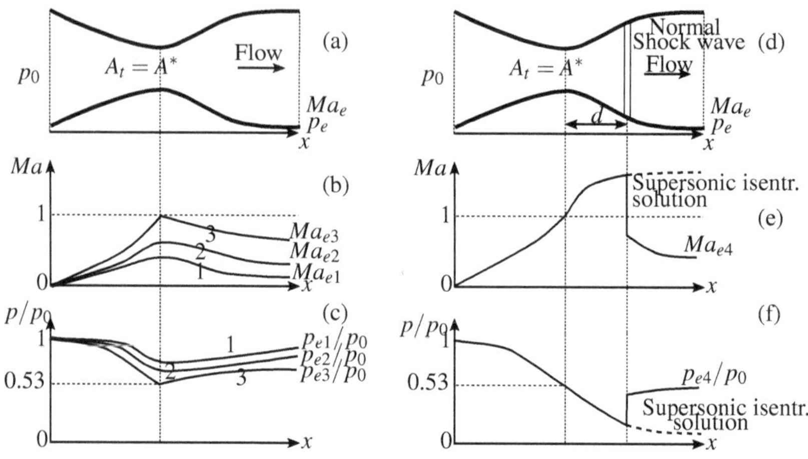

Figure 2.8: Nozzle flow conditions [AJ91].

match the ideal, isentropic value obtained from Eq. 2.91. To this end, let $p_{e,ideal}$ represent the ideal exit pressure for which the flow profiles match the isentropic estimations in Fig. 2.7. If the pressure gradient between nozzle inlet and exit equals zero ($p_e = p_0$), a nozzle flow does not appear. A small reduction of the pressure at the exit of the nozzle to the value $p_e = p_{e,1}$ will produce a low-speed flow from the nozzle inlet towards its exit. Since $p_{e,1}$ is much higher than $p_{e,ideal}$, the flow is still far from reaching sonic conditions at the nozzle throat. Furthermore, because of the relation in Eq. 2.88, the flow experiences a deceleration in the diffuser and Ma drops. This case is illustrated by curve 1 in Fig. 2.8 (b) and (c). A further reduction of the exit pressure to the value $p_e = p_{e,2}$ results in an increase of velocity and Mach number in the nozzle compared to the profiles for $p_e = p_{e,1}$. However, as can be seen from curve 2 in Fig. 2.8 (b) and (c), the Mach number through the nozzle still lies below one and the flow remains subsonic. If the exit pressure is further reduced to the value $p_e = p_{e,3}$, the pressure difference is high enough for $Ma = 1$ to be reached at the nozzle throat. However, the flow downstream of the throat remains subsonic. This case is represented by curve 3 in Fig. 2.8 (b) and (c). From this point on, information like flow disturbances and pressure waves can no longer travel from the diffuser and nozzle exit upstream of the throat, a condition known as *chocked flow*. Thus, the profiles in the convergent section of the nozzle shown in curve 3 of Fig. 2.8 (b) and (c) remain unchanged upon further reduction of p_e.

Let us further reduce the exit pressure to the value $p_{e,4}$, with $p_{e,3} > p_{e,4} > p_{e,ideal}$. In this case, the pressure at the nozzle exit is low enough for a supersonic flow to appear inside the diffuser. However, its value is still too large for the flow to remain supersonic throughout the entire diffuser and a normal shock wave appears at a distance d from the nozzle throat. The flows downstream of the shock wave and upstream of it are both isentropic. However, the entropy increases across the shock wave. Moreover, an abrupt increase of the flow pressure as well as an abrupt drop of velocity and Mach number can be observed across the shock wave and the flow becomes subsonic. This case is illustrated in Fig. 2.8 (d) to (f).

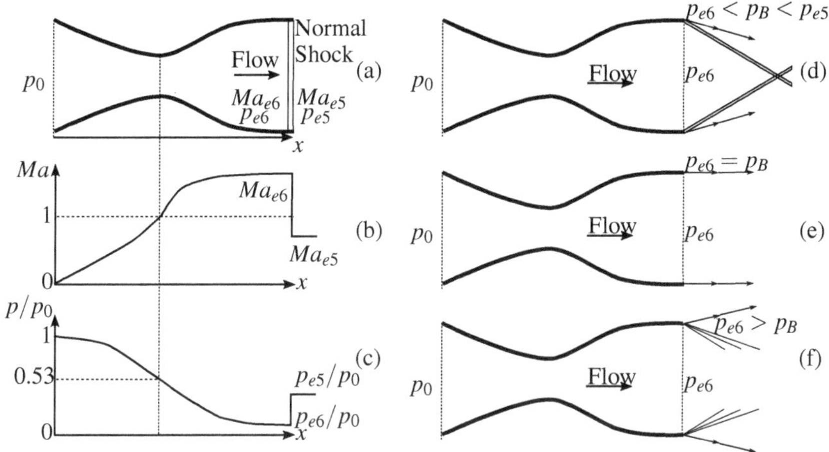

Figure 2.9: Supersonic nozzle flows [AJ91].

As p_e is further reduced, the shock wave moves towards the nozzle exit. At a certain pressure value $p_e = p_{e,5} > p_{e,ideal}$ the shock wave stands precisely at the nozzle exit. In this case, the typical effects of the shock wave manifest exactly at the interface between the nozzle and its environment and the flow entering the surroundings of the nozzle is subsonic. This case is depicted in Fig. 2.9 (a) to (c). Let p_B represent the pressure of the gas surrounding the nozzle (*back pressure*). Note that for a subsonic gas exiting the nozzle, the back pressure equals the exit pressure, i.e., $p_e = p_B$. However, since information cannot propagate upstream of a supersonic flow, this does not necessarily apply if $Ma_e \geq 1$. Now let us further reduce the pressure at the exit so that the value $p_e = p_{e,ideal}$ is reached. In order to examine the influence of p_B, let us consider the case $p_{e,ideal} < p_B < p_{e,5}$. Since $p_{e,ideal} < p_{e,5}$, shock waves are no longer present at the nozzle exit. However, the back pressure p_B is still higher than the ideal value. In order for the flow pressure to reach the back pressure, the flow must be compressed. This process takes place across oblique shock waves attached to the nozzle exit. This case is known as *overexpanded nozzle* and is shown in Fig. 2.9 (d). If p_B is further reduced so that it exactly matches the ideal exit pressure, there is no pressure discrepancy between the flow and the surroundings and the flow exits the nozzle precisely in the direction of the x axis. This corresponds to the ideal case depicted in Fig. 2.9 (e) and whose Mach number, pressure and temperature distributions are shown in Fig. 2.7. If the back pressure is further reduced, i.e., $p_B < p_{e,ideal}$, the nozzle flow must undergo further expansion across *centred expansion waves* attached to the nozzle in order for p_e to match p_B. This case, known as *underexpanded nozzle* is depicted in 2.9 (f). From Fig. 2.9 (d) to (f) it is evident that the non-x components of the exit velocity vector cancel each other out for both overexpanded and underexpanded flows which has a negative impact on the efficiency of the nozzle.

Note that the relations presented in this section can be used to obtain a reasonable prediction of the flow behaviour inside a Laval nozzle. In reality, the nozzle flow is influenced by many additional factors including its three-dimensionality, gas viscosity, the precise nozzle geometry and the characteristics of the nozzle walls [AJ91].

2.8 Fundamentals of plasma

A macroscopically electric neutral medium containing many free charge carriers interacting with one another is known as a *plasma*. In addition to *solid*, *liquid* and *gas*, *plasma* constitutes one of the states of matter. The different states of matter may be classified according to the magnitude of the interaction forces between the atoms or molecules of the substance. The magnitude of these forces depends on the thermal kinetic energy of the individual particles. In the solid state, the intermolecular attractive forces are high. The particles of the substance have essentially no movement freedom and the distance between them is very small. Through heating, the ratio of intermolecular forces to thermal kinetic energy can be decreased. If the kinetic energy of the atoms or molecules is high enough, they become able to move much more freely while still sticking together as a result of cohesive forces. This state of matter is known as liquid. Through further energy addition, the particles reach a point where they move freely and with very high velocities. In this state, known as gas, the distance between the particles is large and intermolecular forces are almost negligible. The transitions between individual states, known as *phase transitions*, take place at substance specific values of pressure and temperature [Bit04].

If enough energy is added to a molecular gas, its individual atoms begin to dissociate through molecular collisions and an atomic gas is produced. If this atomic gas is further heated, some of the atoms may acquire enough thermal kinetic energy to overcome, through collisions, the binding energy of the outermost electrons of the atoms. The substance, now comprised of free electrons, ions and neutral atoms is called a *plasma*. The transition from gas to plasma is a gradual, temperature-dependent process. Hence, it does not fall under the definition of a classic thermodynamic phase transition [Bit04].

A plasma can be created through heating of a gas. In thermodynamic equilibrium, the gas temperature and the ionisation degree are closely related to one another. The exact relation is given by the *Saha-Equation* and will be discussed in detail in Section 2.8.1.4. Plasmas in thermodynamic equilibrium can be found frequently in nature. They are, however, not as common in the laboratory. Two of the most common processes for plasma production are *photoionisation* and *gas discharge*.

In the *photoionisation* process, a plasma is produced by exposure of a gas to photons whose energy is equal to or higher than the ionisation energy of the gas atoms. The ionisation energy of atomic oxygen is, for instance, 13.6 eV. In this case, the amount of energy required to remove the valence electron can be delivered by radiation with a wave length equal to or lower than 91 nm, a value that falls in the range of ultraviolet radiation. An example of a natural plasma produced by photoionisation is the earth ionosphere, which is the consequence of the interaction between molecular and atomic gas in the earth atmosphere and ultraviolet and x-radiation from the sun [Bit04].

In the *gas discharge* process, an electric voltage is applied to a partially ionised gas. The resulting electric field accelerates the free electrons in the plasma until they reach kinetic energy levels equal to or higher than the binding energy of the valence electrons. When a collision between these fast electrons and neutral atoms occurs, valence electrons are released increasing this way the number density of electrons and ions in the plasma and thus, the ionisation degree. One important aspect of this process is that the energy from the applied electric field is transmitted to the light electrons much faster than to the heavy ions. As a consequence, the velocity of the electrons, their thermal kinetic energy and therefore their temperature are often higher than that of the ions [Bit04].

2.8.1 Physical properties of plasma

2.8.1.1 Macroscopic neutrality

A plasma in thermodynamic equilibrium and not influenced by external forces can be assumed to be macroscopic neutral. Macroscopic neutrality applies for a plasma volume large enough for it to contain a large number of charge carriers. At the same time, this plasma volume is smaller than the characteristic length of the system over which gradients of properties like temperature and density are to be expected. Macroscopic neutrality is a direct consequence of the potential energy associated with imbalances in the distribution of charge carriers. The potential energy, in turn, is related to the Coulomb forces appearing if neutrality is not maintained and can be extremely high compared to the thermal kinetic energy of the charge carriers. Let us consider a plasma with a charged particle number density of 10^{20} m^{-3} and an imbalance between the ion n_i and the electron number density n_e of only 1%. The plasma is confined in a spheric volume of radius $r = 10^{-3}$ m. The net charge inside the sphere is given by the equation:

$$q = \frac{4}{3}\pi r^3 (n_i - n_e)e \tag{2.94}$$

where e stands for the elementary charge. Furthermore, the electric potential on the surface of the sphere can be calculated as follows:

$$\phi = \frac{1}{4\pi\varepsilon_0}\frac{q}{r} = \frac{er^2}{3\varepsilon_0}(n_i - n_e) \tag{2.95}$$

with $\varepsilon_0 = 8.854 \times 10^{-12}$ F/m representing the vacuum permittivity. Substitution of the known values into Eq. 2.95 yields a voltage on the sphere surface of approx. 6,000 V. Using the relation $k_B T = 1$ eV$= 1.602 \times 10^{-19}$ J, we can compute the temperature necessary to balance the potential energy:

$$\begin{aligned} k_B T &= 6{,}000\,\text{eV} = 6{,}000 \cdot 1.602 \times 10^{-19}\,\text{J} \\ T &= \frac{6{,}000 \cdot 1.602 \times 10^{-19}\,\text{J}}{1.38 \times 10^{-23}\,\text{J/K}} \approx 6.95 \times 10^7\,\text{K} \end{aligned} \tag{2.96}$$

Hence, a temperature of 6.95×10^7 K is required for the potential and kinetic energy of a plasma with an electric potential of 6,000 V to be balanced. Deviations from the macroscopic neutrality without the influence of external forces are allowed to occur naturally over distances for which the balance between potential and thermal kinetic energy can be guaranteed. This distance is of the order of magnitude of a plasma parameter called *Debye length*. For deviations over larger distances, the resulting Coulomb forces become too strong and the free charges, in particular the light electrons, quickly move towards plasma regions with excess of ions so that macroscopic neutrality is restored [Bit04].

2.8.1.2 Debye shielding and Debye length

The distance over which the electric field produced by a charge carrier influences other charges is called *Debye length* λ_D. In order to define the Debye length, let us examine a general charge carrier inside a plasma. Because of the electric attractive and repulsive forces, the observed charge will have in its surroundings more particles of the opposite charge compared to particles of the same charge. The electrostatic potential and the associated electric field generated by the observed particle is therefore, at a given distance, effectively shielded by the charge carriers in its surroundings. The specific distance at which this occurs corresponds to the Debye length λ_D. For an electrolyte, λ_D is given by the equation:

$$\lambda_D = \left(\frac{\varepsilon_0 k_B T}{n_e e^2}\right)^{(1/2)} \tag{2.97}$$

Consider the test particle Q with positive charge. By using a spherical coordinate system whose origin is located at the position of Q, one can develop an expression for the actual electrostatic potential in the surroundings of the test charge. This equation takes into account the electrostatic potential created by both the test particle Q and the particles in its surroundings and is given by:

$$\phi(r) = \phi_c(r) \exp\left(-\frac{\sqrt{2}r}{\lambda_D}\right) \tag{2.98}$$

where $\phi_c(r)$ corresponds to the electrostatic Coulomb potential, i.e., the potential created by the test charge Q and calculated as follows:

$$\phi_c(r) = \frac{1}{4\pi\varepsilon_0}\frac{Q}{r} \tag{2.99}$$

The comparison between the Debye potential and the Coulomb potential is shown in Fig. 2.10 as a function of the distance from the position of the test charge Q. Here, the y axis is normalized by the Coulomb potential at the position $r = \lambda_D$ and the x axis, by the Debye length λ_D. As evident from Fig. 2.10, the charge distribution in the surroundings of the test charge Q leads to a considerable deviation between the actual and the theoretical Coulomb potential. The potential field created by the positive charge Q is gradually shielded by the excess of negative charges around it. For distances from Q greater than λ_D, the actual electrostatic potential is a lot weaker than the theoretical one. Charge carriers located at distances greater than λ_D are therefore no longer influenced by Q. As a consequence, it can be assumed that a given charge can only interact with other charges inside a sphere with radius $r = \lambda_D$ and whose centre is located at the charge's position. This is known as the *Debye sphere*. Outside of the sphere, the influence of Q on other charges can be neglected. Considering the above, the following criteria regarding the Debye length must be met for a collection of charge carriers to constitute a plasma:

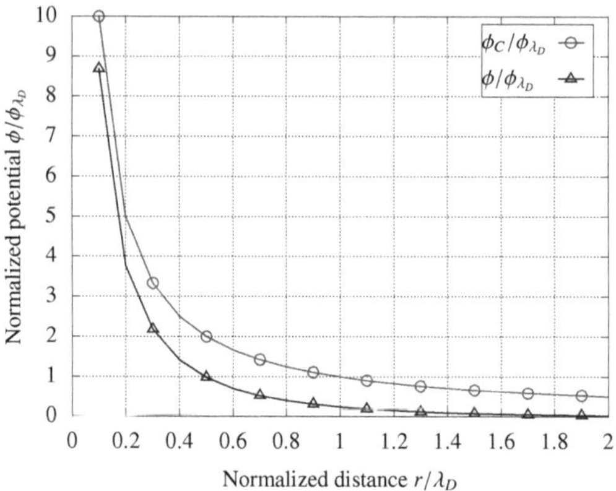

Figure 2.10: Coulomb potential ϕ_C and Debye potential ϕ as a function of the distance r (cp. [Bit04]).

- The geometrical dimensions of the system of particles must be larger than the Debye length. Otherwise, the condition of macroscopic neutrality cannot be fulfilled and the particles collection will not exhibit typical plasma behaviour. Let L represent the characteristic length of the system. The *first criterion* can therefore be expressed as:

$$L >> \lambda_D \tag{2.100}$$

- A charge Q can only influence charges located inside its Debye sphere. Conversely, Q is only influenced by charges inside the Debye sphere and interactions with particles outside it can be neglected. For this condition to apply, a high enough number of particles must be present inside the Debye sphere. Let n_e represent the electron number density. The *second criterion* can then be formulated as follows:

$$n_e \lambda_D^3 >> 1 \tag{2.101}$$

Equation 2.101 implies that the mean distance between the electrons, approximately $n_e^{-1/3}$, must be very small compared to the Debye length λ_D. Eq. 2.101 can be rearranged to give:

$$1 >> \frac{1}{n_e \lambda_D^3} \qquad \text{or} \qquad 1 >> g \tag{2.102}$$

where g is known as the *plasma parameter* and the condition $g << 1$, as the *plasma approximation*.

- The principle of macroscopic neutrality can be derived from Eq. 2.100. Nevertheless, this is sometimes considered as a *third criterion* for the existence of a plasma. Mathematically:

$$n_e = \sum_i n_i \tag{2.103}$$

where n_i represents the number density of positive charge carriers in the system [Bit04].

2.8.1.3 Plasma frequency

A small disturbance in the macroscopic neutrality can be induced by an external force acting on the plasma, leading to the appearance of a region with an excess and a region with a deficit of electrons. As shown schematically in Fig. 2.11, if the external force disappears, the electric forces resulting from the charge imbalance accelerate the electrons from the excess region towards the region with a deficit in an attempt to restore the equilibrium and the macroscopic neutrality.

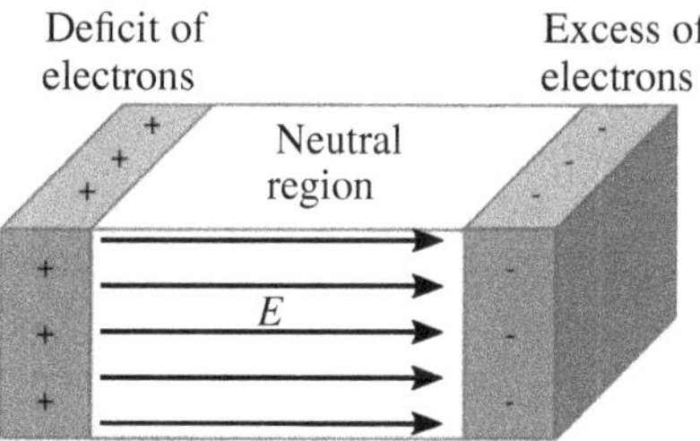

Figure 2.11: Electric field E as a result of the charge imbalance produced by an external force (cp. [Bit04]).

While the electrons are easily accelerated by the electric field, the same does not apply for the heavy ions. Hence, it can be assumed that the ions remain in roughly the same position after the external force disappears. Because of their inertia, the electrons move beyond the equilibrium position in their attempt to restore charge neutrality. The result is the appearance of a new electric field in the opposite direction. This process repeats itself over time and macroscopic neutrality is maintained on a statistical average. Hence, the electrons effectively oscillate around the heavy ions. The oscillation frequency of the electrons, also known as *plasma frequency* ω_{pe}, is given by the equation:

$$\omega_{pe} = \left(\frac{n_e e^2}{m_e \varepsilon_0} \right)^{1/2} \tag{2.104}$$

where m_e stands for the electron mass. During the oscillation process, the electrons collide with neutral particles leading to a loss of kinetic energy and hence, to a damping of the oscillations. For the oscillations to be only slightly damped, the collision frequency ν_{en} must be small when compared to the plasma frequency:

$$\frac{\omega_{pe}}{2\pi} = \nu_{pe} > \nu_{en} \tag{2.105}$$

If the condition in Eq. 2.105 is not fulfilled, the collisions between electrons and neutral particles result in a too strong deceleration of the electrons. In this case, the electrons reach an equilibrium with the neutral particles contained in the plasma and cannot behave in an independent way. Hence, the system does not exhibit typical plasma behaviour and can be treated as a neutral gas. The expression in Eq. 2.105 represents therefore the *fourth criterion* for the existence of a plasma, which can be alternatively formulated as follows:

$$\omega_{pe} \tau > 1 \tag{2.106}$$

where $\tau = 1/\nu_{en}$ stands for the mean time between electron-neutral collisions. According to Eq. 2.106, the mean time between collisions must be large compared to the characteristic time necessary for a change in the plasma properties to be observed. Otherwise, the particles collection will not exhibit typical plasma behaviour [Bit04].

2.8.1.4 Ionisation degree and electron temperature

The ionisation degree of a plasma in thermodynamic equilibrium is strongly dependent of the electron temperature. This relation is given by the *Saha equation*. Let n_a and n_b represent the number densities of two arbitrary species with energies U_a and U_b. Based on statistical mechanics, the ratio of n_a to n_b in thermodynamic equilibrium can be expressed as:

$$\frac{n_a}{n_b} = \frac{g_a}{g_b} \exp\left[-\frac{(U_a - U_b)}{k_B T} \right] \tag{2.107}$$

where g_a and g_b describe the statistical weights associated with the energy states U_a and U_b, i.e., the *degeneracy factors* which represent the number of quantum states with the energies U_a and U_b. If the examined substance has two energy levels U_a and U_b with $U_a > U_b$, the fraction of particles with the higher energy in the system α is given by the equation:

$$\alpha = \frac{n_a}{n_t} = \frac{n_a}{n_a + n_b} = \frac{n_a}{n_b}\left(\frac{n_a}{n_b} + 1\right)^{-1} \tag{2.108}$$

Substitution of Eq. 2.107 into Eq. 2.108 with $U = U_a - U_b$ yields:

$$\alpha = \frac{n_a}{n_b}\left(\frac{n_a}{n_b} + 1\right)^{-1} = \frac{g_a/g_b \exp(-U/k_BT)}{g_a/g_b \exp(-U/k_BT) + 1} \tag{2.109}$$

If the examined substance is a plasma, the energy state a may be assumed to be that of the ionised gas, i.e., the electron-ion pair. Furthermore, the state b corresponds to the ground neutral state and the difference $U = U_a - U_b$, to the ionisation energy. As an example, the ionisation energies of argon and helium are 15.76 eV and 24.59 eV respectively.

The degeneracy factors g_a and g_b in Eq. 2.109 can be determined based on quantum mechanical calculations. If both the small interaction potential between electrons and ions as well as the internal degrees of freedom are neglected, the ratio g_a to g_b is given by:

$$\frac{g_a}{g_b} = \left(\frac{2\pi m_e k_B T}{h^2}\right)^{3/2} \frac{1}{n_i} \tag{2.110}$$

where h stands for the *Planck constant* and n_i, for the number density of the ions. With the temperature T in Kelvin and n_i in m^{-3}, Eq. 2.110 can be simplified as follows:

$$\frac{g_a}{g_b} = 2.405 \times 10^{21}\, T^{3/2} \frac{1}{n_i} \tag{2.111}$$

Substitution of Eq. 2.111 into Eq. 2.107 yields the well-known *Saha equation*:

$$\frac{n_i}{n_n} = 2.405 \times 10^{21}\, T^{3/2} \frac{1}{n_i} \exp\left[-\frac{U}{k_B T}\right] \tag{2.112}$$

where the indexes a and b have been replaced by i for the ionised and n for the neutral state. Furthermore, the ionisation degree α can be expressed as a function of temperature and number density of ionised particles by substituting Eq. 2.111 into Eq. 2.109 yielding:

$$\alpha = \frac{2.405 \times 10^{21}\, T^{3/2} \frac{1}{n_i} \exp(-U/k_BT)}{2.405 \times 10^{21}\, T^{3/2} \frac{1}{n_i} \exp(-U/k_BT) + 1} \tag{2.113}$$

The ionisation degree α as a function of temperature is depicted in Fig. 2.12 for argon with $U = 15.76$ eV and the total number density $n_t = n_i + n_n = 10^{19}$ m^{-3}.

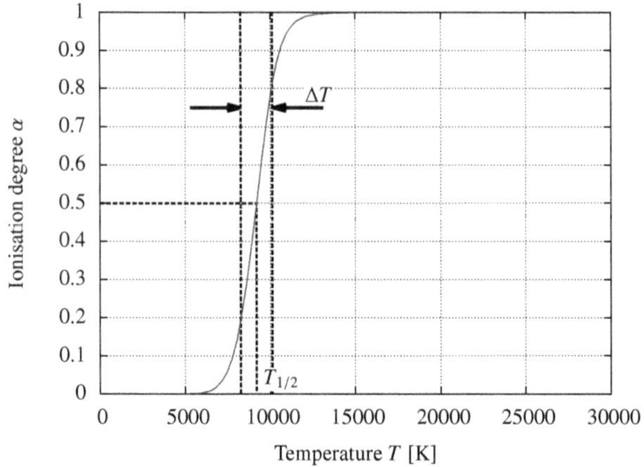

Figure 2.12: Ionisation degree α as a function of temperature T with $n_t = 10^{19}$ m^{-3} and $U = 15.76$ eV (cp. [Bit04]).

As can be seen from Fig. 2.12, the ionisation process takes place over a relatively small temperature range. In order to estimate this temperature range, let us define a straight line whose slope corresponds to that of the actual function for α at the position $\alpha = 1/2$ (s. Fig. 2.12). The slope of this line equals $1/\Delta T$ with ΔT corresponding to the difference $T_{\alpha=1} - T_{\alpha=0}$. On the other hand, the slope can be calculated by taking the derivative of Eq. 2.109 with respect to the temperature with $\alpha = 1/2$. Hence:

$$\frac{1}{\Delta T} = \left(\frac{\mathrm{d}\alpha(T)}{\mathrm{d}T}\right)_{T_{1/2}} = \left[\frac{U\alpha^2}{k_B T^2 (g_a/g_b) \exp(-U/k_B T)}\right]_{T_{1/2}}$$
$$= \frac{U}{4 k_B T_{1/2}^2 (g_a/g_b) \exp(-U k_B T_{1/2})} \tag{2.114}$$

On the other hand, if $\alpha = 1/2$, then $T = T_{1/2}$ and $n_i = n_n$. Substitution into Eq. 2.107 yields:

$$\frac{n_i}{n_n} = 1 = \frac{g_a}{g_b} \exp\left[-\frac{U}{k_B T_{1/2}}\right] \tag{2.115}$$

$$\rightarrow T_{1/2} = \frac{U}{k_B \ln(g_a/g_b)} \tag{2.116}$$

Finally, substitution of Eqs. 2.115 and 2.116 into Eq. 2.114 yields the expression for the estimation of the temperature range ΔT:

$$\frac{1}{\Delta T} = \frac{U}{4k_B T_{1/2}^2} = \frac{k_B[\ln(g_a/g_b)]^2}{4U}$$

$$\rightarrow \Delta T = \frac{4U}{k_B[\ln(g_a/g_b)]^2} \tag{2.117}$$

Note that the ratio g_a/g_b is, in reality, a function of the temperature (s. Eq. 2.110). For this reason, Eqs. 2.116 and 2.117 provide only estimates for $T_{1/2}$ and ΔT. From Eq. 2.117 it can be concluded that an increase of the ratio g_a/g_b leads to a reduction of the temperature interval ΔT in which most of the ionisation process takes place. Hence, the function $\alpha(T)$ becomes steeper. For $g_a >> g_b$, ionisation occurs over a very narrow temperature interval. In this case, a small increase of the gas temperature leads to a substantial increase of the ionisation degree. Since the ionised energy state is much more degenerate than the neutral ground state, g_a is always very large compared to g_b. Therefore, $\alpha(T)$ is usually a steep function [Bit04].

In order to calculate the value of $T_{1/2}$ in a more precise way, Eq. 2.113 can be solved numerically. For instance, with $n_i = \alpha n_t$, the values $n_t = 10^{19}$ m^{-3} and $\alpha = 0.5$ yield the temperature $T_{1/2} = 9,202$ K or 0.7933 eV for argon. The ratio g_a/g_b may now be determined from Eq. 2.111 with $T = T_{1/2} = 9,202$ K and the result can be substituted into Eq. 2.117 to obtain $\Delta T = 1,852$ K. According to this, the largest part of the ionisation process for argon with $n_t = 10^{19}$ m^{-3} takes place in the temperature range $8,276\text{K} < T < 10,129\text{K}$. This interval is depicted in Fig. 2.12.

The ionisation degree as a function of the temperature is depicted in Fig. 2.13 for different values of the total number density n_t. As can be seen from Fig. 2.13, n_t has a strong influence on the ionisation degree. At low number densities, it is possible to achieve relatively high values of α at temperatures far below the actual ionisation energy of the atomic gas (e.g. 15.76 eV for Argon). For a gas with a very low ionisation energy, like caesium ($U = 3.89$ eV), this means that a high ionisation degree is achievable at temperatures as low as a 1,000 K [Bit04].

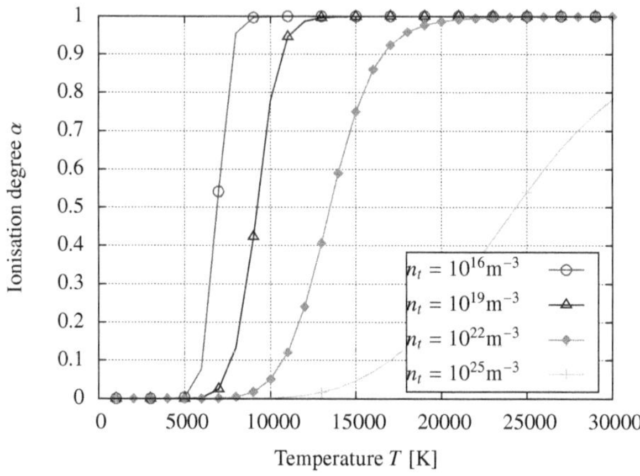

Figure 2.13: Ionisation degree α as a function of temperature T for different values of the number density n_t and with $U = 15.76$ eV.

2.9 Kinetic theory of gases

2.9.1 Fundamental concepts

2.9.1.1 The 1/6 distribution

Let V represent a closed volume containing a large number N of gas particles. The particles are treated as small hard spheres with mass m and diameter σ. The gas in the control volume is macroscopically at rest. Hence, the net gas velocity is zero and both temperature and density are constant. In a real gas, the particles move chaotically in **all directions** and with **different velocities**. This state, known as *Brownian motion*, is described by the Maxwell distribution which is presented in Section 2.9.3. As a simplification of the chaotic Brownian motion, a uniform distribution over six space coordinates can be assumed. In this simple model, known as the 1/6 distribution, the particles are uniformly distributed along six cartesian directions. Furthermore, the particles move with the constant, mean velocity c_0. Although this model does not accurately describe the motion direction and velocities of a real gas, its simplicity makes it useful for the introduction of fundamental concepts and properties associated with the kinetic theory of gases (cp. [Hän04]). The position and momentum spaces for the 1/6 model are depicted in Fig. 2.14.

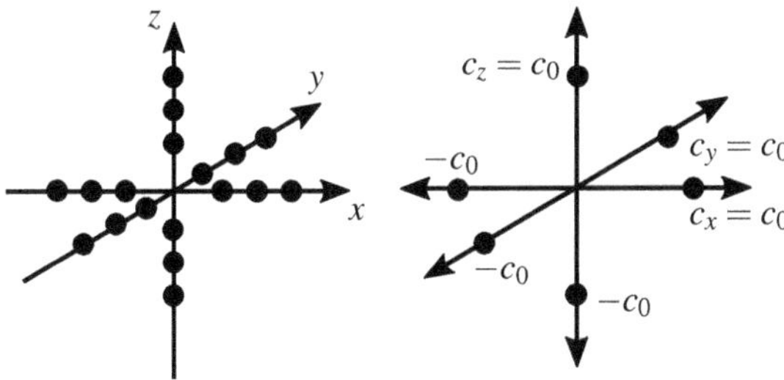

Figure 2.14: Position and momentum spaces for the 1/6 model [Hän04].

2.9.1.2 Thermodynamic and caloric properties

Based on the 1/6 model, the following fundamental thermodynamic and caloric properties of a gas can be defined:

Particle Flux $\dot{n}$: The particle flux $\dot{n}$ describes the number of particles crossing the control area A during the time Δt. Let us consider the flux in the positive x direction $\dot{n}_x$. As depicted in Fig. 2.14, the particles move with the velocity $c_x = c_0$. The number of particles crossing the control area A_x normal to the x direction equals the particles contained in the cylindrical volume $\Delta V = c_0 \Delta t A_x$. This is shown schematically in Fig. 2.15. Since the particles are uniformly distributed in the volume V, the number of particles in the cylinder equals $N/6 \times \Delta V/V$. The particle flux in x direction is therefore given by:

$$\dot{n}_x = \lim_{\Delta t \to 0} \frac{\Delta N}{\Delta t} = \frac{1}{6}\frac{N}{V}\frac{c_0 \Delta t A_x}{\Delta t} = \frac{1}{6} n c_0 A_x \tag{2.118}$$

where n stands for the number density of the gas. The mass flux through the same control area is given by:

$$\dot{m}_x = m \cdot \dot{n}_x \tag{2.119}$$

Equations 2.118 and 2.119 enable the estimation of the particle and mass fluxes across the control area A_x. Note that in reality, particles can cross the control area from any direction and not necessarily along the x axis. This fact is taken into account by the Maxwell distribution, which yields the result $\dot{n}_x = (1/4) n c_0 A_x$ [Hän04].

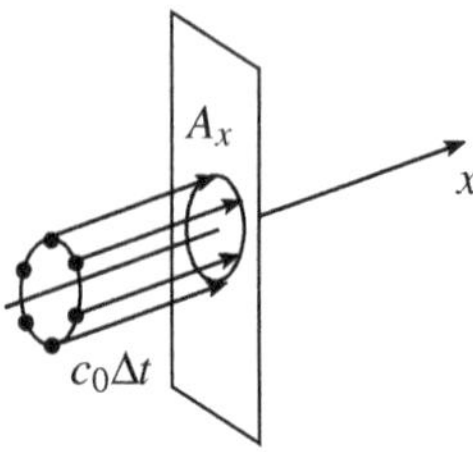

Figure 2.15: Particle flux in the positive x direction $\dot{n}_x$ using the 1/6 model [Hän04].

Pressure p and Temperature T: The pressure force exerted by a gas on a wall of area A can be interpreted as the result of the momentum change of the gas particles after colliding elastically with the wall. The elastic collision process is depicted in Fig. 2.16. If c_0 is the velocity of the particles before the elastic collision and $-c_0$, their velocity after the collision, the change of momentum flux in the x direction $\Delta\dot{I}_x$ can be calculated as follows:

$$\begin{aligned} \Delta\dot{I}_x &= \dot{I}_{x,\text{after}} - \dot{I}_{x,\text{before}} = F_x = -pA_x \\ &= \dot{n}_x(-mc_0) - \dot{n}_x(mc_0) = -pA_x \end{aligned} \tag{2.120}$$

with $\dot{I}_x = \dot{n}_x mc_0$. Eq. 2.120 may be combined with Eq. 2.119 to yield:

$$\begin{aligned} -2m\left(\frac{1}{6}nc_0^2A_x\right) &= -pA_x \\ \rightarrow p &= \frac{1}{3}\rho c_0^2 \end{aligned} \tag{2.121}$$

where $\rho = mn$ stands for the mass density. Furthermore, the temperature can be calculated from the ideal gas relation in Eq. 2.77 as follows (cp. [Hän04]):

$$p = nk_BT \rightarrow T = \frac{p}{nk_B} \tag{2.122}$$

Internal and thermal energy: The definition of pressure in Eq. 2.121 can be expressed in terms of the kinetic energy of the particle as follows:

$$p = \frac{1}{3}nmc_0^2 = \frac{2}{3}n\frac{m}{2}c_0^2 = \frac{2}{3}nE_{kin} \tag{2.123}$$

On the other hand and from a thermodynamic standpoint, a gas at rest has only thermal (internal) energy as energy form. Hence, kinetic and thermal energy must be equivalent.

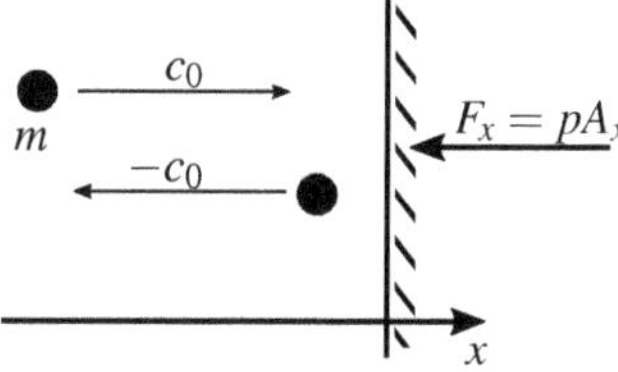

Figure 2.16: Elastic collision of a gas particle with a wall using the 1/6 model [Hän04].

The relation between the kinetic and the thermodynamic definition of thermal energy can be obtained through combination of Eqs. 2.122 and 2.123:

$$E_{kin} = \frac{1}{2} m c_0^2 = \frac{3}{2} \frac{p}{n} = \frac{3}{2} k_B T = E_{therm} \tag{2.124}$$

which can also be interpreted as the kinetic definition of temperature for an atom or molecule [Hän04].

Uniform distribution of energy: Based on Eq. 2.124, the thermal energy of a particle per mass unit can be expressed as:

$$e = \frac{E_{therm}}{m} = \frac{3}{2} \frac{k_B}{m} T = \frac{3}{2} RT \tag{2.125}$$

with R representing the specific gas constant. On the other hand, the velocity vector c of a given atom in translational motion possesses three spatial components (c_x, c_y, c_z). Hence, each of this components contributes to the total kinetic energy of the particle and the atom is said to have three *degrees of freedom*. By comparing this statement to the definition of the internal energy in Eq. 2.125, it can be concluded that each of these translational degrees of freedom contributes to the total energy of a monoatomic gas with the factor $(1/2)RT$. There are, however, more possible degrees of freedom (DF) depending on the nature of the considered gas:

- $DF = 3$ for monoatomic gases with translational motion in three Cartesian coordinates.
- $DF = 2$ or $DF = 3$ for rotational motion of molecules with a finite moment of inertia about two or three Cartesian axes respectively.
- $DF = 2$ for each chemical bond between atoms. These degrees of freedom result from the kinetic and potential energy associated with the vibrational motion of chemical bonds.

Hence, the internal energy of a molecule with fully excited degrees of freedom is given by the equation [Hän04]:

$$e = \frac{DF}{2} RT \tag{2.126}$$

2.9.1.3 Estimation of collision parameters

The relatively simple 1/6 model enables the estimation of important collision parameters based on the assumption of fictive collisions. To this end, let us imagine a test particle moving with the finite velocity c_0 through a cloud of static particles. All the particles in the system are treated as hard spheres of diameter σ. The arrangement is depicted in Fig. 2.17. The test particle either collides or grasps all of the particles contained in a cylinder with diameter $D = 2\sigma$ and length $c_0\Delta t$ in the time interval Δt. The volume of the imaginary cylinder is therefore $\Delta V = \pi\sigma^2 c_0 \Delta t$.

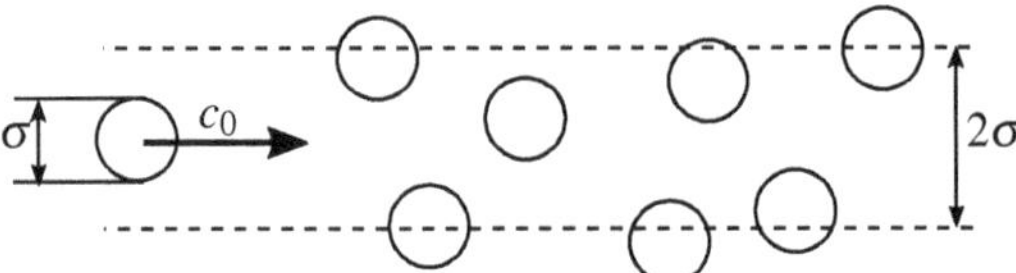

Figure 2.17: Arrangement for estimation of collision parameters [Hän04].

Collision frequency ν_c: The collision frequency equals the number of collisions between the test particle and the static particles in the imaginary cylinder per unit of time. The total number of collisions $\nu_c\Delta t$ is proportional to the volume ΔV as well as to the ratio between the total amount of particles in the system N and the total system volume V. Mathematically:

$$\nu_c \Delta t = \frac{\Delta V}{V} N = \pi\sigma^2 c_0 \Delta t \frac{N}{V} \tag{2.127}$$

Eq. 2.127 can be solved for the collision frequency to obtain:

$$\nu_c = \pi\sigma^2 c_0 n \tag{2.128}$$

Note that Eq. 2.128 is only an estimation of the collision rate which does not take into account the relative velocity between the particles taking part in a collision. The accurate solution obtained using the Maxwell distribution is (s. Section 2.9.3 and [Hän04]):

$$\nu_c = \frac{4}{\sqrt{3\pi}} \pi \sigma^2 c_0 n \tag{2.129}$$

Mean free path λ: The mean free path describes the mean distance covered by a particle or molecule before a collision with a second one takes place. λ can be obtained from the definition of the collision frequency in Eq. 2.128:

$$\lambda = \frac{c_0}{\nu_c} = \frac{1}{\pi \sigma^2 n} = \frac{m}{\pi \sigma^2} \frac{1}{\rho} \tag{2.130}$$

which, as for the collision frequency, represents only an estimation of the mean free path using the 1/6 model. The accurate solution obtained from the Maxwell distribution is [Hän04]:

$$\lambda = \frac{1}{\sqrt{2}} \frac{1}{\pi \sigma^2 n} \tag{2.131}$$

2.9.1.4 Boundary conditions

Boundary conditions for a fluid in the continuum regime can be usually formulated in an accurate manner assuming *complete accommodation* of momentum and energy at the system walls. Mathematically:

$$u = U_W, \quad T = T_W \tag{2.132}$$

where U_W and T_W stand for the wall velocity and temperature respectively. For rarefied conditions, the gas particles collide with the system walls with a much lower frequency, which disrupts the momentum and energy transfer between gas and wall. Hence, the concept of complete accommodation does not describe the gas-wall interaction accurately and a so-called *incomplete accommodation* must be considered. Mathematically, this effect is described through the use of accommodation coefficients for momentum α_p and energy α_E [Hän04].

In the case of momentum transfer, incomplete accommodation manifests as the tendency of a gas to slide along a wall without fully assuming the wall velocity, a phenomenon known as *slip flow*. This is described through the accommodation coefficient for momentum:

$$\alpha_p = \frac{u_i - u_r}{u_i - U_W} \tag{2.133}$$

where u_i stands for the velocity of the incident particles and u_r for the velocity of the reflected particles at the wall. The extreme cases are:

- $\alpha_p = 1 \rightarrow u_r = U_W$ corresponds to the standard no-slip boundary condition.
- $\alpha_p = 0 \rightarrow u_r = u_i$ corresponds to the flow completely sliding along the wall without any momentum transfer.

Similarly, incomplete accommodation for energy implies that the fluid is unable to assimilate the wall temperature and a temperature jump between fluid and wall is observed. The associated accommodation coefficient for temperature is given by:

$$\alpha_E = \frac{T_i - T_r}{T_i - T_W} \tag{2.134}$$

where T_i stands for the temperature of the incident gas particles and T_r, for the temperature of the reflected particles. Here, the following extreme cases apply:

- $\alpha_E = 1 \rightarrow T_r = T_W$ corresponds to a complete thermal accommodation.
- $\alpha_E = 0 \rightarrow T_r = T_i$ corresponds to the flow temperature being independent of the wall temperature (heat flux $q = 0$).

The accommodation coefficients α_p and α_E are influenced by numerous factors like Knudsen number, wall temperature and wall surface characteristics. As such, they are difficult to determine with accuracy [Hän04].

2.9.2 Velocity distribution function and macroscopic properties

2.9.2.1 Distribution function

In Section 2.9.1, important properties of a gas have been introduced by taking advantage of the simple 1/6 kinetic model. Since the atoms and molecules in a real gas move with a wide range of velocities and in all possible directions (*Brownian motion*), the assumptions linked to the 1/6 model are highly unrealistic. In order to accurately describe a gas, let us define a space containing a large amount of particles representing either gas atoms or molecules. The particles can be imagined as material points which, most of the time, move in space in straight lines obeying the laws of classic mechanics. In this case, the state of a given particle is defined by:

- its position in space $r = (x, y, z)$ and
- its absolute velocity in a fixed coordinate system $\xi = (\xi_x, \xi_y, \xi_z)$.

Furthermore, the particle carries during its motion the following properties:

- its mass m,
- its momentum $m\xi$,
- and its kinetic energy $(1/2)m\xi^2$.

If no external force is present, the particle moves in a straight line until it interacts with either a second particle or a wall. By assuming that the interaction with another particle can be modelled as an elastic collision between hard spheres, it is possible to calculate the motion of all particles making up a gas if the positions r and velocities ξ at a given time t are known. The macroscopic properties of the gas can be subsequently computed from the results for r and ξ. The challenges regarding this approach involve the very high number of particles needed to accurately model a system and the high number of collisions between particles and with walls, which are very susceptible to small irregularities and can potentially lead to the modelled process becoming irreversible. Nevertheless, in most cases, the macroscopic properties of the gas rather than the state of each individual particle are of interest. The macroscopic gas properties themselves can be obtained through averaging of particle properties over a large amount of atoms or molecules (cp. [Hän04]).

In order to describe the state of the particles, a statistic function called the *molecular velocity distribution function* $f(r,t,\xi)$ is of vital importance. Let us define a generic volume V containing a large number of particles N. The particles move in all directions and with different velocities ξ. At a given time t we consider all particles:

- contained inside the specific volume $\mathrm{d}V$ located at the position r and
- with a value of velocity lying inside the range ξ to $\xi + \mathrm{d}\xi$.

Furthermore, it is assumed that the number of particles N inside the differential volume $\mathrm{d}V$ moving with velocities lying in the velocity range $\mathrm{d}\xi$ is large enough for the density function to be continuous. The velocity distribution function is then given by:

$$f(r,\xi) = \frac{\mathrm{d}N}{\mathrm{d}V\mathrm{d}\xi} \tag{2.135}$$

and can be interpreted as a deterministic density function describing the number of particles $\mathrm{d}N$ per position and velocity volume as well as a statistical function representing the probability of finding a particle in the space and velocity volume about the specific location r and with the velocity ξ. If a three-dimensional coordinate system is employed, the distribution function is a function of the six variables $\mathrm{d}x$, $\mathrm{d}y$, $\mathrm{d}z$, $\mathrm{d}\xi_x$, $\mathrm{d}\xi_y$, $\mathrm{d}\xi_z$, plus time. The six-dimensional position and velocity space is called the *phase space* [Hän04].

The distribution function in Eq. 2.135 enables the calculation of the number of particles dN contained in the small volume dV located at the position r and moving with velocities ranging from ξ to $\xi + \mathrm{d}\xi$. This is:

$$\mathrm{d}N = f(r,\xi) \cdot \mathrm{d}V\mathrm{d}\xi = f(x,y,z,\xi_x,\xi_y,\xi_z)\mathrm{d}x\mathrm{d}y\mathrm{d}z\mathrm{d}\xi_x\mathrm{d}\xi_y\mathrm{d}\xi_z \tag{2.136}$$

The total number of particles N in the volume V can be obtained through integration of the distribution function over all possible molecular velocities:

$$N(r,t) = \int_{x_1}^{x_2}\int_{y_1}^{y_2}\int_{z_1}^{z_2}\int_{-\infty}^{\infty}\int_{-\infty}^{\infty}\int_{-\infty}^{\infty} f(x,y,z,t,\xi_x,\xi_y,\xi_z)\mathrm{d}x\mathrm{d}y\mathrm{d}z\mathrm{d}\xi_x\mathrm{d}\xi_y\mathrm{d}\xi_z \tag{2.137}$$

or in short form:

$$N(r,t) = \int_V\int_\xi f(r,t,\xi) \cdot \mathrm{d}V\mathrm{d}\xi \tag{2.138}$$

2.9.2.2 Macroscopic properties

If the distribution function $f(r,t,\xi)$ is known, it is possible to calculate all relevant macroscopic properties by multiplication with a function $\Phi(\xi)$ and subsequent integration over the velocity space. The macroscopic properties are also known as the *moments of the distribution function*. Some of the most important ones are introduced in the following.

Density: The number density n corresponds to the number of particles ΔN per volume unit ΔV and is, in general, a function of the position r. The number of particles ΔN is given by:

$$\Delta N(r,t) = \int_{\Delta V}\int_\xi f(r,t,\xi) \cdot \mathrm{d}V\mathrm{d}\xi = \int_\xi f(r,t,\xi)\mathrm{d}\xi \cdot \Delta V \tag{2.139}$$

Hence, the density can be computed as follows:

$$n(r,t) = \lim_{\Delta V \to 0} \frac{\Delta N}{\Delta V} = \int_\xi f(r,t,\xi)\mathrm{d}\xi \tag{2.140}$$

The mass density ρ can be obtained through multiplication of Eq. 2.140 with the molecular mass m.

Thermal and fluid velocity: The collection of particles ΔN contained in the volume ΔV may possess, in average, a non-zero centre of mass velocity in a fixed coordinate system. This velocity corresponds to the mean macroscopic fluid velocity v. The total velocity ξ of a particle contained in the particle collection ΔN can be expressed as the sum of the centre of mass velocity v and a relative (thermal) velocity c:

$$\xi = v + c \tag{2.141}$$

The mean velocity is therefore given by the arithmetic average of the particle velocities ξ:

$$v(r,t) = \frac{1}{n} \int_{\xi} \xi f(r,t,\xi) \mathrm{d}\xi \tag{2.142}$$

The thermal velocity $c = \xi - v$ describes the thermal motion of the particles relative to the mean velocity v and is associated with the thermal (internal) energy, the pressure and the stress tensor of a gas. Since the thermal velocity describes the deviation of the molecular velocity with respect to its mean value, its average equals zero, i.e.,

$$c_{\text{mean}} = \int c f \mathrm{d}c = \int \xi f(\xi) dc - v \int f \mathrm{d}c = 0 \tag{2.143}$$

with $\mathrm{d}c = \mathrm{d}\xi$. For simplicity, the integration limits and the position and time dependency of the distribution function $f(r,t,\xi)$ are omitted in Eq. 2.143 and hereafter (cp. [Hän04]).

Internal energy and temperature: The thermal or internal energy of a gas is defined by the kinetic energy of the thermal velocity c of the particles $E_{\text{kin,th}} = (1/2)mc^2$. The total energy contained in a volume ΔV is obtained through integration over all particles in ΔV:

$$\Delta E_{\text{Tot}} = \int \frac{m}{2} c^2 \mathrm{d}N = \int \int \frac{m}{2} c^2 f(c) \mathrm{d}c \mathrm{d}V = \Delta V \int \frac{m}{2} c^2 f(c) \mathrm{d}c \tag{2.144}$$

Equation 2.144 can be divided by ΔV to obtain the internal energy on a unit volume basis:

$$\rho e_{\text{tr}} = \lim_{\Delta V \to 0} \frac{\Delta E_{\text{Tot}}}{\Delta V} = \frac{m}{2} \int c^2 f(c) \mathrm{d}c \tag{2.145}$$

where $\rho = mn$ stands for the mass density and e_{tr}, for the internal energy per mass unit:

$$e_{\text{tr}} = \frac{1}{\rho} \cdot \frac{m}{2} \int c^2 f(c) \mathrm{d}c = \frac{1}{2n} \cdot \int c^2 f(c) \mathrm{d}c \tag{2.146}$$

The temperature T can be defined based on the internal energy from the expression:

$$e = c_v T \tag{2.147}$$

Monoatomic gases have only translational motion as form of internal energy, i.e., $e = e_{\text{tr}}$. Combination of Eqs. 2.125 and 2.146 yields:

$$e = c_v T = \frac{3}{2} RT = \frac{1}{2n} \cdot \int c^2 f(c) \mathrm{d}c \tag{2.148}$$

Hence, the temperature associated with the translational motion is given by:

$$T_{\text{tr}} = \frac{1}{2c_v n} \cdot \int c^2 f(c) \mathrm{d}c = \frac{1}{3Rn} \cdot \int c^2 f(c) \mathrm{d}c \tag{2.149}$$

The translational temperature T_{tr} in Eq. 2.149 corresponds to the thermodynamic temperature T if the considered gas is monoatomic ($DF = 3$). However, if the particle possesses additional degrees of freedom, further temperature definitions associated with rotational and oscillation energies must be considered [Hän04].

Pressure and stress tensor: The definition of pressure from the velocity distribution function is based on the same approach as for the 1/6 model in Section 2.9.1.2. The pressure force exerted by the gas on a wall corresponds to the rate of change of momentum of the particles as a result of elastic collisions with the wall. Let us define the pressure p_{xx} as a general component of the stress tensor τ_{ij}. Furthermore, p_{xx} corresponds to the force per unit area in the x direction acting on the surface A_x whose normal vector points in the x direction. If the particles fly towards the wall with an arbitrary velocity c_x and are reflected with the velocity $-c_x$, then the following relation can be formulated:

$$\mathrm{d}\dot{n}_x(-mc_x) - \mathrm{d}\dot{n}_x(mc_x) = -\mathrm{d}p_{xx} A_x \tag{2.150}$$

which corresponds to Eq. 2.120 in differential form. The particle flux $\dot{n}_x$, units [1/s], describes the number of particles per unit time flying with a velocity ranging between c_x and $c_x + \mathrm{d}c_x$ towards the surface A_x. Mathematically:

$$\mathrm{d}\dot{n}_x = c_x \left(\int \int f(c) \mathrm{d}c_y \mathrm{d}c_z \right) \cdot \mathrm{d}c_x A_x \tag{2.151}$$

The integration over the velocities c_y and c_z in Eq. 2.151 reflects the fact that the rate of particles interacting with the wall is independent of their velocities in y and z directions. Substitution of the definition of $\mathrm{d}\dot{n}_x$ into Eq. 2.150 and subsequent division by A_x yields:

$$\mathrm{d}p_{xx} = 2mc_x^2 \left(\int \int f(c) \mathrm{d}c_y \mathrm{d}c_z \right) \cdot \mathrm{d}c_x \tag{2.152}$$

which can be integrated to obtain:

$$p_{xx} = 2m \int_{c_x=0}^{\infty} \left(\int_{-\infty}^{\infty} \int_{-\infty}^{\infty} c_x^2 f(c) \mathrm{d}c_y \mathrm{d}c_z \right) \cdot \mathrm{d}c_x \tag{2.153}$$

where the limits of the integral over c_x reflect the fact that only particles with a positive velocity in the x direction move towards the wall. Since it is assumed that the particles are elastically reflected, then the transformation,

$$\int_0^{\infty} ...\mathrm{d}c_x = \frac{1}{2} \int_{-\infty}^{\infty} ...\mathrm{d}c_x \tag{2.154}$$

is valid, which combined with Eq. 2.153 produces:

$$p_{xx} = m \int c_x^2 f(c) \mathrm{d}c \tag{2.155}$$

The general form of the normal stresses is therefore given by:

$$p_{ii} = m \int c_i^2 f(c) \mathrm{d}c \tag{2.156}$$

The scalar pressure p, as used in fluid mechanics, is defined as an isotropic property corresponding to the average of all the normal stresses produced by the gas. Hence,

$$\begin{aligned} p = \frac{1}{3}(p_{xx} + p_{yy} + p_{zz}) &= \frac{m}{3} \int (c_x^2 + c_y^2 + c_z^2) f(c) \mathrm{d}c \\ &= \frac{m}{3} \int c^2 f(c) \mathrm{d}c \end{aligned} \tag{2.157}$$

Equation 2.157 may be combined with the definition of temperature in Eq. 2.149 to obtain the ideal gas law:

$$p = \frac{m}{3} \int c^2 f(c) \mathrm{d}c = mnRT = \rho RT \tag{2.158}$$

The distribution function is symmetric for a gas in thermodynamic equilibrium. In this case, the contributions of the shear stresses disappear and the normal stresses are equivalent, i.e., $p = p_{xx} = p_{yy} = p_{zz}$ [Hän04]. However, if the gas is not in thermodynamic equilibrium and macroscopic velocity gradients are present, the normal stress p_{ij} consists of the mean pressure as defined in Eq. 2.157 and a component for shear stress. The components of the shear stress are associated with the rate of change of the tangential momentum. A particle interacting with a surface A_x, whose normal vector points in the positive x direction, carries the tangential momentum components mc_y and mc_z. The shear stresses can be defined in a similar way as the normal (pressure) stresses as follows:

$$\tau_{xy} = m \int c_x c_y f(c) \mathrm{d}c \qquad \text{and} \qquad \tau_{xz} = m \int c_x c_z f(c) \mathrm{d}c \tag{2.159}$$

where τ_{xy} describes the shear stress in the y direction acting on the surface A_x and τ_{xz}, the shear stress in the z direction acting on the same surface (s. Fig. 2.2). The components of the shear stresses acting on the surfaces A_y and A_z can be formulated in a similar way. The general stress tensor is therefore given by:

- Tangential or shear stresses:

$$\tau_{ij} = m \int c_i c_j f(c) \mathrm{d}c \qquad i \neq j \tag{2.160}$$

- Normal stresses:

$$\tau_{ii} = m \int c_i^2 f(c) \mathrm{d}c \tag{2.161}$$

2.9.3 Maxwell distribution

In thermodynamic equilibrium, the velocity distribution of the gas particles exhibits all possible directions and magnitudes but does not possess a dominant velocity direction. The resulting velocity distribution can therefore be assumed to be spherically symmetric. This velocity distribution, known as the *Maxwell distribution* $F(x,c,t)$, can be used to calculate the macroscopic and molecular properties in thermodynamic equilibrium [Bit04].

The derivation of the Maxwell distribution is based on the calculation of the number of particles contained in the spherical shell in velocity space constrained by velocities with magnitudes in the range between $|c|$ and $|c + \mathrm{d}c|$. In the case of thermodynamic equilibrium without external forces, the following assumptions are valid (cp. [Kre10]):

- The number of particles $\mathrm{d}N_i$ with velocity components ranging from c_i to $c_i + \mathrm{d}c_i$ is proportional to the differential thickness of the spherical shell $\mathrm{d}c_i$. If $g(c_i)$ represents the probability of finding a particle in the given velocity range, then:

$$\mathrm{d}N_i = g(c_i)\mathrm{d}c_i \tag{2.162}$$

- Because of the symmetry of the distribution function, the probabilities $g(c_i)$ are independent from one another. Hence, the total number of particles in the velocity interval between c_i and $c_i + \mathrm{d}c_i$ is obtained via multiplication of the three components:

$$\mathrm{d}N = \mathrm{d}N_x \cdot \mathrm{d}N_y \cdot \mathrm{d}N_z = g(c_x)\mathrm{d}c_x \cdot g(c_y)\mathrm{d}c_y \cdot g(c_z)\mathrm{d}c_z \tag{2.163}$$

- Since the velocity is spherically symmetric, the function $F(c)$ is dependent of the magnitude of the velocity $|c|$ and independent of the single velocity components. The number of particles in the spherical shell with velocities in the interval between $|c|$ and $|c+\mathrm{d}c|$ is therefore:

$$\mathrm{d}N = F(|c|)\mathrm{d}c_x\mathrm{d}c_y\mathrm{d}c_z \tag{2.164}$$

  where the function $F(|c|)$ stands for the molecular or *Maxwell velocity distribution* in thermodynamic equilibrium and describes the number of particles per velocity volume unit and space volume unit in the six-dimensional phase space.

Through combination of Eqs. 2.163 and 2.164, the following expression for $F(|c|)$ can be formulated:

$$F(|c|) = g(c_x)g(c_y)g(c_z) \tag{2.165}$$

Taking the logarithm of Eq. 2.165 leads to:

$$\ln F(|c|) = \ln g(c_x) + \ln g(c_y) + \ln g(c_z) \tag{2.166}$$

which can be differentiated successively with respect to c_x, c_y and c_z to yield:

$$\frac{1}{|c|}\frac{\mathrm{d}\ln F(|c|)}{\mathrm{d}c} = \frac{1}{c_x}\frac{\mathrm{d}\ln g(c_x)}{\mathrm{d}c_x} = \frac{1}{c_y}\frac{\mathrm{d}\ln g(c_y)}{\mathrm{d}c_y} = \frac{1}{c_z}\frac{\mathrm{d}\ln g(c_z)}{\mathrm{d}c_z} \tag{2.167}$$

Equation 2.167 can only be fulfilled if all four terms are constant. Let us define the constant term as $-2\beta^2$. Based on Eq. 2.167, the following expression is therefore valid:

$$\frac{1}{c_i}\frac{\mathrm{d}\ln g(c_i)}{\mathrm{d}c_i} = -2\beta^2 \tag{2.168}$$

which can be integrated and rearranged to yield (cp. [Fas11]):

$$\begin{aligned} \frac{\mathrm{d}\ln g(c_i)}{\mathrm{d}c_i} &= -2\beta^2 c_i \\ \ln g(c_i) &= -\beta^2 c_i^2 + a \\ g(c_i) &= e^{-\beta^2 c_i^2 + a} \\ g(c_i) &= e^{-\beta^2 c_i^2} e^a \\ g(c_i) &= \sqrt[3]{A} e^{-\beta^2 c_i^2} \end{aligned} \tag{2.169}$$

Hence, the distribution function is given by:

$$F(|c|) = Ae^{-\beta^2 c^2} \tag{2.170}$$

The unknown constants A and β can be determined by taking advantage of the definition of the number density n in Eq. 2.140, the internal energy ρe_{tr} in Eq. 2.145 and the spherical symmetry of the velocity distribution. This yields (cp. [Hän04]):

$$A = \frac{n}{(2\pi RT)^{3/2}} \qquad \text{and} \qquad \beta^2 = \frac{1}{2RT} \tag{2.171}$$

Therefore, the Maxwell distribution function for a gas at rest and in thermodynamic equilibrium is given by:

$$F(|c|) = \frac{n}{(2\pi RT)^{3/2}} \exp\left[-\frac{1}{2}\frac{c^2}{RT}\right] \tag{2.172}$$

On the other hand, if the gas in thermodynamic equilibrium moves with a uniform net flow velocity v, the thermal velocity c in Eq. 2.172 can be replaced by the difference between the absolute ξ and the centre of mass velocity v using the expression in Eq. 2.141. This leads to:

$$F(|c|) = \frac{n}{(2\pi RT)^{3/2}} \exp\left[-\frac{1}{2}\frac{(\xi - v)^2}{RT}\right] \tag{2.173}$$

The properties of the Maxwell distribution can be examined based on a one-dimensional velocity distribution. To this end, the distribution function in Eq. 2.172 is split into its three independent velocity components, which after division by the number density n yields the following expression:

$$\frac{F(|c|)}{n} = g'(c_x)g'(c_y)g'(c_z) \tag{2.174}$$

where the function,

$$g'(c_i) = \frac{1}{(2\pi RT)^{3/2}} \exp\left[-\frac{1}{2}\frac{c_i^2}{RT}\right] \tag{2.175}$$

represents the normal or Gaussian distribution shown in Fig. 2.18 for two different temperature values $T_1 = 300$ K and $T_2 = 1{,}200$ K. The one-dimensional Maxwell distribution can hence be interpreted as the probability of finding particles with the velocity value indicated in the x-axis of the function in Fig. 2.18. Furthermore, the area under the curve is the same for both temperature values and equals the number density of the gas. As evident from the figure, the wideness of the distribution is an indicator of the temperature of the gas. Moreover, if the temperature approaches zero, the distribution takes the

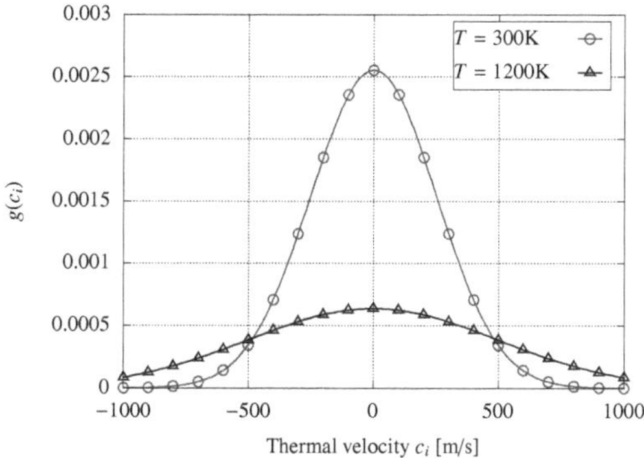

Figure 2.18: Graphical representation of the one-dimensional Maxwell distribution for argon at two temperature values $T_1 = 300$ K and $T_2 = 1,200$ K (cp. [Hän04]).

form of the Dirac delta function. In Fig. 2.18, the mean velocity of the distribution is zero and the gas has no net macroscopic velocity. For a gas macroscopically in motion, the Maxwell distribution slides to the right or the left side of the graph depending on the direction of the macroscopic flow velocity.

2.9.4 Boltzmann equation

2.9.4.1 Derivation

In the previous sections, the assumptions of a 1/6 distribution function or of a gas in thermodynamic equilibrium enabled the definition of simple forms of the distribution function $f(\xi,r,t)$. However, the obtained functions are not valid for the more general case of a gas in thermodynamic non-equilibrium. Based on fundamental definitions of the kinetic theory of gases, it is possible to derive a general equation which allows the determination of a distribution function for both thermodynamic equilibrium and non-equilibrium cases. The result is known as the *Boltzmann equation.* Some of the main properties of the Boltzmann equation are ([Hän04]):

- The Boltzmann equation is an equation for the determination of the molecular velocity distribution function $f(\xi,r,t)$ in terms of position and time and it is valid for general thermodynamic non-equilibrium process.
- The Boltzmann equation is the most general equation for the determination of the

distribution function $f(\xi, r, t)$ for gas dynamic processes.

- The Boltzmann equation covers gas problems in all possible flow regimes from continuum regime $Kn << 1$ to free molecular flows $Kn >> 1$.
- The Boltzmann equation is not only valid for gases in non-equilibrium but also for problems in thermodynamic equilibrium. One of the possible solutions of the Boltzmann equation is the Maxwell distribution.
- For small deviations of thermodynamic equilibrium, transport equations for flows in the continuum regime can be obtained from the Boltzmann equation through a Chapman-Enskog Expansion. The best-known equations obtainable with this approach are the Navier-Stokes equations presented in Section 2.4.
- The Boltzmann equation is a very complex integro-differential equation. As such, only a handful of analytic approximations for some simple, specific problems are possible. General solutions can, however, be obtained via numerical methods such as the *Direct Simulation Monte Carlo (DSMC)* and the *Lattice Boltzmann Methods*.

In order to derive the Boltzmann equation, let us consider a phase volume which includes the three-dimensional molecular velocity space and the three-dimensional position space. The balance of particles in this phase volume can be expressed in terms of a transport and a collision term as:

$$\frac{Df}{Dt}|_{\text{Transport}} = \frac{Df}{Dt}|_{\text{Collision}} \tag{2.176}$$

The transport term in Eq. 2.176 includes the temporal change of the number of particles in the observed phase volume, the change of the number of particles due to molecular transport associated with the particles velocities and the transport due to external forces. On the other hand, the collision term describes the gain or loss of particles in the observed phase volume as a result of molecular collisions. The molecular collisions themselves are assumed to be accurately described by reversible elastic collisions between hard spheres [Hän04].

The derivation of the Boltzmann equation involves a balance of the number of particles in the control volume $\Delta V \cdot \xi$ in the phase space about the position vector r and the velocity vector ξ as depicted in Fig. 2.19. The control volume consists of a space volume (s. Fig. 2.19 LHS):

$$\Delta V = \Delta x \cdot \Delta y \cdot \Delta z \qquad \text{about the vector} \qquad r = (x, y, z) \tag{2.177}$$

and a velocity volume (s. Fig. 2.19 RHS):

$$\Delta\xi = \Delta\xi_x \cdot \Delta\xi_y \cdot \Delta\xi_z \qquad \text{about the vector} \qquad \xi = (\xi_x, \xi_y, \xi_z) \tag{2.178}$$

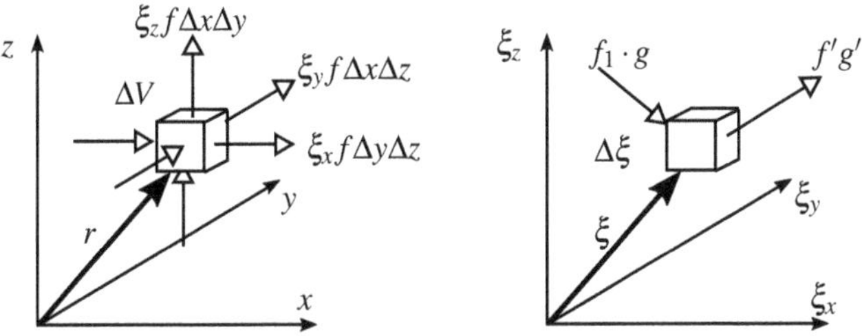

Figure 2.19: Position and velocity volume in the phase space $\Delta V \cdot \Delta\xi$ [Hän04].

The total number of particles in the phase volume is given by:

$$\Delta N(\xi, r, t) = f(\xi, r, t) \cdot \Delta V \cdot \xi \tag{2.179}$$

where $f(\xi, r, t)$ represents an arbitrary velocity distribution function valid in thermodynamic non-equilibrium and to be determined from the Boltzmann equation. As defined in Eq. 2.176, the balance of particles in the phase volume can be expressed in terms of a transport and a collision term. In an expanded form, this relation is given by (cp. [Hän04]):

$$(\Delta N_t + \Delta N_r + \Delta N_\xi)_{\text{Transport}} = (\Delta N_G - \Delta N_L)_{\text{Collision}} \tag{2.180}$$

The individual components of Eq. 2.180 are derived in the following.

Transport term

The temporal change ΔN_t of the number of particles ΔN in the control volume ΔV can be expressed as:

$$\begin{aligned} \Delta N_t &= \Delta N(\xi, r, t + \Delta t) - \Delta N(\xi, r, t) \\ &= (f(\xi, r, t + \Delta t) - f(\xi, r, t)) \cdot \Delta V \Delta\xi \end{aligned} \tag{2.181}$$

By taking advantage of a Taylor series for small values of Δt:

$$f(\xi, r, t + \Delta t) = f(\xi, r, t) + \frac{\partial f(\xi, r, t)}{\partial t} \Delta t + \ldots \tag{2.182}$$

the temporal change ΔN_t can be formulated as follows:

$$\Delta N_t = \frac{\partial f}{\partial t} \cdot \Delta V \cdot \Delta\xi \cdot \Delta t \tag{2.183}$$

The spatial change ΔN_r of the number of particles ΔN in the control volume ΔV is the consequence of variations in the particle fluxes through the surfaces of the control volume. This term can be divided into the individual contributions in each spatial coordinate, yielding:

$$\Delta N_r = \Delta N_x + \Delta N_y + \Delta N_z \tag{2.184}$$

Let us consider the balance of fluxes in the x direction. According to Fig. 2.19, this can be expressed as the difference between the flux exiting the control volume and the flux entering it through the surface $\Delta y \Delta z$. Hence:

$$\Delta N_x = \xi_x f(\xi, x+\Delta x, y, z, t) \cdot \Delta y \Delta z \Delta t \Delta \xi - \xi_x f(\xi, x, y, z, t) \cdot \Delta y \Delta z \Delta t \Delta \xi \tag{2.185}$$

Based on a Taylor series for small values of x:

$$f(\xi, x+\Delta x, y, z, t) = f(\xi, r, t) + \frac{\partial f(\xi, r, t)}{\partial x} \Delta x + \ldots \tag{2.186}$$

the spatial change in x direction ΔN_x is defined by the expression:

$$\Delta N_x = \xi_x \frac{\partial f}{\partial x} \cdot \Delta V \cdot \Delta \xi \cdot \Delta t \tag{2.187}$$

which expanded for all spatial directions yields:

$$\Delta N_r = \left(\xi_x \frac{\partial f}{\partial x} + \xi_y \frac{\partial f}{\partial y} + \xi_z \frac{\partial f}{\partial z} \right) \cdot \Delta V \cdot \Delta \xi \cdot \Delta t \tag{2.188}$$

The change ΔN_ξ describes the change of the number of particles in the molecular velocity volume $\Delta \xi$ as a result of a change in velocity produced by the acceleration $\mathrm{d}\xi/\mathrm{d}t$. The acceleration is, in turn, the product of an external force F described by Newton's second law of motion:

$$F = m \cdot \frac{\mathrm{d}\xi}{\mathrm{d}t} \tag{2.189}$$

This acceleration causes the particles to exit the velocity space $\Delta \xi$ about the vector ξ or, in the opposite case, to enter it. Similar to the spatial change ΔN_r, the term ΔN_ξ can be expressed in terms of its individual components:

$$\Delta N_\xi = \Delta N_{\xi,x} + \Delta N_{\xi,y} + \Delta N_{\xi,z} \tag{2.190}$$

The balance in x direction is given by:

$$\Delta N_{\xi,x} = \left(f(\xi_x + \Delta \xi_x, \xi_y, \xi_z, r, t) - f(\xi_x, \xi_y, \xi_z, r, t) \right) \cdot \Delta V \cdot \Delta \xi \tag{2.191}$$

By taking advantage of a Taylor series for small values of $\Delta\xi_x$:

$$f(\xi_x+\Delta\xi_x,\xi_y,\xi_z,r,t)=f(\xi_x,\xi_y,\xi_z,r,t)+\frac{\partial f(\xi_x,\xi_y,\xi_z,r,t)}{\partial \xi_x}\cdot\Delta\xi_x+\dots \tag{2.192}$$

and of the definition of the acceleration produced by an external force:

$$\Delta\xi_x=\frac{\mathrm{d}\xi_x}{\mathrm{d}t}\Delta t \tag{2.193}$$

the change $\Delta N_{\xi,x}$ can be defined as follows:

$$\Delta N_{\xi,x}=\frac{F_x}{m}\cdot\frac{\partial f}{\partial \xi_x}\cdot\Delta V\cdot\delta\xi\cdot\Delta t \tag{2.194}$$

which expanded in all three directions yields the following expression for ΔN_ξ:

$$\Delta N_\xi=\left(\frac{F_x}{m}\cdot\frac{\partial f}{\partial \xi_x}+\frac{F_y}{m}\cdot\frac{\partial f}{\partial \xi_y}+\frac{F_z}{m}\cdot\frac{\partial f}{\partial \xi_z}\right)\cdot\Delta V\cdot\delta\xi\cdot\Delta t \tag{2.195}$$

Collision term

Collisions inside or outside the observed volume in phase space lead to a change of the molecular velocities. The number of particles in the volume $\Delta V\Delta\xi$ can therefore increase or decrease as a result of collisional processes. This behaviour is described through the gain term ΔN_G and the loss term ΔN_L.

The loss term ΔN_L is associated with collisions between particles with velocities between ξ and $\xi+\Delta\xi$ contained in the examined phase volume with particles of a second phase volume $\Delta\xi_1$ about the velocity value ξ_1. Let us consider the collision between a particle with velocity ξ and a particle with velocity ξ_1. The post-collision velocities are ξ' and ξ_1' respectively and they lie outside the original velocity spaces, i.e., the collision induces a loss of particles for the phase space $\Delta V\cdot\Delta\xi$. The total number of particles ΔN_L exiting the observed velocity space is proportional to the collision frequency Z_c, which in turn corresponds to the number of collisions per time and volume unit with the relative velocity:

$$g=\xi_1-\xi \tag{2.196}$$

For particles with the velocities ξ and ξ_1 the collision frequency Z_c is given by (cp. [Hän04]):

$$Z_c=\int_{A_c}\int_{\xi}\int_{\xi_1} gf(\xi,r,t)f(\xi_1,r,t)\mathrm{d}\xi_1\mathrm{d}\xi\mathrm{d}A_c \tag{2.197}$$

where dA_c represents the differential collision cross section. Similarly, the total number of particles ΔN_L contained in the velocity volume $\Delta\xi$ which collide with particles of velocity ξ_1 in the time interval Δt is given by (cp. [Hän04]):

$$\begin{aligned} \Delta N_L &= \int_{\xi_1}\int_{A_c} f(\xi,r,t)\Delta\xi\Delta V \cdot g \cdot f(\xi_1,r,t)\mathrm{d}\xi_1 \mathrm{d}A_c \cdot \Delta t \\ &= \int_{\xi_1}\int_{A_c} f f_1 g \mathrm{d}A_c \mathrm{d}\xi_1 \cdot \Delta\xi \cdot \Delta V \Delta t \end{aligned} \tag{2.198}$$

which at the same time corresponds to the number of particles exiting the velocity volume $\Delta\xi$ as a result of molecular collisions.

The gain term ΔN_G is associated with particles entering the observed velocity space $\Delta\xi$ as a product of collisions between particles in other phase spaces. The collision between two particles with pre-collision velocities ξ' and ξ_1' leads to a change of their velocities to the post-collision values ξ and ξ_1. Since elastic collisions are reversible, this process can be regarded as the inversion of the collision described in the definition of the loss term ΔN_L. Hence, the variables ξ, ξ_1 and g in Eq. 2.198 can be replaced by ξ', ξ_1' and g' in order to formulate the expression of the gain term. ΔN_G is therefore given by:

$$\Delta N_G = \int_{\xi_1'}\int_{A_c} f' f_1' g' \mathrm{d}A_c \mathrm{d}\xi_1' \cdot \Delta\xi' \cdot \Delta V \Delta t \tag{2.199}$$

By taking advantage of collision mechanics, the gain term can be further rearranged so it can be written inside the same integral as the loss term (s. [Hän04]). The rearranged gain term can be written as follows:

$$\Delta N_G = \int_{\xi_1}\int_{A_c} f' f_1' g \mathrm{d}A_c \mathrm{d}\xi_1 \cdot \Delta\xi \cdot \Delta V \Delta t \tag{2.200}$$

which combined with the loss term gives the following expression for the collision term:

$$(\Delta N_G - \Delta N_L)_{\text{Collision}} = \int_{\xi_1}\int_{A_c} (f' f_1' - f f_1) g \mathrm{d}A_c \mathrm{d}\xi_1 \cdot \Delta\xi \cdot \Delta V \Delta t \tag{2.201}$$

Boltzmann Equation

Substitution of Eq. 2.183, 2.188, 2.195 and 2.201 into 2.180 and subsequent division by the phase volume $\Delta\xi \cdot \Delta V$ and the time interval Δt produces the **Boltzmann equation**

[Hän04]:

$$\frac{\partial f}{\partial t}+\sum_{i=1}^{3}\xi_i\frac{\partial f}{\partial x_i}+\sum_{i=1}^{3}\frac{F_i}{m}\frac{\partial f}{\partial \xi_i}=\int_{\xi_1}\int_{A_c}(f'f_1'-ff_1)g\mathrm{d}A_c\mathrm{d}\xi_1 \tag{2.202}$$

which can be also expressed in **tensor notation** as:

$$\frac{\partial f}{\partial t}+\xi_i\frac{\partial f}{\partial x_i}+\frac{F_i}{m}\frac{\partial f}{\partial \xi_i}=\int_{\xi_1}\int_{A_c}(f'f_1'-ff_1)g\mathrm{d}A_c\mathrm{d}\xi_1 \tag{2.203}$$

and in **vector notation** as:

$$\frac{\partial f}{\partial t}+\xi\cdot\nabla_r f+\frac{F_i}{m}\nabla_\xi f=\int_{\xi_1}\int_{A_c}(f'f_1'-ff_1)g\mathrm{d}A_c\mathrm{d}\xi_1 \tag{2.204}$$

2.9.4.2 Asymptotic solutions of the Boltzmann equation

Analytic solutions of the Boltzmann equation are extremely complex. For this reason, only asymptotic solutions for the extreme cases of a free molecular flow and a flow in the continuum regime are described in this section. These cases are defined by the dimensionless Knudsen number $Kn=\lambda/l$ introduced in Section 2.1. According to the definition of the flow regimes in Table 2.1, a free molecular flow can be represented by $Kn\rightarrow\infty$ while for a continuum flow, $Kn\rightarrow 0$. In order to analyse the Boltzmann equation for these two cases, characteristic parameters of the system must be introduced. These are the characteristic length of the system l, the mean free path of the particles λ, a reference number density n and a typical molecular velocity magnitude c_0. The variables in the Boltzmann equation can then be replaced by the following relations (cp. [Hän04]):

$$\begin{aligned}&x_i=\bar{x}_i\cdot l \qquad t=\bar{t}\cdot c_0/l \qquad \xi=\bar{\xi}\cdot c_0 \qquad g=\bar{g}\cdot c_0\\&f=\bar{f}\cdot n/c_0^3 \qquad \mathrm{d}A_c=\mathrm{d}\bar{A}_c\cdot 1/(n\lambda) \qquad F_i/m=\bar{F}_i/\bar{m}\cdot c_0^2/l\end{aligned} \tag{2.205}$$

yielding the following dimensionless form of the Boltzmann equation:

$$\frac{\partial \bar{f}}{\partial \bar{t}}+\bar{\xi}_i\frac{\partial \bar{f}}{\partial \bar{x}_i}+\frac{\bar{F}_i}{\bar{m}}\frac{\partial \bar{f}}{\partial \bar{\xi}_i}=\frac{1}{Kn}\int_{\bar{\xi}_1}\int_{\bar{A}_c}(\bar{f}'\bar{f}_1'-\bar{f}\bar{f}_1)\bar{g}\mathrm{d}\bar{A}_c\mathrm{d}\bar{\xi}_1 \tag{2.206}$$

where the bars denote dimensionless quantities. $Kn\rightarrow\infty$ represents the case of a **free molecular flow** and the right-hand side of Eq. 2.206 disappears. The resulting form of the Boltzmann equation is, in dimensional form, given by:

$$\frac{\partial f}{\partial t}+\xi_i\frac{\partial f}{\partial x_i}+\frac{F_i}{m}\frac{\partial f}{\partial \xi_i}=0 \tag{2.207}$$

Equation 2.207 corresponds to a transport equation for the distribution function f. The characteristic solution of this equation, neglecting external forces, represents the transport of f along the trajectory of the particles and the expression:

$$r-\xi\cdot t=\text{const.} \tag{2.208}$$

is valid. Therefore, the distribution function f is, in the case of a free molecular flow, constant along a particle trajectory. This results from the fact that, without intermolecular collisions, property exchanges cannot take place and the distribution function can only be altered by interaction of the particles with the system walls.

Multiplication of Eq. 2.206 with Kn and subsequent definition of $Kn\rightarrow 0$ leads to the disappearance of the left-hand side of the Boltzmann equation. The resulting expression in dimensional form is given by:

$$0=\int_{\xi_1}\int_{A_c}(f'f_1'-ff_1)g\mathrm{d}A_c\mathrm{d}\xi_1 \tag{2.209}$$

in which only the collisional term is present. Eq. 2.209 is fulfilled if its integrand equals zero. Hence,

$$f'f_1'-ff_1=0 \qquad \text{or} \qquad f'f_1'=ff_1 \tag{2.210}$$

Equation 2.210 implies that the number of particles exiting the phase space as a result of intermolecular collisions equals the number of particles entering it as a result of collisions taking place in other phase spaces. Hence, for each collision taking place in the observed phase space, an opposite collision outside the observed space also occurs, restoring this way the equilibrium. Therefore, this case corresponds to the condition for thermodynamic equilibrium in the limit case of a continuum flow and, as mentioned in Section 2.9.3, is described by the **Maxwell distribution function**. In order to demonstrate this, let us take the logarithm of Eq. 2.210, which yields:

$$\ln f+\ln f_1=\ln f'+\ln f_1' \tag{2.211}$$

Equation 2.211 clearly corresponds to a conservation equation for generic properties of two particle types before and after a collision. This condition is fulfilled in the case of elastic, reversible collisions by the collision invariants mass m, momentum mc and energy $(1/2)mc^2$. Mathematically:

$$\begin{aligned} m+m_1 &= m'+m'_1 \\ mc+m_1c_1 &= m'c'+m'_1c'_1 \\ \frac{m}{2}c^2+\frac{m_1}{2}{c_1}^2 &= \frac{m'}{2}c'^2+\frac{m'_1}{2}{c'_1}^2 \end{aligned} \tag{2.212}$$

where the apostrophes denote post-collision properties. A linear combination of the collision invariants in Eq. 2.212 must fulfil the conservation of the logarithmised distribution function in Eq. 2.211. Hence, the following expression can be assumed for $\ln f$:

$$\ln f = Am + Bmc + C\frac{m}{2}c^2 \tag{2.213}$$

The resulting distribution function is therefore given by:

$$f = \exp(Am)\cdot\exp(Bmc)\cdot\exp(C\frac{m}{2}c^2)$$

As mentioned in Section 2.9.3, the constants A, B and C can be determined through the definition of the number density n and the internal energy ρe_{tr}. This yields the Maxwell distribution presented in Eq. 2.172:

$$F(|c|) = \frac{n}{(2\pi RT)^{3/2}}\exp\left[-\frac{1}{2}\frac{c^2}{RT}\right]$$

Based on this derivation of the Maxwell distribution from the Boltzmann equation, the following conclusions can be drawn (cp. [Hän04]):

- The derivation of the Maxwell distribution does not require assumptions regarding intermolecular forces.
- The condition $f'f'_1 = ff_1$ is both sufficient and necessary for the existence of a thermodynamic equilibrium.
- The condition $f'f'_1 = ff_1$ implies that, on average, for each particle exiting the phase space through a collision, another particle enters it as a consequence of a collision in other phase spaces.
- $Kn \to 0$ implies that the mean free path is very small compared to the characteristic macroscopic length l. The intermolecular collisions occur with a very high frequency and thermodynamic equilibrium is always reached.

2.10 Summary

In this chapter, the main theoretical principles employed in the frame of the present work have been presented. Starting from the definition of the Knudsen number, different flow regimes have been identified and their characteristics, discussed. For flows in the continuum regime, the main analytic expressions include different sets of partial differential equations describing the conservation of mass, momentum and energy. After a brief discussion of ideal gases and the application case of a Laval nozzle, fundamental concepts of plasma physics and the kinetic theory of gases are explored in order to round out the basic concepts handled in the present work. After several key computational methods are introduced in Chapter 3, these concepts are applied to the modelling of an arcjet thruster in cold-gas operating mode and the development of a kinetic model for plasma behaviour during hot-operating modes in Chapters 4 to 6.

Chapter 3

Computational Methods

In this chapter, the computational methods employed for the numerical modelling of the arcjet thruster INGA III in cold-gas operation as well as for the development of a plasma model for hot-gas operation conditions are introduced. In cold-gas operation and for high mass flow rates $\dot{m}$, the characteristic Knudsen numbers of the system are low and the flow can be accurately modelled by the compressible, transient form of the mass continuity equation, the Navier-Stokes-Equations and an energy conservation equation. To this end, the set of partial differential equations is discretized with the *Finite Volume Method* (FVM) described in Section 3.1.2. On the other hand, for low values of $\dot{m}$, the Knudsen numbers of the studied cold-gas system increase and the flows reach the slip and transition regimes. Hence, the results of the continuum-based equations lose accuracy. In order to study the cold-gas flows in these conditions, the kinetic *Direct Simulation Monte Carlo* (DSMC) approach, described in Section 3.2, is employed. Finally, in hot-gas operation, the propellant gas is partially ionised resulting in a medium consisting of free electrons, positive ions and neutral gas atoms, a *plasma*. The model developed for this case is based on the Particle-In-Cell (PIC) Method presented in Section 3.3.

3.1 Methods based on transport equations

3.1.1 Finite Difference Method

The *Finite Difference Method* (FDM) is considered to be one of the easiest approaches for the mathematical modelling of flows in simple geometries. The approach is applied on the differential form of a transport equation with its derivatives being replaced by algebraic expressions. Furthermore, the geometrical domain is discretized using a numerical grid consisting of nodal points as shown in Fig. 3.1. Let us consider a continuous differentiable function $\varphi(x)$ representing an arbitrary flow property. In order to derive expressions for the first and second derivatives often present in transport equations, a Taylor series in the vicinity of x_i can be formulated:

$$\varphi(x) = \varphi(x_i) + (x - x_i)\left(\frac{\partial \varphi}{\partial x}\right)_i + \frac{(x-x_i)^2}{2!}\left(\frac{\partial^2 \varphi}{\partial x^2}\right)_i + \frac{(x-x_i)^3}{3!}\left(\frac{\partial^3 \varphi}{\partial x^3}\right)_i + H \quad (3.1)$$

with H denoting higher-order terms. Based on Eq. 3.1, it is possible to define expressions for the first and second derivatives. Replacing x by x_{i+1} yields for the first derivative:

$$\left(\frac{\partial \varphi}{\partial x}\right)_i = \frac{\varphi_{i+1} - \varphi_i}{x_{i+1} - x_i} - \frac{x_{i+1} - x_i}{2}\left(\frac{\partial^2 \varphi}{\partial x^2}\right)_i - \frac{(x_{i+1}-x_i)^2}{6}\left(\frac{\partial^3 \varphi}{\partial x^3}\right)_i - H \quad (3.2)$$

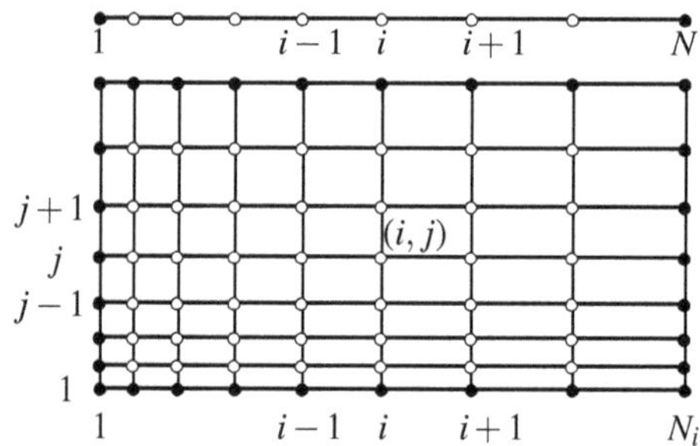

Figure 3.1: 1D (top) and 2D representation (bottom) of a Cartesian grid in the Finite Difference Method. Full circles represent boundaries and open circles, computational nodes (cp. [FP02]).

If the space between the nodal points is small, the second and higher order derivatives on the RHS of Eq. 3.2 are also small. Hence, the first derivative can be approximated by the expression:

$$\left(\frac{\partial \varphi}{\partial x}\right)_i \approx \frac{\varphi_{i+1} - \varphi_i}{x_{i+1} - x_i} \quad (3.3)$$

Equation 3.3 is known as the *forward difference scheme* (FDS) because the derivative is computed based on the value of φ at the node i and the downstream node $i+1$. Similarly, substitution of x by x_{i-1} in Eq. 3.1 yields the *backward difference scheme* (BDS):

$$\left(\frac{\partial \varphi}{\partial x}\right)_i \approx \frac{\varphi_i - \varphi_{i-1}}{x_i - x_{i-1}} \tag{3.4}$$

while the use of the values of φ at the positions x_{i-1} and x_{i+1} produces the *central difference scheme* (CDS):

$$\left(\frac{\partial \varphi}{\partial x}\right)_i \approx \frac{\varphi_{i+1} - \varphi_{i-1}}{x_{i+1} - x_{i-1}} \tag{3.5}$$

The second derivative can be obtained by taking advantage of the already defined approximations for the first derivative. A forward difference approach for the second derivative yields:

$$\left(\frac{\partial^2 \varphi}{\partial x^2}\right)_i \approx \frac{\left(\frac{\partial \varphi}{\partial x}\right)_{i+1} - \left(\frac{\partial \varphi}{\partial x}\right)_i}{x_{i+1} - x_i} \tag{3.6}$$

The first order derivatives in Eq. 3.6 might be computed with any of the first derivative schemes introduced above. For equidistant grid points and a BDS or CDS approach for the first derivatives, the approximation for the second derivative becomes:

$$\left(\frac{\partial^2 \varphi}{\partial x^2}\right)_i \approx \frac{\varphi_{i+1} + \varphi_{i-1} - 2\varphi_i}{(\Delta x)^2} \tag{3.7}$$

This expression requires the values of the variable φ at three different nodal points. Note that there are many other approaches for the approximation of the first and second derivatives. The FDM is both easy to employ and accurate for simple geometries. However, it loses precision for complex flow problems. One additional disadvantage is that the approach does not automatically guarantee conservation of fluid properties [FP02].

3.1.2 Finite Volume Method

The *Finite Volume Method* (FVM) takes advantage of the integral form of the conservation equations for mass, momentum and energy presented in Chapter 2. The flow domain is divided into small, contiguous control volumes (CVs) on which the transport equations are applied. The flow properties are calculated at the centre of each CV, a location also known as *computational node*. The flow variables at the boundary surfaces of each CV are determined through interpolation of the nodal values. The surface and volume integrals making up the transport equations are approximated using appropriate quadrature expressions [FP02]. In the following sections, the main aspects of the FVM are introduced.

3.1.2.1 Discretization of transport equations

A compressible flow in the continuum regime can be accurately described by a set of partial differential equations consisting of the continuity equation (Eq. 2.14),

$$\frac{\partial \rho}{\partial t} + \frac{\partial \rho u_j}{\partial x_j} = 0 \tag{3.8}$$

the Navier-Stokes Equations (Eq. 2.44),

$$\rho \frac{\partial u_i}{\partial t} + \rho u_j \frac{\partial u_i}{\partial x_j} = \rho g_i - \frac{\partial p}{\partial x_i} + \frac{\partial}{\partial x_j}\left[\mu \left(\frac{\partial u_i}{\partial x_j} + \frac{\partial u_j}{\partial x_i} - \frac{2}{3}\delta_{ij}\frac{\partial u_k}{\partial x_k}\right)\right] \tag{3.9}$$

and one form of the energy equation, for instance, in terms of internal energy (Eq. 2.63).

$$\rho\left(\frac{De}{Dt}\right) = \nabla \cdot (\kappa \nabla T) + \tau_{ij}\frac{\partial u_i}{\partial x_j} \tag{3.10}$$

With one Navier-Stokes equation per Cartesian coordinate plus the ideal gas law in Eq. 2.74, a system consisting of six equations and the six variables u_x, u_y, u_z, p, T and ρ is available. In order to describe the discretization strategies applied on these equations, let us define a general transport equation for the arbitrary flow property φ:

$$\underbrace{\frac{\partial}{\partial x_j}(\rho \varphi u_j)}_{\text{convection}} = \underbrace{\frac{\partial}{\partial x_j}\left(\Gamma \frac{\partial \varphi}{\partial x_j}\right)}_{\text{diffusion}} + \underbrace{q_\varphi}_{\text{source}} \tag{3.11}$$

which can be expressed either in terms of volume integrals:

$$\int_V \frac{\partial}{\partial x_j}(\rho \varphi u_j)\,\mathrm{d}V = \int_V \frac{\partial}{\partial x_j}\left(\Gamma \frac{\partial \varphi}{\partial x_j}\right)\mathrm{d}V + \int_V q_\varphi \,\mathrm{d}V \tag{3.12}$$

or by taking advantage of the divergence theorem (s. Appendix A), in terms of surface integrals:

$$\underbrace{\int_{\delta V} \rho \varphi (u \cdot n)\,\mathrm{d}S}_{\text{convection}} = \underbrace{\int_{\delta V} \Gamma(\nabla \varphi) \cdot n\,\mathrm{d}S}_{\text{diffusion}} + \underbrace{\int_V q_\varphi \,\mathrm{d}V}_{\text{source}} \tag{3.13}$$

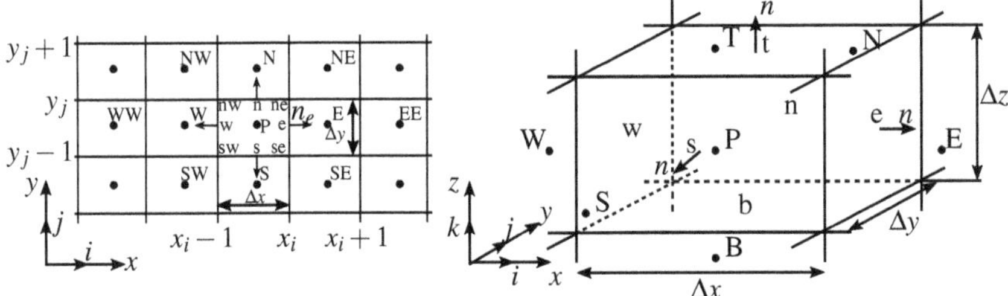

Figure 3.2: Schematic representation of the control volumes in the Finite Volume Method for a 2D (left) and a 3D domain (right) [FP02].

As already mentioned, the flow domain is divided into small contiguous control volumes (CVs). Fig. 3.2 shows a general CV for both a 2D and a 3D problem as well as the nomenclature typically used in the FVM. The node of the depicted CV is called P. In the 2D case, the CV is surrounded by four neighbour CVs whose nodes are denoted as *east* E, *west* W, *north* N and *south* S. The boundary surfaces of the CV separating it from its neighbours are defined in a similar way as e, w, n, and s. In the three-dimensional case, the additional neighbour CVs *top* T and *bottom* B and the associated faces t and b must be considered. The convective and diffusive surface integrals in Eq. 3.13 can be approximated as the sum over all the surfaces of the individual surface integrals:

$$\int_{\delta V} f \,\mathrm{d}S = \sum_k \int_{S_k} f \,\mathrm{d}S \tag{3.14}$$

where S_k stands for the area of the kth surface, $f = \rho\varphi(u \cdot n)$ for the convective integrand and $f = \Gamma(\nabla\varphi) \cdot n$, for the diffusive integrand. Note that f corresponds to the component of the convective or diffusive flux vector of the arbitrary property φ in the direction normal to the considered CV face. Furthermore, if the velocity field and the fluid properties are assumed to be known, the only unknown quantity is φ. Now let us consider the surface integral over the face "e" in Fig. 3.2 (left). In order to calculate the exact value of the integral, the values of the function f must be known at all locations over the surface. However, since the variable φ is computed only on the nodal points of the grid, this information is not available. Hence, an approximation scheme based on the available information must be developed. To this end, the surface integral is first approximated as a function of one or more values of f on the surface "e". Moreover, the surface values of f must be estimated based on the available nodal values.

The simplest approximation approach for the surface integral is the so-called *midpoint rule*. In this approach of second-order accuracy, the integral is computed as the product of the cell-face area S_e and the value of the integrand at the cell-face centre f_e, which in turn corresponds to an approximation of the mean value $\bar{f}_e$ over the face. Hence:

$$F_e = \int_{S_e} f \,\mathrm{d}S = \bar{f}_e S_e \approx f_e S_e \tag{3.15}$$

Another approach of second-order accuracy is the so-called trapezoid rule. Here, the integral is approximated based on values of the function f in the corners ne and se of the cell (s. Fig. 3.2, left) which yields:

$$F_e = \int_{S_e} f \,\mathrm{d}S \approx \frac{S_e}{2}(f_{ne} + f_{se}) \tag{3.16}$$

After the surface integrals have been defined, the source term in Eq. 3.13 must be discretized. This term corresponds to a volume integral which is also to be approximated by a function based on the available information in the numerical grid. In the simplest approach of second-order accuracy, the volume integral is estimated as the product of the cell's volume and the mean value of the function f in the CV. Hence, the mean value is assumed to be equal to the value in the cell centre which gives:

$$Q_P = \int_V q_\varphi \,\mathrm{d}V = \bar{q_\varphi} \Delta V \approx q_{\varphi,P} \Delta V \tag{3.17}$$

where $q_{\varphi,P}$ corresponds to the value of the source term q_φ at the node P and interpolation schemes are not needed. Approximation approaches of higher-order require values of q_φ at more locations than the cell's centre P. These can be estimated either via interpolation or by employing shape functions [FP02].

3.1.2.2 Interpolation schemes

As mentioned in the previous section, the computation of the surface integrals for the convective flux $f^c = \rho\varphi(u \cdot n)$ and the diffusive flux $f^d = \Gamma(\nabla\varphi) \cdot n$ requires an estimation of the values of f on the faces of the control volume. Let us assume that the velocity field and the fluid properties ρ and Γ are known at all locations. In this case, the values of the arbitrary property φ and of its gradient $\nabla\varphi$ must be estimated at the cell surfaces from the available nodal information. There are plenty of interpolation schemes with different orders of accuracy available. In this work, three of the most commonly used approaches are briefly described. They are the upwind, linear and linear upwind schemes [FP02].

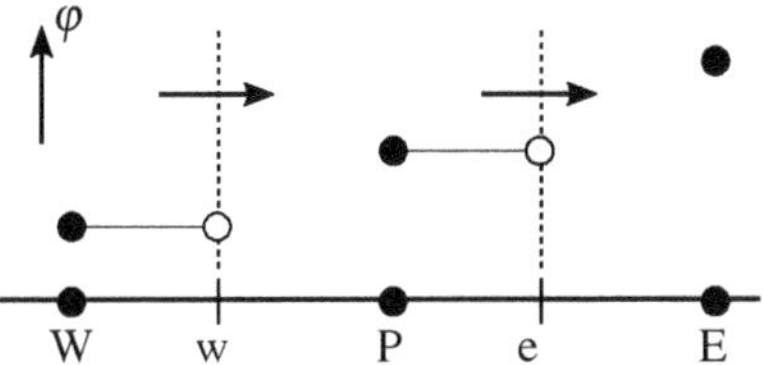

Figure 3.3: Schematic representation of the upwind interpolation scheme [Nol93].

Upwind interpolation scheme (UDS) The upwind interpolation scheme, depicted in Fig. 3.3, defines the value of φ in the cell face e as its value in the node upstream of the surface. Hence,

$$\varphi_e = \begin{cases} \varphi_P & \text{if } (u \cdot n)_e > 0 \\ \varphi_E & \text{if } (u \cdot n)_e < 0 \end{cases}, \tag{3.18}$$

This approach is also known as *upwind differencing scheme* (UDS) because it is equivalent to a backward (BDS) or forward-difference scheme (FDS) for the first derivative in the Finite Difference Method defined in Eq. 3.3 as follows (cp. [FP02]):

$$\left(\frac{\partial \varphi}{\partial x}\right) \approx \frac{\varphi_{i+1} - \varphi_i}{x_{i+1} - x_i} \qquad \text{FDS} \tag{3.19}$$

The similarity is evident when the expression in Eq. 3.19 is compared to the Taylor series expansion of φ_e about P for Cartesian coordinates and $(u \cdot n)_e > 0$:

$$\varphi_e = \varphi_P + (x_e - X_P)\left(\frac{\partial \varphi}{\partial x}\right)_P + \frac{(x_e - x_P)^2}{2}\left(\frac{\partial^2 \varphi}{\partial x^2}\right)_P + H \tag{3.20}$$

where H stands for higher-order terms. The upwind differencing scheme retains the first term on the RHS of Eq. 3.20 and the approach is therefore of first-order accuracy. The leading truncation error is given by:

$$f_e^d = \Gamma_e \left(\frac{\partial \varphi}{\partial x}\right)_P \tag{3.21}$$

The expression in Eq. 3.21 is comparable to a diffusive term with the so-called *coefficient of numerical or false diffusion* $\Gamma_e^{num} = (\rho u)_e \Delta x/2$. The truncation error is therefore linked to an artificial numerical diffusion in the system, which in turn leads to a reduction of peaks in the flow variables and the subsequent loss of resolution. As a consequence of this behaviour and the inherent low order of the scheme, the use of very high resolution grids is necessary in order to obtain accurate solutions [FP02].

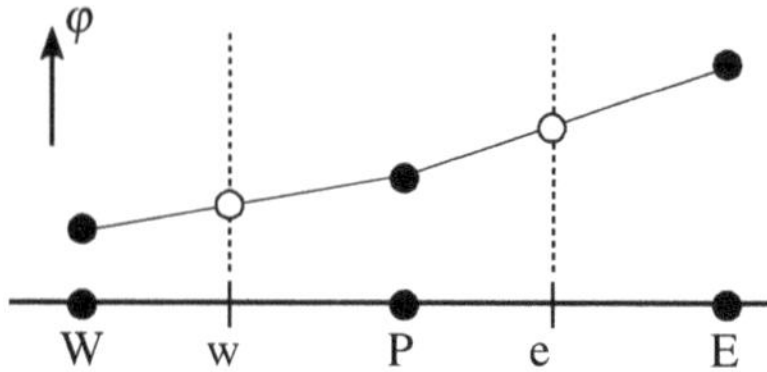

Figure 3.4: Schematic representation of the linear interpolation scheme [Nol93].

Linear interpolation scheme (CDS) The simplest second-order scheme results from the linear interpolation of the variable φ between its nearest nodal points as depicted in Fig. 3.4. With this approach, the value of φ on the face e is given by:

$$\varphi_e = \varphi_E \lambda_e + \varphi_P (1 - \lambda_e) \tag{3.22}$$

where λ_e, the linear interpolation factor, describes the relative distance between the nodal points and the cell's boundary:

$$\lambda_e = \frac{x_e - x_P}{x_E - x_P} \tag{3.23}$$

By developing the Taylor series of φ_E about the point x_P and using the result to eliminate the first derivative in Eq. 3.20, it can be shown that:

$$\varphi_e = \varphi_E \lambda_e + \varphi_P (1 - \lambda_e) - \frac{(x_e - x_P)(x_E - x_e)}{2} \left(\frac{\partial^2 \varphi}{\partial x^2} \right)_P + H \tag{3.24}$$

The first and second terms on the RHS of Eq. 3.24 correspond to the result of the linear interpolation in Eq. 3.22. The leading truncation error is therefore proportional to the square of the grid spacing and the scheme is, as already mentioned, of second-order accuracy. As a scheme of order higher than one, this approach can produce, in some cases, oscillatory results. On the other hand, it is the simplest and most widely used second-order approach. Note that this scheme is equivalent to the central-difference approximation of the first derivative in the Finite Difference Method defined in Eq. 3.5:

$$\left(\frac{\partial \varphi}{\partial x} \right) \approx \frac{\varphi_{i+1} - \varphi_{i-1}}{x_{i+1} - x_{i-1}} \qquad \text{CDS} \tag{3.25}$$

Hence, the linear interpolation scheme is also known as *central-difference scheme* (CDS). Based on the assumption of a linear profile between neighbour nodes, the gradient of the arbitrary property φ on the face surface, necessary for the computation of diffusive fluxes, can be estimated in a simple way:

$$\left(\frac{\partial \varphi}{\partial x} \right)_e \approx \frac{\varphi_E - \varphi_P}{x_E - x_P} \tag{3.26}$$

The truncation error associated with this approximation is given by:

$$\varepsilon_\tau = \frac{(x_e - x_P)^2 - (x_E - x_e)^2}{2(x_E - x_P)} \left(\frac{\partial^2 \varphi}{\partial x^2}\right) - \frac{(x_e - x_P)^3 + (x_E - x_e)^3}{6(x_E - x_P)} \left(\frac{\partial^3 \varphi}{\partial x^3}\right) + H \tag{3.27}$$

If the location of the face e lies midway between the nodes P and E, then $x_e - x_P = x_E - x_e$ and the first term on the RHS of Eq. 3.27 vanishes. Hence, the approximation of the gradient on the face surface presented in Eq. 3.26 is of second-order accuracy and its error is proportional to $(\Delta x)^2$ [FP02].

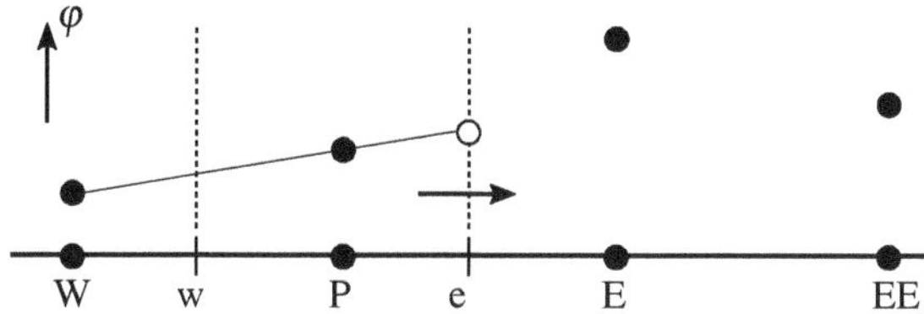

Figure 3.5: Schematic representation of the linear upwind interpolation scheme (LUDS) [Nol93].

Linear upwind interpolation scheme (LUDS) Another second-order scheme is based on the extrapolation of the value of φ at two upstream nodes and onto the cell face e. The approach, known as *linear upwind differencing scheme* (LUDS), is depicted in Fig. 3.5. The value of φ on the face e for $(u \cdot n)_e > 0$ is therefore given by:

$$\varphi_e = \varphi_W \lambda_w + \varphi_P (1 - \lambda_w) \tag{3.28}$$

where λ_w, the linear interpolation factor, describes the relative distance between the nodal points W and P and the cell's boundary:

$$\lambda_w = \frac{x_P - x_e}{x_P - x_W} \tag{3.29}$$

In general, the LUDS scheme is more complex than CDS and may produce unbounded solutions. Specially in regions with strong gradients, LUDS can lead to overshooting or undershooting of the solution [Nol93]. As a consequence, CDS is generally preferred [FP02].

Other Schemes Other interpolation schemes involve the use of more interpolation points, leading to approaches of order of accuracy higher than two. One example is the *quadratic upwind interpolation scheme for convective mathematics* (QUICK.) In this approach, the profile of the arbitrary property φ is described by a parabola rather than a straight line and therefore, a third interpolation point is required. Furthermore, because of the nature of convection phenomena, the additional point is selected on the upstream side of the flow. With $(u \cdot n)_e > 0$ and on a uniform Cartesian grid, the value φ_e is given by:

$$\varphi_e = \frac{6}{8}\varphi_P + \frac{3}{8}\varphi_E - \frac{1}{8}\varphi_W \tag{3.30}$$

The QUICK scheme is of third-order accuracy and hence, slightly more accurate than CDS. However, compared to CDS, it is also more complex as it requires information from one additional node. In addition, the overall accuracy of a computational approach is also a function of the chosen approximation for the surface integrals. Hence, if QUICK is used in combination with the second-order *midpoint rule* for the surface integrals, the overall approach will retain second-order accuracy and the difference to CDS will be small. For the same reason, interpolation schemes of order higher than third are only practical if the integrals are also approximated with higher-order formulae and if the grid is fine enough to capture the details of the solution. As an example, an interpolation scheme of fourth-order of accuracy would involve the use of a polynomial of third degree, which is to be constructed from the variable values at four nodes (two on either side of the cell face e). This approach is sometimes called *fourth-order* CDS.

Other approaches involve the combination of different methods in an attempt to benefit from the advantages of two or more approximation schemes. Some examples are schemes with automatic switching between UDS and CDS as a function of the local value of the Peclet number and hybrid schemes obtained through blending of approximations of different order of accuracy with the goal of reducing non-physical oscillations or computational requirements [FP02].

3.1.3 Methods for unsteady problems

In unsteady flow problems, the time coordinate must also be discretized. This additional coordinate may be regarded as a succession of either discrete points in time (finite difference view) or of contiguous *time volumes* (finite volume view). The main difference between the spatial and time coordinates is, of course, the direction of influence. A force applied on a specific fluid location at a specific point in time might affect the flow variables anywhere in the space grid but only in the future in the time coordinate. It follows that, barring conditions at the flow boundaries, no further conditions can be imposed on the solution after initialisation of a computation. In the same spirit, solution methods for unsteady problems advance in time in a step-by-step fashion. In order to introduce fundamental concepts of time discretization, four relatively simple procedures for initial value problems for ordinary differential equations (ODEs) are presented in the following. Let us consider a first-order ODE with an initial condition:

$$\frac{\mathrm{d}\varphi(t)}{\mathrm{d}t} = f(t,\varphi(t)); \qquad \varphi(t_0) = \varphi^0 \tag{3.31}$$

where φ represents an arbitrary flow variable. In order to advance the solutions of φ in time, a new solution at $t_1 = t_0 + \Delta t$ depending on the initial condition φ^0 should be determined. Furthermore, this new solution may be taken as a new initial value for the subsequent time step $t_2 = t_1 + \Delta t$ and the procedure is repeated until the final time is reached. In order to advance the problem in time, Eq. 3.31 can be integrated from t_n to $t_{n+1} = t_n + \Delta t$ to give:

$$\int_{t_n}^{t_{n+1}} \frac{\mathrm{d}\varphi(t)}{\mathrm{d}t}\,\mathrm{d}t = \varphi^{n+1} - \varphi^n = \int_{t_n}^{t_{n+1}} f(t,\varphi(t))\,\mathrm{d}t \tag{3.32}$$

where the notation $\varphi^{n+1} = \varphi_(t_{n+1})$ has been used for simplicity. Although Eq. 3.32 is exact, the RHS cannot be evaluated without knowing the exact solution $\varphi(t)$ of the ODE. In theory, one could take advantage of the mean value theorem of calculus which states that at the specific point $t = \tau$, the integral on the RHS equals the area of the rectangle $f(\tau,\varphi(\tau)) \cdot (\tau - t_n)$. However, τ is also unknown and an approximation is therefore necessary. One option is to estimate the RHS of Eq. 3.32 by employing the value of the integrand at the initial point t_n. This approach, known as the *explicit* or *forward Euler method* yields:

$$\varphi^{n+1} = \varphi^n + f(t_n,\varphi^n)\Delta t \tag{3.33}$$

A second approach, known as the *implicit* or *backward Euler method* uses the final point t_{n+1} in order to estimate the integral yielding the following equation:

$$\varphi^{n+1} = \varphi^n + f(t_{n+1},\varphi^{n+1})\Delta t \tag{3.34}$$

Furthermore, it is possible to estimate the integral by taking advantage of the midpoint of the interval, an approach known as the *midpoint rule* and closely related to the *leapfrog method*, popular for solving partial differential equations:

$$\varphi^{n+1} = \varphi^n + f(t_{n+\frac{1}{2}},\varphi^{n+\frac{1}{2}})\Delta t \tag{3.35}$$

A final important strategy is based on the approximation of the integral by a straight line between the first and the final point. This is known as the *trapezoid rule* and is closely related to the widely used Crank-Nicolson method for solving partial differential equations:

$$\varphi^{n+1} = \varphi^n + \frac{1}{2}[f(t_n, \varphi^n) + f(t_{n+1}, \varphi^{n+1})]\Delta t \tag{3.36}$$

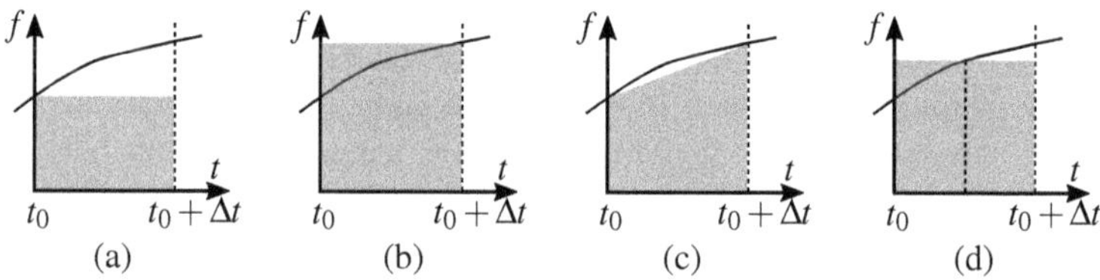

Figure 3.6: Estimation of the integral $f(t)$ over the interval Δt. (a) explicit Euler; (b) implicit Euler; (c) trapezoidal rule; (d) midpoint rule (cp. [FP02]).

The four methods described above are depicted schematically in Fig. 3.6. Since they rely on values of the unknown variable at two different times, they are known as two-level methods. Furthermore, the last three approaches involve values of φ at times at which the solution is, in fact, unknown. Hence, they require additional approximation or iterations and are therefore called *implicit*. In contrast, the forward Euler strategy falls into the category of the *explicit* methods.

Although all the discussed methods are stable for small values of Δt, it is important to establish their behaviour for large time steps. This is specially important for many problems in fluid mechanics involving multiple phenomena with large differences of time scales for which the solution of the slow, long term time scale often constitutes the focus of interest. In such problems, known as *stiff*, large time steps may be necessary and the stability of the employed time discretization scheme becomes crucial. One possible definition of stability, among several in the literature, judges a method for its ability to produce a bounded solution if the solution of the underlying differential equation is also known to be bounded. From this definition, it follows that the explicit Euler method is stable if the condition:

$$\left| 1 + \Delta t \frac{\partial f(t, \varphi)}{\partial \varphi} \right| < 1 \tag{3.37}$$

is fulfilled. If the generic function $f(t,\varphi)$ is allowed to assume complex values, which is the case for higher order systems with complex eigenvalues, the above condition requires the term $\Delta t \partial f(t,\varphi)/\partial\varphi)$ to be located inside a unit circle with centre at -1. The reason for this condition is that only values of this term with zero or negative real part lead to bounded solutions. Furthermore, if $f(t,\varphi)$ has only real values, the condition in Eq. 3.37 becomes:

$$\left|\Delta t \frac{\partial f(t,\varphi)}{\partial \varphi}\right| < 2 \tag{3.38}$$

Hence, the explicit Euler method falls into the category of *conditionally stable* methods. The other three methods discussed in this section are *unconditionally stable* because they produce bounded solutions regardless of the time step, provided that $\partial f(t,\varphi)/\partial\varphi$ < 0. There are, however, some differences in the behaviour of these three approaches for large values of Δt. The implicit Euler method, for instance, usually produces smooth solutions even for very large values of Δt, which makes it a good approach even for non-linear problems. On the other hand, the trapezoid rule tends to yield solutions with oscillations which might lead to instability in systems with high non-linearity.

Regarding their accuracy, both the explicit and implicit Euler approaches are first-order methods, meaning that their errors are proportional to Δt. On the other hand, the trapezoid and midpoint rule methods are of second-order accuracy with errors proportional to Δt^2. Note that the order of a method describes only the rate at which the error decreases with smaller values of Δt after the time step has already become *small enough*. Since the definition of a *small enough* Δt depends on both problem and method, the order of accuracy should be regarded as only one of several indicators of accuracy.

The application of the time discretization methods for a generic transport equation like the one introduced in Eq. 3.11 is discussed in the following. After adding the transient term to Eq. 3.11 and rearranging, one obtains:

$$\frac{\partial(\rho\varphi)}{\partial t} = -\frac{\partial(\rho\varphi u_j)}{\partial x_j} + \frac{\partial}{\partial x_j}\left(\Gamma\frac{\partial\varphi}{\partial x_j}\right) + q_\varphi = f(t,\varphi(t)) \tag{3.39}$$

Equation 3.39 closely resembles the ODE in Eq. 3.31 and any of the time discretization methods introduced above may be employed. The generic term $f(t,\varphi(t))$ contains the convective, diffusive and source terms which are to be discretized in space using either the Finite Difference Method (s. Section 3.1.1) or the Finite Volume Method (s. Section 3.1.2) at one or more time values. Several important characteristics of the time discretization methods applied to a generic transport equation will be examined in the following sections.

3.1.3.1 Explicit Euler method

In the easiest approach, the explicit Euler method, the values of the arbitrary property φ at the new time level t_{n+1} are evaluated based on known values at the earlier time t_n. Furthermore, the values at the neighbour nodal points of a given CV are all evaluated at earlier time levels. Therefore, the estimation of the value φ_i^{n+1} is done in an explicit fashion. Let us introduce a simplified, 1D version of the generic transport equation Eq. 3.39 assuming constant velocity, constant fluid properties and no source terms:

$$\frac{\partial \varphi}{\partial t} = -u\frac{\partial \varphi}{\partial x} + \frac{\Gamma}{\rho}\frac{\partial^2 \varphi}{\partial x^2} \tag{3.40}$$

Note that Eq. 3.40 implies that the important balance in the considered unsteady flow is the one between the convective and the diffusive terms. Although this behaviour is not usually seen in real flows, the simplified expression may be used to examine the time discretization methods. With the spatial derivatives discretized using the CDS scheme and with a uniform spatial grid in the x-direction, the value of the studied variable at the new time level is given by the algebraic expression:

$$\begin{aligned}\varphi_i^{n+1} &= \varphi_i^n + \left[-u\frac{\varphi_{i+1}^n - \varphi_{i-1}^n}{2\Delta x} + \frac{\Gamma}{\rho}\frac{\varphi_{i+1}^n + \varphi_{i-1}^n - 2\varphi_i^n}{(\Delta x)^2}\right]\Delta t \\ &= (1-2d)\varphi_i^n + (d - \frac{Co}{2})\varphi_{i+1}^n + (d + \frac{Co}{2})\varphi_{i-1}^n \end{aligned} \tag{3.41}$$

with the dimensionless parameters,

$$d = \frac{\Gamma \Delta t}{\rho (\Delta x)^2} \tag{3.42}$$

and

$$Co = \frac{u\Delta t}{\Delta x} \tag{3.43}$$

Here, the parameter d represents the ratio of time step Δt to the time required by a disturbance to be transported via diffusion over the distance Δx. The dimensionless number Co, known as the *Courant number*, equals the ratio of time step Δt to the time necessary for a disturbance to be transmitted via convection over the distance Δx and is a parameter of great importance in Computational Fluid Dynamics.

In order to examine the stability of the algebraic expression in Eq. 3.41, let us introduce the norm ε which represents the amount of change of the solution between consecutive time steps:

$$\varepsilon = \|\varphi^n - \varphi^{n-1}\| = \sqrt{\sum_i (\varphi_i^n - \varphi_i^{n-1})^2} \tag{3.44}$$

In order for the differential equation to reach steady-state solution, the parameter ε should, because of dissipation, decrease with time. However, from Eq. 3.41 it is evident that the coefficients of the terms φ_i^n and φ_{i+1}^n on the RHS might lead to an increasing value of ε and thus, to numerical instability if they become negative. Therefore, the restrictions $d < 0.5$ and $Co < 2d$ are necessary in order to guarantee stability. The resulting condition for the time step is given by:

$$\Delta t < \frac{\rho(\Delta x)^2}{2\Gamma} \tag{3.45}$$

while the relation between the convection and diffusion coefficients is given by:

$$\frac{\rho u \Delta x}{\Gamma} = Pe < 2 \tag{3.46}$$

where Pe stands for the *Peclet number.* As a combination of the explicit Euler scheme for ODEs and the CDS method, the approach presented here inherits the accuracy of both schemes and is therefore of first-order accuracy in time and second-order in space. As can be seen from Eq. 3.42, the condition $d < 0.5$ requires the time step to be reduced by a factor of four if the grid resolution in the x-direction is doubled. This leads to considerable computational requirements for problems approaching steady state or with large time scales. One possible solution to this issue is to apply UDS instead of CDS for the spatial discretization of the convective term in problems dominated by convection. The approach leads to the following expression:

$$\begin{aligned} \varphi_i^{n+1} &= \varphi_i^n + \left[-u \frac{\varphi_i^n - \varphi_{i-1}^n}{\Delta x} + \frac{\Gamma}{\rho} \frac{\varphi_{i+1}^n + \varphi_{i-1}^n - 2\varphi_i^n}{(\Delta x)^2} \right] \Delta t \\ &= (1 - 2d - Co)\varphi_i^n + d\varphi_{i+1}^n + (d + Co)\varphi_{i-1}^n \end{aligned} \tag{3.47}$$

In this case, only the coefficient of the nodal value φ_i^n can become negative and lead to instability. Hence, the stability requirement with this modified approach is given by:

$$\Delta t < \frac{1}{\dfrac{2\Gamma}{\rho(\Delta x)^2} + \dfrac{u}{\Delta x}} \tag{3.48}$$

If convection effects are negligible, the term $u/\Delta x$ in Eq. 3.48 becomes small and the stability criterion for the time step equals the one on Eq. 3.45. On the other hand, for problems with dominant convection, the criterion becomes:

$$\Delta t < \frac{\Delta x}{u} \qquad \text{or} \qquad Co < 1 \tag{3.49}$$

If both diffusion and convection are important for an accurate flow description, the stability requirement becomes more complex. One simple and widely used approach is to guarantee that both criteria are satisfied simultaneously. Although this approach is somehow restrictive, it is both easy to apply and robust. Furthermore, because of the introduction of UDS, the approach is now of first-order accuracy in both time and space. Therefore, small values of Δt and Δx are required.

3.1.3.2 Implicit Euler method

As already described, the implicit Euler method applied on ODEs produces smooth results even for large values of Δt, making it a good approach for problems with high non-linearity. In this approach, all of the fluxes and source terms on the RHS of the transport equation Eq. 3.40 are evaluated in terms of the values of the arbitrary variable φ at the new time level. If the spatial derivatives are discretized using CDS, the application of the implicit Euler method on the generic simplified transport equation yields:

$$\varphi_i^{n+1} = \varphi_i^n + \left[-u \frac{\varphi_{i+1}^{n+1} - \varphi_{i-1}^{n+1}}{2\Delta x} + \frac{\Gamma}{\rho} \frac{\varphi_{i+1}^{n+1} + \varphi_{i-1}^{n+1} - 2\varphi_i^{n+1}}{(\Delta x)^2} \right] \Delta t \tag{3.50}$$

which can be rearranged to obtain:

$$(1+2d)\varphi_i^{n+1} + \left(\frac{Co}{2} - d \right) \varphi_{i+1}^{n+1} + \left(-\frac{Co}{2} - d \right) \varphi_{i-1}^{n+1} = \varphi_i^n \tag{3.51}$$

where the parameters d and Co correspond to the already defined dimensionless numbers in Eqs. 3.42 and 3.43. Although the method remains stable with arbitrary values of Δt, oscillation issues might arise if CDS is applied on a coarse grid. Additional general disadvantages of the implicit Euler method include its first-order accuracy in time and the increase in computational cost as a result of the higher complexity of the system of algebraic equations to be solved. As already mentioned, the main advantage of this approach resides in the possible use of very high values of Δt without loss of stability, a factor that can, in many cases, outweigh the described disadvantages [FP02].

3.1.3.3 Crank-Nicolson method

The Crank-Nicolson method is based on the second-order trapezoid rule for initial value problems. Its application to the generic simplified transport equation with CDS for the spatial derivation produces the expression:

$$\varphi_i^{n+1} = \varphi_i^n + \frac{\Delta t}{2}\left[-u\frac{\varphi_{i+1}^{n+1} - \varphi_{i-1}^{n+1}}{2\Delta x} + \frac{\Gamma}{\rho}\frac{\varphi_{i+1}^{n+1} + \varphi_{i-1}^{n+1} - 2\varphi_i^{n+1}}{(\Delta x)^2}\right] + \frac{\Delta t}{2}\left[-u\frac{\varphi_{i+1}^{n} - \varphi_{i-1}^{n}}{2\Delta x} + \frac{\Gamma}{\rho}\frac{\varphi_{i+1}^{n} + \varphi_{i-1}^{n} - 2\varphi_i^{n}}{(\Delta x)^2}\right] \tag{3.52}$$

As evident from Eq. 3.52, this approach corresponds to a combination or blending of the implicit and the explicit Euler methods. Since the values of the variable φ at the new time level are unknown, the Crank-Nicolson method is implicit. Moreover, the resulting system of algebraic equations is very similar to the one obtained with the implicit Euler scheme and the increase in computational cost is therefore minimal. It should be noted that the expression in Eq. 3.52 corresponds to the equal blending of both methods. In this form, the Crank-Nicolson scheme is, like the trapezoid rule, of second-order accuracy and the approach becomes a good alternative when time accuracy is key. On the other hand, it is also possible to alter the blending factor based on the particular problem at hand. For instance, a higher factor for the implicit contribution would lead to an increase in stability, albeit at the expense of accuracy. Indeed, second-order accuracy is only obtained with equal contributions of both the implicit and explicit Euler approaches. For any other blending factor, the accuracy falls back to first-order [FP02].

3.1.3.4 Other methods

A variety of additional methods for the time discretization of a generic transport equation are available. They include the *Leapfrog method*, an explicit scheme based on the midpoint rule commonly used in applications like meteorology and oceanography and described in Section 3.3.1. Another alternative is the *Three time level method*, which is obtained from a quadratic backward approximation in time and is hence, fully implicit and of second-order accuracy. For detailed information on these methods, the reader is referred to [FP02].

3.1.4 Solution algorithms for the Navier-Stokes equations

In Sections 3.1.1, 3.1.2 and 3.1.3, the discretization methods for the spatial and temporal terms of a generic transport equation were introduced. In this section, computational algorithms for solving the system of equations describing mass and momentum conservation are described in detail. Comparison between the momentum equations Eq. 2.43 and the generic transport equation Eq. 3.39 with $\varphi = u_i$ shows that the convective terms in both equations are equal. Hence, when working with the Navier-Stokes equations, the convective term can be treated with either the FDM or the FVM plus any of the interpolation and temporal schemes described in the prior sections. The same applies for the transient term. On the other hand, the viscous contribution, represented in the generic transport equation by the diffusive term, requires a closer examination. Specifically, the viscous term of the compressible Navier-Stokes equations includes contributions from both the bulk viscosity and the spatial variability of the viscosity. For simplicity, let us consider an incompressible Newtonian fluid. In this case, the viscous term is given by:

$$\frac{\partial \tau_{ij}}{\partial x_j} = \frac{\partial}{\partial x_j}\left[\mu\left(\frac{\partial u_i}{\partial x_j} + \frac{\partial u_j}{\partial x_i}\right)\right] \tag{3.53}$$

The term present in both the momentum equations and the generic transport equation with $\Gamma = \mu$ is:

$$\frac{\partial}{\partial x_j}\left(\mu\frac{\partial u_i}{\partial x_j}\right) \tag{3.54}$$

This term can be discretized with any of the approaches already introduced. Furthermore, the term not present in the generic equation is:

$$\frac{\partial}{\partial x_j}\left(\mu\frac{\partial u_j}{\partial x_i}\right) \tag{3.55}$$

where summation over j applies. This additional term can essentially be discretized as the term in Eq. 3.54 using the methods described in the past sections. Since the term equals zero for constant viscosity, it is often assumed to be small in comparison with Eq. 3.54 and frequently treated in an explicit manner, even if implicit solution schemes are employed.

One key difference between the generic transport equation and the Navier-Stokes equations is the presence of the pressure term. Note that strictly speaking, this term lacks its own conservation equation. Furthermore, the mass conservation equation for an incompressible flow basically imposes a restriction over the flow velocity field without directly enabling the computation of any additional variable. One way to circumvent this, is to construct a pressure field in such a way that it satisfies the continuity equation. On the other hand, for compressible flows, the continuity equation may be used to compute the density and through an equation of state, the pressure. This approach is, however, not appropriate for incompressible flows or low values of the Mach number. Two of the most well-known and widely used algorithms to achieve coupling between the momentum and the continuity equation as well as to calculate the pressure field for incompressible flows are shown in the next sections. They are known as the *SIMPLE* and *PISO* algorithms.

3.1.4.1 The SIMPLE and PISO algorithms

Let us begin by considering the implicitly discretized general transport equation Eq. 3.56:

$$(1+2d)\varphi_i^{n+1} + \left(\frac{Co}{2} - d\right)\varphi_{i+1}^{n+1} + \left(-\frac{Co}{2} - d\right)\varphi_{i-1}^{n+1} = \varphi_i^n \tag{3.56}$$

With $\varphi = u_i$ and using the cell notation $P = i$, $E = i-1$ and $W = i+1$, the following discretized momentum equation can be formulated:

$$\underbrace{(1+2d)}_{A_P^{u_i}} u_{i,P}^{n+1} + \underbrace{\left(\frac{Co}{2} - d\right)}_{A_E^{u_i}} u_{i,E}^{n+1} + \underbrace{\left(-\frac{Co}{2} - d\right)}_{A_W^{u_i}} u_{i,W}^{n+1} = Q_{u_i}^{n+1} - \left(\frac{\partial p^{n+1}}{\partial x_i}\right)_P \tag{3.57}$$

where the pressure term has been written as a separate term and the source term $Q_{u_i}^{n+1}$ includes all terms that may be explicitly calculated as a function of the velocity value at the old time level u_i^n, as well as linearized body forces that depended on the velocity at the new time level u_i^{n+1}. Eq. 3.57 may be written in the following condensed manner:

$$A_P^{u_i} u_{i,P}^{n+1} = \sum_l A_l^{u_i} u_{i,l}^{n+1} = Q_{u_i}^{n+1} - \left(\frac{\partial p^{n+1}}{\partial x_i}\right)_P \tag{3.58}$$

where the index l stands for the neighbour points. Since the A coefficients and sometimes the source term are a function of the unknown velocity u_i^{n+1}, this equation is nonlinear. Hence, an iterative method should be employed. During the flow computation, two types of iterations are performed. In the so-called *outer iterations*, the coefficients of the matrix and source terms are updated. During the *inner iterations*, the coefficients are kept constant and the linear systems are solved until a given tolerance is reached. The equations solved during an outer iteration are:

$$A_P^{u_i} u_{i,P}^{m*} + \sum_l A_l^{u_i} u_{i,l}^{m*} = Q_{u_i}^{m-1} - \left(\frac{\partial p^{m-1}}{\partial x_i}\right)_P \tag{3.59}$$

where the outer iteration counter m has been introduced instead of the time step index $n+1$. Therefore, u_i^m stands for the current estimate of the velocity u_i^{n+1}. Furthermore, at the beginning of each outer iteration, the RHS of Eq. 3.59 is evaluated from the values of the previous outer iteration $m-1$. For this reason, the velocities estimated with Eq. 3.59 do not necessarily satisfy the continuity equation and are hence marked with an asterisk (*). Eq. 3.59 can be rewritten as follows:

$$u_{i,P}^{m*} = -\frac{1}{A_P^{u_i}}\left(\frac{\partial p^{m-1}}{\partial x_i}\right)_P + \frac{1}{A_P^{u_i}}\left(Q_{u_i}^{m-1} - \sum_l A_l^{u_i} u_{i,l}^{m*}\right) \tag{3.60}$$

In the next step, a pressure p^m or a pressure correction $p' = p^m - p^{m-1}$ satisfying the continuity equation should be computed. The result is subsequently used to correct the velocity. The relation between the pressure value and the corrected velocity can be expressed as follows:

$$u_{i,P}^{m} = -\frac{1}{A_P^{u_i}}\left(\frac{\partial p^m}{\partial x_i}\right)_P + \frac{1}{A_P^{u_i}}\left(Q_{u_i}^{m-1} - \sum_l A_l^{u_i} u_{i,l}^{m}\right) \tag{3.61}$$

while the velocity correction $u'_{i,P} = u_{i,P}^{m} - u_{i,P}^{m*}$ is given by,

$$u_{i,P}^{m} - u_{i,P}^{m*} = -\frac{1}{A_P^{u_i}}\left(\frac{\partial (p^m - p^{m-1})}{\partial x_i}\right)_P - \frac{1}{A_P^{u_i}}\underbrace{\left(\sum_l A_l^{u_i}(u_{i,l}^{m} - u_{i,l}^{m*})\right)}_{\approx 0} \tag{3.62}$$

The last term on the RHS of Eq. 3.62 is unknown and hence, assumed to be zero. Since this assumption is hardly accurate, it is considered to be the main reason for the general slow convergence of the algorithm. By taking the divergence of Eq. 3.62 after multiplication with ρ, the following expression is obtained:

$$\underbrace{\frac{\partial \rho u_{i,P}^{m}}{\partial x_i}}_{=0} - \frac{\partial \rho u_{i,P}^{m*}}{\partial x_i} = -\frac{\partial}{\partial x_i}\left[\frac{\rho}{A_P^{u_i}}\left(\frac{\partial p'}{\partial x_i}\right)\right] \tag{3.63}$$

The first term on the LHS of Eq. 3.63 corresponds to the continuity equation and equals zero. The remaining expression corresponds to an implicit equation for the pressure correction in terms of the estimated value of $u_{i,P}^{m*}$. The pressure equation might be discretized and solved with any of the schemes introduced in prior sections. The solution is a pressure correction or pressure value that satisfies the continuity equation. In the last step, the velocity is corrected through substitution of p' into Eq. 3.62. This yields:

$$u_{i,P}^{m} = u_{i,P}^{m*} - \frac{1}{A_P^{u_i}}\left(\frac{\partial p'}{\partial x_i}\right)_P \tag{3.64}$$

This widely used algorithm is known as the *Semi-Implicit Method for Pressure-Linked Equations* (SIMPLE) and was developed by Prof. Brian Spalding and his student Suhas Patankar in the 1970s [PS72]. An alternative procedure can be developed by using the SIMPLE algorithm as an intermediate step while considering the pressure at the time levels $m-1$ and the intermediate time level $m-1/2$. In this case, the first estimate of the velocity is given by:

$$u_{i,P}^{m*} = -\frac{1}{A_P^{u_i}}\left(\frac{\partial p^{m-1}}{\partial x_i}\right)_P + \frac{1}{A_P^{u_i}}\left(Q_{u_i}^{m-1} - \sum_l A_l^{u_i} u_{i,l}^{m*}\right) \tag{3.65}$$

while the relation between the second estimate of the velocity $u_{i,P}^{m**}$ and the pressure at the intermediate time level $p^{m-1/2}$ can be expressed as follows:

$$u_{i,P}^{m**} = -\frac{1}{A_P^{u_i}}\left(\frac{\partial p^{m-1/2}}{\partial x_i}\right)_P + \frac{1}{A_P^{u_i}}\left(Q_{u_i}^{m-1} - \sum_l A_l^{u_i} u_{i,l}^{m**}\right) \tag{3.66}$$

The velocity correction $u'_{i,P} = u_{i,P}^{m**} - u_{i,P}^{m*}$ is hence given by,

$$u'_{i,P} = u_{i,P}^{m**} - u_{i,P}^{m*} = -\frac{1}{A_P^{u_i}}\left(\frac{\partial(p^{m-1/2} - p^{m-1})}{\partial x_i}\right)_P - \frac{1}{A_P^{u_i}}\underbrace{\left(\sum_l A_l^{u_i}(u_{i,l}^{m**} - u_{i,l}^{m*})\right)}_{\approx 0} \tag{3.67}$$

where the last term on the RHS of Eq. 3.67 is assumed to equal zero. Up to this point, the procedure is the same as the SIMPLE algorithm. In the next step, the relation between the velocity $u_{i,P}^{m**}$ and the pressure $p^{m-1/2}$ is redefined as follows:

$$u_{i,P}^{m**} = -\frac{1}{A_P^{u_i}}\left(\frac{\partial p^{m-1/2}}{\partial x_i}\right)_P + \frac{1}{A_P^{u_i}}\left(Q_{u_i}^{m-1} - \sum_l A_l^{u_i} u_{i,l}^{m*}\right) \tag{3.68}$$

while the corrected velocity is given by:

$$u_{i,P}^{m} = -\frac{1}{A_P^{u_i}}\left(\frac{\partial p^{m}}{\partial x_i}\right)_P + \frac{1}{A_P^{u_i}}\left(Q_{u_i}^{m-1} - \sum_l A_l^{u_i} u_{i,l}^{m**}\right) \tag{3.69}$$

Note that the last terms on the RHS of Eqs. 3.68 and 3.69 (diagonal elements in the coefficient matrix) are now defined in terms of the velocities of the prior intermediate step rather than the current one. The velocity correction $u''_{i,P} = u_{i,P}^{m} - u_{i,P}^{m**}$ for the second sub-step is given by:

$$u_{i,P}^{m} - u_{i,P}^{m**} = -\frac{1}{A_P^{u_i}}\left(\frac{\partial(p^{m} - p^{m-1/2})}{\partial x_i}\right)_P - \frac{1}{A_P^{u_i}}\underbrace{\left(\sum_l A_l^{u_i}(u_{i,l}^{m**} - u_{i,l}^{m*})\right)}_{u'_{i,P}\text{ from SIMPLE step}} \tag{3.70}$$

Because of the manner in which the velocities for the second sub-step were defined, neglecting of the last term in Eq. 3.70 is no longer necessary as $u'_{i,P}$ is estimated in the prior intermediate SIMPLE step. In order to correct the velocities and obtain a pressure which satisfies the continuity equation, Eq. 3.67 is multiplied with the density and the divergence of the resulting expression is taken. This yields:

$$\underbrace{\frac{\partial \rho u_{i,P}^{m**}}{\partial x_i}}_{=0} - \frac{\partial \rho u_{i,P}^{m*}}{\partial x_i} = -\frac{\partial}{\partial x_i}\left[\frac{\rho}{A_P^{u_i}}\left(\frac{\partial p'}{\partial x_i}\right)\right] \tag{3.71}$$

with $p' = p^{m-1/2} - p^{m-1}$. This expression corresponds to an implicit equation for the first pressure correction in terms of the known value of $u^{m*}_{i,P}$. The pressure equation might be discretized and solved with any of the schemes introduced in prior sections. Substitution of p' into Eq. 3.67 yields the first correction of the velocity:

$$u^{m**}_{i,P} = u^{m*}_{i,P} \underbrace{- \frac{1}{A^{u_i}_P} \left(\frac{\partial p'}{\partial x_i} \right)_P}_{u'_{i,P}} \tag{3.72}$$

In a similar way, by taking the divergence of Eq. 3.70 after multiplication with ρ, an implicit equation for the second pressure correction $p'' = p^m - p^{m-1/2}$ is obtained:

$$\underbrace{\frac{\partial \rho u^m_{i,P}}{\partial x_i}}_{=0} - \frac{\partial \rho u^{m**}_{i,P}}{\partial x_i} = - \frac{\partial}{\partial x_i} \left[\frac{\rho}{A^{u_i}_P} \left(\frac{\partial p''}{\partial x_i} + \sum_l A^{u_i}_l \underbrace{(u^{m**}_{i,l} - u^{m*}_{i,l})}_{u'_i} \right) \right] \tag{3.73}$$

Substitution of p'' into Eq. 3.70 yields the final correction of the velocity:

$$u^m_{i,P} = u^{m**}_{i,P} \underbrace{- \frac{1}{A^{u_i}_P} \left[\left(\frac{\partial p''}{\partial x_i} \right)_P + \sum_l A^{u_i}_l u'_{i,l} \right]}_{u''_{i,P}} \tag{3.74}$$

while the final pressure is given by $p^m = p^{m-1/2} + p''$ and $p^{m-1/2} = p^{m-1} + p'$ respectively. This procedure is known as the PISO algorithm (Pressure Implicit with Splitting of Operator) and was developed by Dr. Raad Issa in 1986 [Iss86]. Although the computational requirements of the PISO algorithm are higher than those of the SIMPLE procedure, the PISO algorithm has been shown to have positive effects on the stability and convergence time of numerical computations. The steps involved in the numerical flow simulation using these or similar methods can be summarized as follows:

1. Use the latest solutions at the prior time level u^n_i and p^n as starting values for the new time level u^{n+1}_i and p^{n+1}.
2. Obtain the first estimate of the velocity in the new time level u^{m*}_i from the linearized algebraic momentum equations.
3. Obtain p' from the pressure-correction equation.
4. With p', calculate the corrected velocities u^m_i and the new pressure field p^m both satisfying the continuity equation. If the PISO algorithm is employed, solve the second pressure equation and use the result to correct both velocity and pressure one additional time.

5. Since the corrected values now satisfy the continuity equation but not necessarily the momentum equation, the procedure must be repeated from step 2 with u_i^m and p^m as new estimates of u_i^{n+1} and p^{n+1}. The steps are repeated until the desired tolerances are reached.
6. Proceed to the next time step.

Both the SIMPLE and PISO algorithms as well as many other related methods display a good efficiency when dealing with steady state problems. Because of its stability and greater overall accuracy, the PISO method is also a good choice for transient problems. On the other hand, if time accuracy is not of interest, the SIMPLE algorithm is a good approach because of its overall lower computational requirements (cp. [FP02]).

3.2 Direct Simulation Monte Carlo (DSMC)

The Direct Simulation Monte Carlo Method (DSMC) developed by Prof. Graeme Bird attempts to solve the Boltzmann-Equation in a probabilistic manner and is based on the kinetic theory of gases. A gas is modelled as a collection of spherical test particles or molecules, each representing a given number of real gas molecules. Test particles move in a numerical, discretized spatial grid and possess a variety of properties like mass, diameter, position and velocity. Since each test particle represents a large number of real molecules, the properties of the individual test particles should be considered as mean values over a large number of molecules. Since the method attempts to solve the Boltzmann-Equation, it is considered to be very general, adequate for flows in thermodynamic non-equilibrium and essentially applicable across all flow regimes. However, because of the relatively high computational requirements at low values of the Knudsen number Kn, DSMC is more adequate and hence widely used for rarefied gas flows. At low values of Kn, classic CFD approaches like FVM are much more efficient.

In the DSMC method, the particles motion and their collisions with each other and with the domain boundaries are handled separately. This approach is accurate if the time step Δt used to advance the simulation in time is smaller than the characteristic time between collisions. Furthermore, only binary collisions are considered. For continuum flows with high number densities, the molecules collide very often with each other and the assumption of exclusive binary collisions becomes inaccurate. Although this can be mitigated using very low values of Δt, the steep increase in computational requirements renders the approach usually unpractical for continuum flows. The main steps of a typical DSMC simulation involve the collision-less transport or motion of the test particles, the inter-molecular collisions and the interaction between molecules and walls. These aspects will be examined in the following sections.

In order to introduce the main characteristics of a typical DSMC approach, some basic definitions should be presented first. Let the *weight* W equal the number of real atoms or molecules represented by a single DSMC particle N_i. If the weight is the same for all test molecules, the number density of a given species in an arbitrary control volume V is given by:

$$n = \frac{W}{V} \sum N_i \tag{3.75}$$

Hence, the number density can be obtained through a simple arithmetic summation of the DSMC particles multiplied with their weight and divided by the cell's volume. At the beginning of a DSMC simulation, test particles are usually generated by assuming a gas in thermodynamic equilibrium with known number density n, temperature T and macroscopic fluid velocity v. Since thermodynamic equilibrium is assumed, the individual particles are created in the domain according to the Maxwell distribution described in Section 2.9.3. Particles and their properties like position r, particle's velocity c, mass m, diameter d and degrees of freedom are stored in computational memory. If the simulation does not involve chemical reactions, most of these properties remain constant over the course of a simulation and only the particles positions and velocities must be calculated at each time step. Furthermore, the accurate computation of the positions and velocities is crucial for the estimation of the macroscopic properties, which are in most cases the primary objective of a DSMC simulation. Because of the discrete handling of gas particles and the usually very high values of W, the DSMC method is from a statistical standpoint rather noisy. Hence, some kind of time averaging is usually applied before the estimation of macroscopic properties.

3.2.1 Molecular transport

Let us consider a DSMC particle with velocity c_{old}, position r_{old} and mass m located in an arbitrary spatial domain. The motion of the particle is easily described by Newton's second law of motion. In the presence of an external force F, the position of the particle after the time Δt is given by:

$$r_{new} = r_{old} + \Delta t c_{old} + \frac{1}{2}(\Delta t)^2 \frac{F}{m} \tag{3.76}$$

Possible external forces in Eq. 3.76 are the gravitational force mg or, for charged particles, the electric force. The case of charge carriers is described in Section 3.3. Since Δt in Eq. 3.76 is a small quantity, the gravitational force term with factor Δt^2 is usually much smaller than the first and second terms and is often neglected (cp. [Fas11]).

As long as the test molecule does not encounter a domain boundary, the new particle's position is obtained from Eq. 3.76. On the other hand, there are several approaches for the handling of molecules on domain boundaries. In the case of an inlet or outlet boundary, molecules are usually just deleted from memory. Although this approach is adequate for an outlet boundary, it can lead to deviations if the mass flow rate entering the domain through the inlet is to be defined in an accurate manner. An alternative is to handle the inlet as an open boundary for molecules entering the domain and as a wall for particles attempting to exit through the inlet. This approach enables the precise definition of the mass flow rate entering the simulation domain and its implementation has been described in [Fas11]. If the test particle encounters a wall during the motion procedure, one needs to determine the alternative time step $\Delta t'$ at which the interaction with the wall takes place and move the particle only until this point in time. The particle then bounces off the wall and flies back into the domain with a new post-collision velocity. The particle's motion is then continued until the global time step Δt is covered. Two widely used approaches for the estimation of the particle's velocity after a wall interaction are *specular reflection* and *diffusive reflection*. They are described briefly in Section 3.2.3.

3.2.2 Molecular collisions

After the motion procedure is completed, the next step in a typical DSMC simulation involves the computation of the intermolecular collisions and the subsequent update of the molecular velocities c. The considered collection of particles is handled in a cell-wise manner, while the actual collisions take place between molecules inside sub-cells. To achieve this, the total number of collisions in the considered cell must be estimated so that the actual collision rate of the gas is modelled accurately. The collision rate ν_c was introduced in Eq. 2.128 for a moving particle in a static background gas as follows:

$$\nu_c = \pi d^2 c_0 n$$

This expression can be rewritten for a homogeneous collection of moving particles to give:

$$\nu_c = \pi d^2 \overline{c_r} n = n \overline{\sigma_T c_r} \tag{3.77}$$

where $c_r = |c_r|$ represents the magnitude of the relative velocity between two given particles in the cell and the bar indicates the mean cell value of the product $\sigma_T c_r$. Furthermore, σ_T stands for the total collision cross section πd^2. The correct modelling of the collision diameter d is discussed in detail later in this section. Nonetheless, it should be noted that d may be a function of the particles velocities. For this reason, d is assumed to be variable over the particles collection, which explains the bar over the term σ_T in Eq. 3.77. The total number of collisions per unit volume in the time interval Δt can be expressed as follows:

$$N_c \Delta t = \frac{1}{2} n \nu_c = \frac{1}{2} n^2 \overline{\sigma_T c_r} \tag{3.78}$$

where N_c stands for the number of collisions per unit volume and unit time. The collision probability P for a given particle pair is proportional to the product of their total collision cross sections σ_T and their relative velocities c_r. One straightforward approach to obtain the right number of collisions in the cell involves the definition of random collision pairs, the computation of their respective collision probabilities $\sigma_T c_r$ and the subsequent estimation of the maximum value in the cell $(\sigma_T c_r)_{\max}$. Next, the ratio $\sigma_T c_r / (\sigma_T c_r)_{\max}$ for each particle pair can be compared to a random number between zero and one in order to select the pairs involved in actual collisions. This procedure is known as the *acceptance-rejection method.* However, if N stands for the total number of test particles in the cell, this approach involves very high computational requirements of the order of the total number of possible pair combinations $N(N-1)/2$ (cp. [Bir94]). For this reason, a variety of alternative methods has been proposed. One of the most widely used schemes is the *no time counter* (NTC) method developed by Bird ([Bir89]) and described in the following. The probability P of a collision between two DSMC particles can be defined as:

$$P = W \sigma_T c_r \Delta t / V \tag{3.79}$$

where W stands for the weight factor of the simulated particles and V, for the volume of the considered cell. According to Eq. 3.79, the collision probability of a particle pair equals the ratio of total volume covered by the particles in the time interval Δt to the total cell's volume. Note that for a given cell, the relative velocity c_r depends on the selected particle's pair. Furthermore, the total collision cross section σ_T is, as already mentioned, usually a function of c_r. The remaining variables in Eq. 3.79 are constant. In theory, one could now select all possible particle pairs $N(N-1)/2$ and compute the collisions with probability P. Considering the possible high number of particle combinations and the fact that P is usually very small, this approach is not particularly efficient. In the NTC scheme, the probability P is increased through division by the factor P_{max}, with:

$$P_{max} = W (\sigma_T c_r)_{max} \Delta t / V \tag{3.80}$$

Furthermore, instead of selecting all possible pairs, only the fraction $P_{max} N(N-1)/2$ is considered as collision candidates, which reduces the computational cost. The method can therefore be summarized as follows. The number of collision candidates is reduced by the factor P_{max}. To compensate this, the collision probability is increased by the same factor. The selected pairs to be considered as collision partners is then given by:

$$Candidates = \frac{1}{2}N(N-1)W(\sigma_T c_r)_{max}\frac{\Delta t}{V} \tag{3.81}$$

and the selected pairs collide with the probability:

$$P_{collision} = \frac{\sigma_T c_r}{(\sigma_T c_r)_{max}} \tag{3.82}$$

In this scheme, the quantity $(\sigma_T c_r)_{max}$ must be stored in computational memory for each cell in the domain. Note that this factor appears in the numerator of Eq. 3.81 and the denominator of Eq. 3.82 and as a consequence, it does not affect the resulting collision rate (cp. [Bir94]). The collisions between the considered particles might be elastic (no exchange of translational or internal energy) or inelastic. In this section, only elastic collisions are considered.

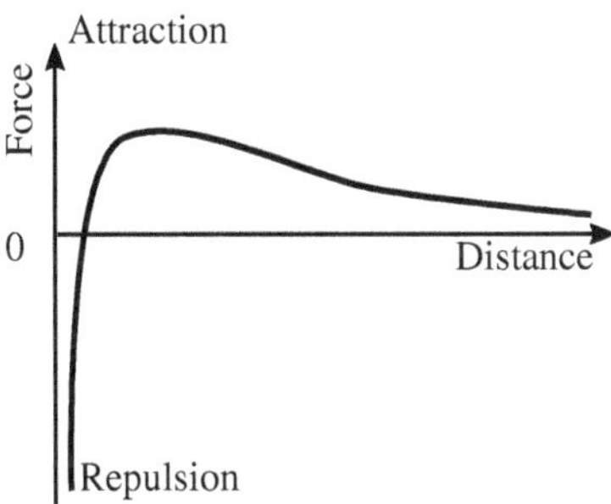

Figure 3.7: Molecular interaction profile as a function of molecular distance (cp. [Bir94]).

Let us now shift focus to the estimation of the total collision cross section σ_T. Real gas particles interact with each other in either an attractive or a repulsive way over a range of distances as shown in Fig. 3.7. At large distances, the interaction force approaches zero. If the distance decreases, the force becomes first moderate attractive and at short distances, strongly repulsive. This means that the total collision cross section σ_T is, in reality, infinite. However, most of the interactions at large distances correspond to glancing collisions with very small deflection angles. Quantum mechanical considerations have shown that these glancing interactions cannot be defined in an exact manner and hence, the use of a finite collision cross section is not only practical but also necessary (cp. [Bir94]). In the simplest approach, the *hard sphere model*, gas molecules are handled as hard spheres with constant diameter d as described in Section 2.9.1. The distance at which a binary interaction occurs is given by:

$$r = \frac{1}{2}(d_1 + d_2) = d_{12} \tag{3.83}$$

where d_1 and d_2 stand for the collisional diameter of the respective particles. Note that the collisional diameter d is not the atomic diameter but a phenomenological quantity strongly related to the viscosity of the gas. For a hard sphere gas with the coefficient of viscosity μ_{ref} at the temperature T_{ref}, the collisional diameter can be estimated as follows:

$$d = \left(\frac{(5/16)(mk_B T_{\mathrm{ref}}/\pi)^{1/2}}{\mu_{\mathrm{ref}}} \right)^{1/2} \tag{3.84}$$

where m represents the molecular mass, k_B the Boltzmann constant and the subscript $_{\mathrm{ref}}$ stands for reference quantities. The total collision cross section for a hard sphere gas is given by:

$$\sigma_T = \pi d_{12}^2 \tag{3.85}$$

The hard sphere model leads to isotropic particle scattering in the centre of mass frame of reference. Although this scattering law is not realistic, changes in this approach have shown to have only a very small influence in analytic and numerical studies of gas flows. On the other hand, this model effectively assumes that the particles interact with each other only if the distance between their centres decreases to d_{12} and d_{12} is assumed to be independent of the relative velocity of the particles c_r. In reality, σ_T decreases with higher values of c_r, an effect which is related to the temperature dependency of the viscosity coefficient μ. To take this into account, the variable hard sphere (VHS) model was proposed by Bird in 1981 [Bir81]. In this approach, the relation between the collisional diameter and the relative velocity is usually defined through an inverse power law of the quantity ν:

$$d = d_{\mathrm{ref}}(c_{r,\mathrm{ref}}/c_r)^{\nu} \tag{3.86}$$

or in a more explicit form,

$$d = d_{\mathrm{ref}} \left[\left\{ 2k_B T_{\mathrm{ref}}/(m_r c_r^2) \right\}^{\omega - 1/2} \Big/ \Gamma(5/2 - \omega) \right]^{1/2} \tag{3.87}$$

with $\nu = \omega - 1/2$. Furthermore, ω stands for the *viscosity index*, $m_r = m_1 m_2/(m_1 + m_2)$ for the *reduced mass* of the colliding molecules and Γ, for the gamma function. The viscosity index describes the temperature dependency of the viscosity in the following way:

$$\mu \propto T^{\omega} \tag{3.88}$$

which can be rewritten if a reference viscosity at a reference temperature is known, yielding:

$$\mu = \mu_{\text{ref}} \left(\frac{T}{T_{\text{ref}}} \right)^{\omega} \tag{3.89}$$

If values of μ as a function of T are known, Eq. 3.89 may be used to estimate the viscosity index. Values of the viscosity index for commonly used gases can also be found in the literature ([CC70], [Bir94]). The reference diameter for a VHS molecule in Eqs. 3.86 and 3.87 is given by:

$$d_{\text{ref}} = \left(\frac{5(\alpha+1)(\alpha+2)(m k_B T_{\text{ref}}/\pi)^{1/2}}{4\alpha(5-2\omega)(7-2\omega)\mu_{\text{ref}}} \right)^{1/2} \tag{3.90}$$

Where $\alpha = 1$. Note that $\alpha \neq 1$ corresponds to the *variable soft sphere* (VSS) model which may be employed if an accurate prediction of the Schmidt number, essentially describing the ratio of viscosity to diffusion, is important for the problem at hand. For details in this model, the reader is referred to [Bir94] and [Bir13]. The effective total collision cross section for the VHS model is $\sigma_T = \pi d_{12}^2$ with d from Eq. 3.87. As in the hard sphere model, the scattering of the particles through collisions in the VHS model is isotropic. This means that the post-collision relative velocity can take all directions with equal probability. For an specific collision in both the hard sphere and VHS models, the *deflection angle* χ of the relative velocity vector is given by:

$$\chi = 2cos^{-1}(b/d) \tag{3.91}$$

where b stands for the distance of closest approach of the undisturbed trajectories as shown in Fig. 3.8. Hence, for a head-on collision, the deflection angle is π, while the limiting case $b = d$ leads, as expected, to a deflection angle of zero. Since the scattering in the hard sphere and VHS models is isotropic, all values in the range $0 < b < d$ are equally likely. This greatly simplifies the calculation of the post-collision velocities as will be shown in the following.

If it has been established through the NTC method that a collision between two particles takes place, the next step involves the estimation of their post-collision velocities.

Let c_1 and c_2 denote the pre-collision velocities and c_1^* and c_2^* the post-collision velocities of the particles involved in an elastic binary collision. The determination of the post-collision velocities is based on geometrical considerations, on the conservation of momentum,

$$m_1 c_1 + m_2 c_2 = m_1 c_1^* + m_2 c_2^* = (m_1 + m_2) c_m \tag{3.92}$$

and the conservation of internal energy,

$$m_1 c_1^2 + m_2 c_2^2 = m_1 c_1^{*2} + m_2 c_2^{*2} \tag{3.93}$$

where m denotes the mass of the particles involved in the collision and c_m, the velocity of the centre of mass. From these considerations, it can be easily shown that the magnitude of the relative velocity remains unchanged after a collision. Hence:

$$c_r^* = c_r \tag{3.94}$$

The pre-collision velocities might be expressed in terms of the centre of mass and pre-collision relative velocities as follows:

$$c_1 = c_m + \frac{m_2}{m_1 + m_2} c_r \tag{3.95}$$

$$c_2 = c_m - \frac{m_1}{m_1 + m_2} c_r \tag{3.96}$$

and similarly, the post-collision velocities are given by:

$$c_1^* = c_m + \frac{m_2}{m_1 + m_2} c_r^* \tag{3.97}$$

$$c_2^* = c_m - \frac{m_1}{m_1 + m_2} c_r^* \tag{3.98}$$

Note that the velocity of the centre of mass c_m remains unchanged after a collision. As already mentioned, because of the isotropic scattering in the hard sphere and VHS models, the post-collision relative velocity can take all directions with equal probability. Hence, the determination of the post-collision velocities c_1^* and c_2^* involves the generation of the random vector c_r^* with magnitude c_r and subsequent substitution of c_r^* into Eqs. 3.97 and 3.98. For more details on the mechanics of elastic collisions the reader is referred to [Bir94].

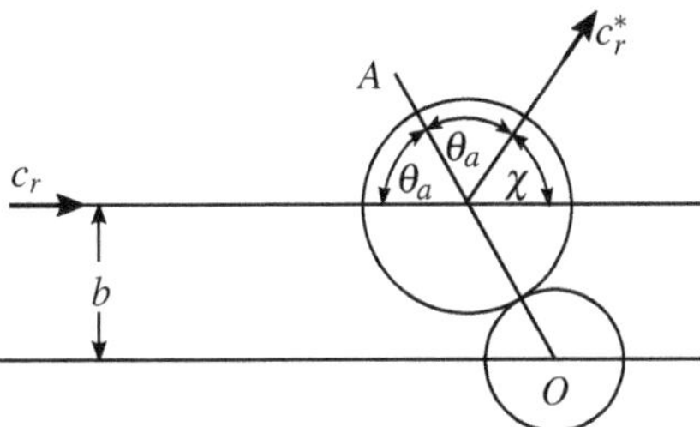

Figure 3.8: Geometry of a binary collision (cp. [Bir94]).

3.2.3 Implementation of boundary conditions

Molecules moving in the computational domain might reach, at a certain point, the domain boundaries. Furthermore, if the domain is not completely enclosed by walls, molecules might also be generated or erased at boundaries. The case of open computational boundaries such as inlet and outlet has been briefly discussed in Section 3.2.1. At an outlet boundary, it is sufficient to delete the particles from memory and hence, from the domain. The case of an inlet boundary, at which molecules are also generated, requires additional consideration. Let us examine the approach for particle generation at the inlet. First, the number of particles to be initialised at a given iteration corresponding to the time t should be defined. One common approach involves the determination of the flux of DSMC particles crossing the inlet boundary of area A while assuming a gas in thermodynamic equilibrium upstream of the inlet boundary surface. In this case, the specification of the gas temperature and number density upstream of the inlet is necessary. If the normal vector of the inlet surface agrees with the positive x direction of the coordinate system, the particle flux $\dot{N}$ across the inlet per unit area is given by:

$$\frac{\dot{N}}{A} = n \int_{-\infty}^{\infty} \int_{-\infty}^{\infty} \int_{0}^{\infty} u f \, \mathrm{d}u \mathrm{d}v \mathrm{d}w \tag{3.99}$$

where n stands for the number density of the equilibrium gas, $u = \{u, v, w\}$ for the particle velocities and f for the Maxwell distribution function. Note that the limits of the integral over the velocity in x direction are defined in a way that only positive velocities (molecules entering the domain) are considered. The integrals in Eq. 3.99 can be solved to obtain the total number of particles N_{part} that enter the domain in the time interval Δt (cp. [Bir94]):

$$N_{\text{part}} = \frac{A \Delta t n \bar{c}}{2\pi^{1/2}} \Big(\exp(-s^2 \cos^2 \theta) + \pi^{1/2} s \cos\theta [1 + \mathrm{erf}(scos\theta)] \Big) \tag{3.100}$$

where $\bar{c} = (2RT)^{1/2}$ stands for the most probable velocity obtained from the temperature value T defined at the inlet and erf corresponds to the *error function*. Furthermore, θ describes the angle between the unit normal vector of the boundary element and the defined boundary velocity or *free stream velocity* c_0. The parameter s, known as the *molecular speed ratio*, is given by:

$$s = \frac{c_0}{\bar{c}} = \frac{c_0}{(2RT)^{1/2}} \tag{3.101}$$

If the result of Eq. 3.100 shows that particles enter the domain, they are randomly generated at the boundary surface based on the specified inlet temperature and the free stream velocity c_0. Although widely used, this particle generation approach has some important disadvantages. Specifically, the procedure assumes that temperature and number density (or pressure) at the inlet boundary are known quantities and the mass flow rate through the inlet is effectively the result of the boundary conditions for T and n. There are, however, numerous applications in which the mass flow rate at the inlet is fixed and the inlet pressure is the desired output of the computation. For this reason, the alternative approach presented in [Fas11] and based on a mass flow rate boundary condition is introduced in the following. Let us define the number of particles to be created as follows:

$$N_{\text{part}} = A \Delta t \dot{N} \tag{3.102}$$

where $\dot{N}$ represents the particle flux across the inlet per unit area [$\text{s}^{-1}\ \text{m}^{-2}$] which can be easily calculated from the mass flow rate $\dot{m}$, the particle mass m and the area of the inlet surface A. This approach allows the exact specification of a mass flow rate across the inlet and into the domain. In general, particles already present in the computational domain can reach the inlet and be subsequently erased from the system. Although this approach is technically correct, it can lead to deviations of the mass flow rate boundary condition. To counteract this, the number of particles exiting the domain through the inlet might be monitored and these particles could be added to the result of Eq. 3.102 in subsequent time steps. An alternative approach involves the handling of the inlet as a wall for particles reaching it from the computational domain. The particles bounce off the inlet back into the computational domain and the mass flow rate boundary condition is satisfied.

There is a number of different approaches for the handling of particle-wall interactions. In general, the particles can be scattered by walls in either a specular (mirror-like) or a diffusive way. Both cases are depicted in Fig. 3.9. In the easiest case, specular reflection, a particle colliding with a wall retains its tangential velocity while its wall-normal velocity is inverted. During this process, the particle transfers momentum to

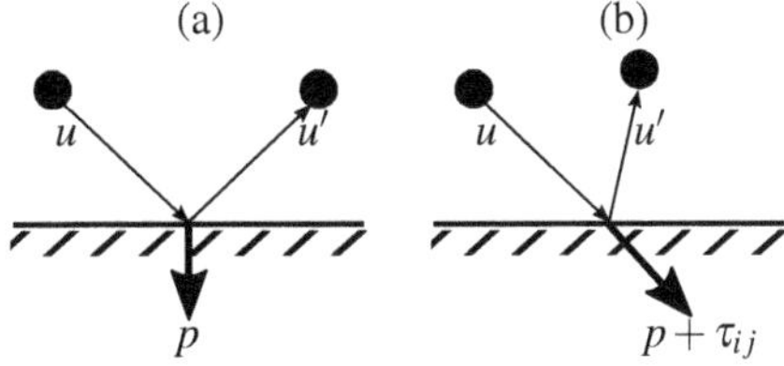

Figure 3.9: Specular (a) and diffusive particle reflection (b).

the wall which manifests as pressure. The case of diffusive reflection, in which the molecule is scattered by the wall in a random direction, requires closer consideration. The pre-collision particle's velocity can be decomposed in a wall normal component u_n and a wall tangential component u_t. Furthermore, let n_w represent the wall normal unit vector. The tangential unit vector in the flow direction is given by $t_{w1} = u_t/|u_t|$ and a second tangential unit vector t_{w2} is defined as the vector orthogonal to both n_w and t_{w1}. The post-collision velocity u' for diffusive scattering is computed as follows:

$$u' = \sqrt{\frac{k_B T}{m}} \left(G_1 t_{w1} + G_2 t_{w2} - \sqrt{-2\log(1-C)} n_w \right) \tag{3.103}$$

where G_1 and G_2 stand for random scalars obtained from the normal Gaussian distribution with mean value of zero and unity variance and C represents a random scalar between 0 and 1 with a uniform probability distribution. Furthermore, T represents the wall temperature and m, the mass of the molecule. From Eq. 3.103, it is evident that the resulting post-collision tangential velocity can have, in principle, any direction in the plane containing the vectors t_{w1} and t_{w2}. However, since the mean value of the Gaussian distribution used to generate G_1 and G_2 is zero, the average tangential post-collision velocity over a large number of particle-wall interactions also equals zero. Furthermore, the normal post-collision velocity is scaled with the factor $\sqrt{\log(1-C)}$. Over a large number of collisions, this approach equals a Maxwell distribution with an expected value of zero and a standard deviation of $\sqrt{k_B T/m} = \sqrt{RT}$ (cp. [Fas11]). Furthermore, both pressure and shear stresses are applied to the wall. Since the expected value of the resulting velocity distribution function in wall proximity is zero, the approach is analogue to the widely used no-slip boundary condition and the gas represented by the DSMC particles fully adopts the wall velocity. This leads to the formation of a boundary layer which, for rarefied gases, is also known as the Knudsen boundary layer. Note that the magnitude of the post-collision velocity in Eq. 3.103 is roughly equal to $(2RT)^{1/2}$ and hence, only dependent of the wall temperature T.

3.2.4 Macroscopic properties

After a DSMC iteration is completed, a list of positions and velocities for the test particles in the computational domain is available. This list is, in most cases, of little use. However, based on this data and on the physical properties of the modelled gas, it is possible to use this information to calculate macroscopic properties of interest. This concept has been briefly discussed in Section 2.9.2.2 for a known velocity distribution function. Here, expressions required for the computation of typical macroscopic properties based on a discrete particle distribution are introduced. For a monoatomic gas with no internal degrees of freedom, the number density n, mass density ρ, linear kinetic energy per unit volume E_{kin} and momentum per unit volume p in a control volume V are given by:

$$n = \frac{W}{V}\sum_i N_i \tag{3.104}$$

$$\rho = mn \tag{3.105}$$

$$E_{kin} = \frac{W}{V}\frac{1}{2}m\sum_i u_i \cdot u_i \tag{3.106}$$

$$p = \frac{W}{V}m\sum_i u_i \tag{3.107}$$

where W stands for the numerical weight or number of real particles represented by each DSMC molecule and m for the molecular mass (66.3×10^{-27} kg for argon). Furthermore, the term $\sum_i$ represents the arithmetic sum over the particles present in the control volume. Note that the momentum is obtained from the arithmetic sum of the particle velocities. If only two particles with same velocity magnitudes but opposite directions are present, the momentum p equals zero. The linear kinetic energy E_{kin} is, however, the sum of the squared velocities and hence, positive. This simplified case corresponds therefore to a cell with no net macroscopic flow velocity (stationary gas). The fluid velocity u, temperature T and pressure p are obtained as follows:

$$u = \frac{p}{\rho} \tag{3.108}$$

$$T = \frac{2}{3k_B n}\left(E_{kin} - \frac{1}{2}\rho(u \cdot u)\right) \tag{3.109}$$

$$p = k_B n T \tag{3.110}$$

Because of the statistical nature of the Monte Carlo method, the macroscopic properties should be averaged over a large number of time steps. Only then, the results are representative of the behaviour of a real gas flow.

3.3 Particle-In-Cell Method (PIC)

The Particle-In-Cell method (PIC) and related approaches like the Particle-In-Cell method with Monte Carlo Collisions (PIC-MCC) [Bir91] are widely used procedures for the modelling of plasma phenomena on a kinetic level. The PIC method has many similarities with the already described DSMC approach. A gas consisting of neutral atoms and charge carriers is modelled in a numerical grid as a collection of individual particles, each representing a large number of real molecules. Like in the DSMC method, the particles motion and interactions are decoupled and hence, computed in separated steps of the procedure. The main difference with respect to the DSMC method lies in the nature of the interactions between particles. In a neutral gas, particles interact or collide with each other only when the distance separating them becomes small enough. This assumption is clearly not adequate for charge carriers. As described in Section 2.8.1.2, charged particles interact with each other over large distances of the magnitude of the Debye length λ_D. These interactions are electrostatic in nature and are the consequence of the Lorentz forces generated by the charged particles. Moreover, the forces are closely related to the spatial distribution and velocities of the particles.

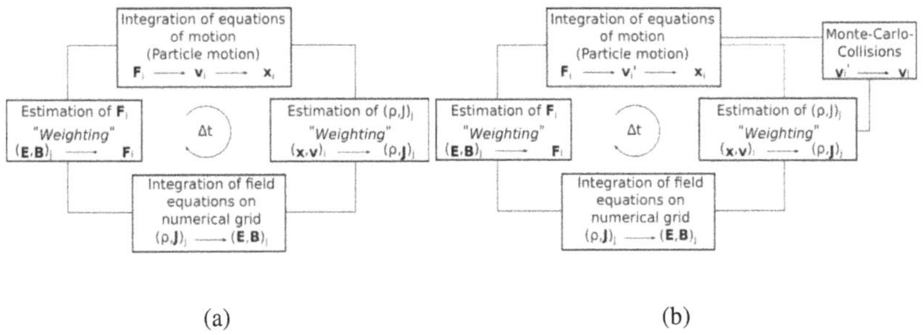

Figure 3.10: Schematic representation of the PIC algorithm (a) and the PIC-MCC algorithm (b) (cp. [Bir91]).

A typical PIC algorithm is shown in Fig. 3.10 (a). As can be seen, the procedure consists of four main steps. The *particle motion* is performed in a similar way as in the DSMC method. However, the motion equations now include a force term corresponding to the Lorentz force. Hence, not only the particles positions but also their velocities are updated and the step is sometimes referred to as *Lorentz solver*. In the next step, known as *weighting*, the particles distribution and velocities are used to calculate the charge density ρ and current density J. Based on these fields, the Maxwell equations are solved in the next step known as *field or Maxwell solver*. In order to achieve this, a variety of approaches may be employed to discretize the differential equations in the numerical grid. They include the Finite Volume Method (FVM) (s. [HLO08]), the Finite Element Mcthod (FEM) (s. [ADH+93]) and the Discontinuous Galerkin method (DG) (s. [MAKF+14, Sti15], cp. [Bir91]). The obtained electric field E and magnetic field B are used in the next step, also called *weighting*, to update the Lorentz force acting on each charge carrier. With this information, the particles velocities and positions are updated in the next time step and the simulation is progressed in time.

In a classic PIC procedure, the test particles inside a given grid cell are identified and used to compute macroscopic quantities like E and B. The necessary deposition of charge in the corresponding cell is usually performed with a so-called *form function*. A known disadvantage of this approach is that the Coulomb force between two similarly charged carriers present in the same cell is strongly underpredicted and, as shown in Fig. 3.11, approaches zero instead of infinite when the distance becomes very small [Ver05]. Hence, a classic PIC procedure can be considered as a collision-free approach with only long-range interactions modelled accurately. In order to take into account short-range effects between charged particles, a Monte Carlo collision approach, as described in Section 3.2, can be introduced into the classic PIC procedure. The resulting algorithm, known as the Particle-In-Cell method with Monte Carlo Collisions (PIC-MCC) [Bir91] is schematically shown in Fig. 3.10 (b). The approach is hence useful if both long-range interactions (Lorentz force) and short-range plasma phenomena (Coulomb interactions, ionisation) are of interest.

In the following, the main aspects of the PIC method will be briefly discussed. For detailed information on the PIC procedure, the reader is referred to [BL04, Tsk08] and [Ver05].

3.3.1 Particle motion - Lorentz solver

In order to model the particles motion inside a plasma, the equations of motion must be integrated in time in the step known as *Lorentz solver* using an adequate time integration

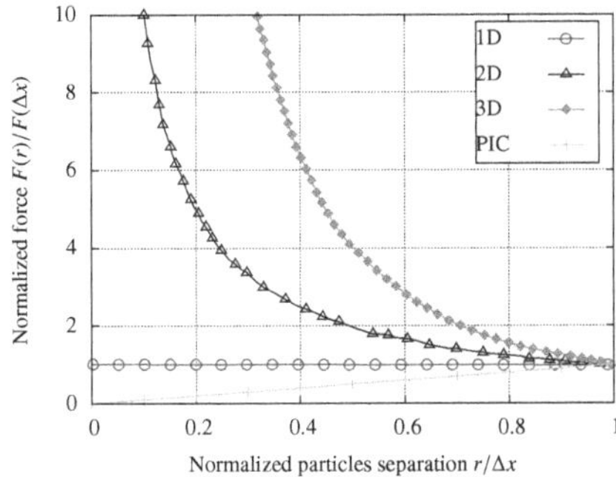

Figure 3.11: Intermolecular force between two charge carriers as a function of their separation calculated with the PIC method and the Coulomb's law in one-, two- and three-dimensional form (cp. [Ver05]).

approach. Some typical methods have already been introduced in Section 3.1.3. The first-order differential equations describing the particle motion in a one-dimensional case are:

$$m\frac{\mathrm{d}v}{\mathrm{d}t} = F \tag{3.111}$$

$$\frac{\mathrm{d}x}{\mathrm{d}t} = v \tag{3.112}$$

where m, x and v stand for the particle's mass, position and velocity respectively and F, for the Lorentz force acting on the particle. One particular characteristic of this set of equations is the fact that they are coupled. As a result of the force F, the particle's velocity changes from v_{old} to v_{new}. In turn, the velocity produces a change of the particle's position from x_{old} to x_{new} where a different force field might apply. From this observation it is clear that a variety of approaches can be employed leading to slightly different results. Specifically, one could update the particle's velocity first using F_{old} at the location x_{old} and then use the new velocity v_{new} to advance the particle in space. Alternatively, one could move the particle with v_{old} first and then perform the velocity update at the position x_{new}. This issue becomes less problematic if very small values of Δt are used. However, the computational requirements increase should not be ignored. A better approach is to solve Eqs. 3.111 and 3.112 in an "unsynchronized" way in order to achieve time-centring during the particle motion step of the PIC simulation. Equations 3.111 and 3.112 can be discretized with the Finite Difference approach to give:

$$m\frac{v_{new} - v_{old}}{\Delta t} = F_{old} \tag{3.113}$$

and

$$\frac{x_{new} - x_{old}}{\Delta t} = v_{new} \tag{3.114}$$

One widely used approach to achieve time-centring is the so-called *leapfrog method* depicted in Fig. 3.12. As can be seen, Eqs. 3.113 and 3.114 are not solved simultaneously but rather in a shifted manner. The velocity x_{old} at the time t is updated using the velocity v_{old} defined at the time level $t - \Delta t/2$. On the other hand, the velocity v_{old} is advanced to the time level $t + \Delta t/2$ using the force value at the time level t. Hence, the position update through Eq. 3.114 is performed with time-centring of v while the velocity update is achieved with a time-centred F.

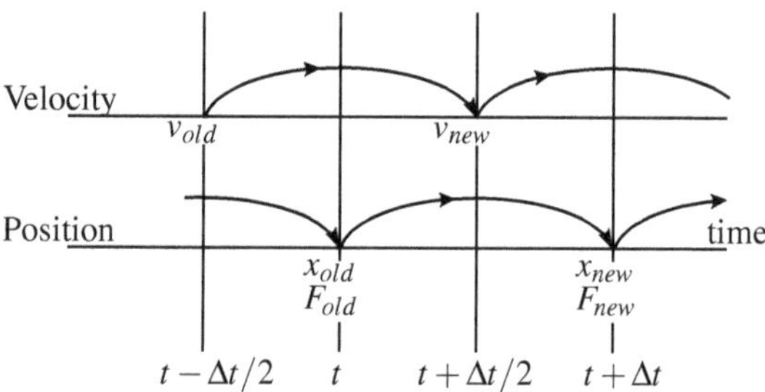

Figure 3.12: Schematic representation of the leapfrog method (cp. [BL04]).

From Fig. 3.12 it is also evident that x and v are not known at the same time level. This must be taken into account while performing the computation of macroscopic properties and writing down data for post-processing. Furthermore, the initial particles velocities v_0 must be shifted back to the time level $-\Delta t/2$ at the beginning of the simulation or when being initialised during the course of the computation (e.g. particles creation through ionisation) in order to retain the time-centred nature shown in Fig. 3.12. At the start of the simulation, this is achieved through Eq. 3.113 using $-\Delta t/2$ and the force at the time $t = 0$. The leapfrog method is a widely used approach because of both its simplicity and high accuracy [BL04].

3.3.2 Field equations - Maxwell solver

The force F acting on a particle with charge q moving with the velocity v can be divided into an electric and a magnetic component as follows:

$$\begin{aligned} F &= F_{electric} + F_{magnetic} \\ &= qE + q(v \times B) \end{aligned} \tag{3.115}$$

Hence, in order to determine F, the calculation of the electric field E and magnetic field B must be performed first. Starting from a known distribution of particles and their velocities in the numerical grid, the initial step involves the determination of the charge density ρ and current density J at the grid points. In the easiest approach, the charge density (units Asm^{-3}) is obtained in the same manner as the density calculation in the DSMC method by performing a simple summation of the particles present in a given cell (s. Section 3.2.4) followed by multiplication with the respective electric charge. More advanced approaches will be presented in Section 3.3.3. On the other hand, the current density J (units Am^{-2}) is obtained from the particles drift or net velocity u as follows:

$$J = \rho u \tag{3.116}$$

Note that the drift velocity u is analogue to the macroscopic flow velocity defined in Section 3.2.4 for the DSMC method. If ρ and J are known at all cell grid points, they can be used in the next step, known as the *Maxwell or field solver*, to determine the electric field E and magnetic field B from the Maxwell equations. They are, in differential form:

$$\nabla \cdot E = \frac{\rho}{\varepsilon_0} \qquad \text{Gauss's law for electric fields.} \tag{3.117}$$

$$\nabla \cdot B = 0 \qquad \text{Gauss's law for magnetic fields.} \tag{3.118}$$

$$\nabla \times E = -\frac{\partial B}{\partial t} \qquad \text{Faraday's law.} \tag{3.119}$$

$$\nabla \times B = \mu_0 \left(J + \varepsilon_0 \frac{\partial E}{\partial t} \right) \qquad \text{Ampere-Maxwell law.} \tag{3.120}$$

where $\varepsilon_0 = 8.8542 \times 10^{-12}$ Fm^{-1} stands for the vacuum permittivity and μ_0, for the vacuum permeability or magnetic constant and with values $\mu_0 = 4\pi \times 10^{-7}$ Hm^{-1} or approximately $1.2566 \times 10^{-6} \mathrm{NA}^{-2}$.

If the problem is assumed to be electrostatic, $\nabla \times E = -\partial B / \partial t \approx 0$ and the relation between the electric field and the electric potential ϕ is given by,

$$E = -\nabla \phi \tag{3.121}$$

which combined with Gauss's law for electric fields produces Poisson's equation:

$$\nabla^2 \phi = -\frac{\rho}{\varepsilon_0} \tag{3.122}$$

The problem is now reduced to the computation of ϕ from the charge density through Eq. 3.122 and subsequent calculation of E from Eq. 3.121. Eq. 3.122 can be discretized using a variety of approaches like the Finite Difference Method (s. Section 3.1.1), the Finite Volume Method (s. Section 3.1.2) or the Discontinuous Galerkin method (s. [MAKF+14, Sti15]). One additional commonly employed approach applicable for periodic systems is based on the use of a discrete Fourier series for the grid quantities. Since this method is not applied in this work, the reader is referred to [BL04] for additional information.

3.3.3 Particle and force weighting

The two remaining steps in a typical PIC algorithm are particle and force weighting. In general, they correspond to the estimation of quantities at the cells grid points based on the discrete particles distribution and velocities (particle weighting) as well as the extrapolation of field quantities available at the grid points, like the electric field, onto the particles positions (force weighting). Weighting can be achieved through a variety of approaches with different accuracies and computational requirements. Note that it is recommended to use the same procedure in both weighting steps in order to avoid a numerical phenomenon in which a particle accelerates itself (self-force).

In order to introduce two common weighting approaches, let us focus on the particle weighting step and in particular, on the determination of the number density based on the particle distribution. The easiest method is known as *zero-order weighting* or *nearest-grid-point* (NGP) and is schematically shown in Fig. 3.13 for a one-dimensional grid. Here, the density in a grid cell is obtained through simple summation of the number of test particles present inside the cell's volume. This approach is therefore analogue to the one described in Section 3.2.4 for the DSMC method. As can be seen in Fig. 3.13 (a), a single particle located at the arbitrary location x_i inside a cell leads to a constant density value across the whole cell. Thus, the exact particle's position inside the cell is irrelevant. As the particle moves from left to right, the density remains constant until the particle exits the origin cell and enters the neighbour one. In this example,

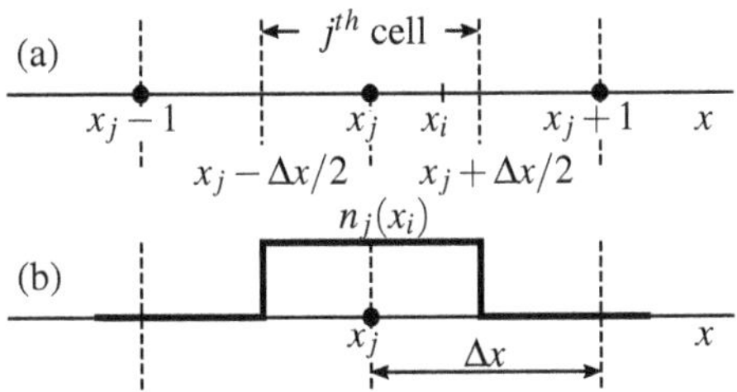

Figure 3.13: Schematic representation of the zero-order weighting or NGP approach. (a) One-dimensional grid with a particle located at x_i; (b) Effective particle shape (cp. [BL04]).

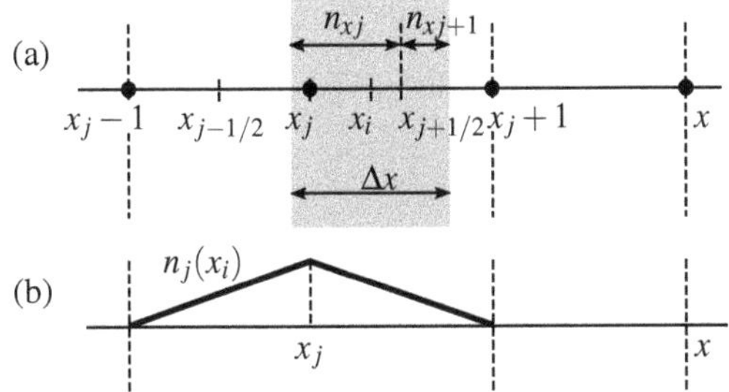

Figure 3.14: Schematic representation of the first-order weighting or CIC approach. (a) One-dimensional grid with a particle located at x_i; (b) Effective particle shape (cp. [BL04]).

the number density in the origin cell falls instantaneously from n to zero while the value at the neighbour cell jumps from zero to n. Therefore, the effective form of the particle resembles a **rectangular shape** as shown in Fig. 3.13 (b). Force weighting in the NGP approach is performed in a similar way, i.e., the force acting on a particle inside a given cell corresponds to the force value at the grid point regardless of the exact particle's position. The NGP approach is computationally very efficient because particles must be accounted for only in the considered cell. The main disadvantage of the NGP approach lies in the possible appearance of noise resulting from abrupt jumps of cell quantities like n, E and B. This noise can be reduced by employing higher-order approaches. Let us imagine a nominal rectangular charge shape with uniform density n and length Δx as shown in Fig. 3.14 (a). The centre of the nominal shape is located at the particle's position x_i. In the *first-order weighting* approach, the total charge of the particle in Fig. 3.14 is assigned to two grid points, x_j and x_{j+1}. The distribution of the charge in both grid points is straightforward. The fraction contained between the cell boundaries $x_{j-1/2}$ and $x_{j+1/2}$ is assigned to the grid point x_j. The remaining charge is assigned to the cell x_{j+1}. Hence, this approach corresponds to the independent application of the NGP approach to each of the charge fractions in Fig. 3.14 (a). If one focuses on the arbitrary position x_i while the particle moves from left to right, the number density increases linearly, reaches a maximum when the particle's position equals x_i and decreases again in a linear manner as the particle moves further away. Therefore, the particle has effectively a triangular shape as shown in Fig. 3.14 (b). This approach is also known as *cloud-in-cell* or CIC model. Higher-order approaches involve the use of quadratic or cubic functions to better capture the actual particle distribution in the numerical grid and reduce noise in the plasma quantities. However, such approaches have high computational requirements and are used less frequently than the NGP and CIC weighting schemes [BL04].

3.4 Summary

In this chapter, the main numerical approaches employed for the simulation of the arcjet thruster INGA III in cold-gas operation mode as well as for the development of a solver for the modelling of plasma phenomena in hot-gas operation conditions have been introduced. The discussed numerical schemes and algorithms build upon the theoretical principles underlined in Chapter 2 and enable their application to a large number of engineering and scientific fluid-related problems. The approaches based on transport equations and applicable to flows in the continuum regime are employed in the present work for the numerical modelling of the arcjet thruster cold-gas operation described in Chapter 4. Because of the low mass flow rates and associated high values of the Knudsen number, the DSMC method is also employed for selected cold-gas operating conditions. The PIC method presented in the final section of this chapter serves as basis for the development of a plasma model aiming to accurately describe basic plasma phenomena as expected to take place during hot-gas operation modes of the INGA III arcjet thruster. The developed solver is described in Chapter 5 and its validation is discussed in Chapter 6.

Chapter 4

Transonic Gas Flows Across Multiple Flow Regimes

In this chapter, the cold gas expansion of the noble gases xenon, argon, krypton and neon through the Laval nozzle of the arcjet thruster INGA III is studied experimentally and numerically. The pressurized gases are accelerated through the nozzle into a vacuum chamber in an attempt to simulate the operating conditions of a cold-gas thruster for attitude control of a micro-satellite. The gases are evaluated at several mass flow rates ranging between 0.178 mg/s and 3.568 mg/s. The Reynolds numbers Re are low (8 to 256) and the estimated values of the Knudsen number Kn lie between 0.33 and 0.02. Hence, flows in both the transition and slip-flow regime are part of the study. DSMC and continuum-based simulations with a no-slip boundary condition are performed and compared with the experimentally obtained data in order to better understand the influence of the Knudsen number Kn and of the gas collision cross sections on the gas behaviour during the expansion process. Furthermore, the thrust produced by the INGA III thruster in cold-gas operation as well as the associated specific impulse are estimated based on the numerical results and potential thrust optimisation approaches are discussed.

4.1 State of the art and previous studies

The influence of the Knudsen number on the behaviour of gases is a key aspect of microscale and nanoscale flows. Here, the Knudsen number reaches values higher than the continuum boundary $Kn = 1 \times 10^{-2}$ (s. Section 2.1). Hence, the continuum assumption, on which the derivation of the Navier-Stokes equations is based, is not fulfilled. This problem is usually dealt with through modification of the flow boundary conditions for the Navier-Stokes equations. One of the simplest and best-known approaches was first proposed by James Clerk Maxwell in 1879 [Max79] and is based on the formulation of first-order gas-slip boundary conditions for rarefied flows. In the following years, several additional theoretical and heuristic approaches have been developed in an effort to explain and correct the observed deviations between experimental data and theoretical results obtained from the Navier-Stokes equations with no-slip boundary conditions at high values of Kn [DSD09]. The majority of these approaches focuses on the *Knudsen layer*, a local thermodynamically non-equilibrium region of thickness $\sim O(\lambda)$ [DZR11] in which modified boundary conditions for the near-wall velocity profile are applied. They include the already mentioned first-order as well as second or higher-order slip-flow boundary conditions [Dei64, MTJW03]. While the simplest form of the slip-flow boundary conditions assumes constant values of the molecular mean free path λ, approaches in which λ is a function of the wall distance have also been proposed. Examples for channel geometries include methods in which λ is calculated from the classic probability distribution function [Sto70] or from a power law probability distribution function [DZR11]. While accuracy has been shown to improve, these methods are based on adjustable coefficients which must be calibrated based on geometrical and Knudsen-related considerations [DSD09]. Further approaches avoid the formulation of modified velocity boundary conditions and are based instead on the modification of the Navier-Stokes equations to include additional diffusive terms such as *volume diffusion* [Bre05, Bre06] or *self-mass diffusion* [DGS06]. As shown in [DSD09] the results of the latter approach are in good agreement with experimental data, theoretical results from linearized Boltzmann equations [Son02] as well as numerical simulations using the Direct Simulation Monte Carlo method (DSMC) [Bir94] across a great range of Kn numbers.

The computational modelling of flows with high Kn numbers also presents plenty of challenges. The standard Maxwell slip condition for Navier-Stokes based models is, in theory, not applicable for curved surfaces [OLRE07, LREB04], which compromises the quality of the numerical results for complex three-dimensional geometries. Special approaches such as the formulation of wall functions for the Knudsen layer, similar to the wall functions used in turbulence modelling, have disadvantages associated with their very restrictive assumptions (low Mach number, relatively low Knudsen numbers among others) [OLRE07]. Several sets of higher-order continuum equations have also been proposed [Bal04, ZMC93, JS01, ST03]. However, as pointed out in [OLRE07], higher order equations often present disadvantages including their numerical instability, high non-linearity and the need for new boundary conditions. An alternative approach proposed in [LR08] and based on the formulation of a new model for slip boundary conditions and near-wall scaling of the Navier-Stokes constitutive relations has shown an increase in accuracy compared to conventional second-order slip boundary conditions. However, some aspects such as stability at high values of Kn, the effect of non-parallel wall interactions and applicability for non-isothermal flows require further investigation. The most accurate results for complex geometries and three-dimensional flows are obtained via DSMC simulations. The DSMC method is, however, usually very expensive from a computational point of view. This is a particular challenge when dealing with problems with mixed conditions, such as hypersonic micronozzles flows where high gradients of pressure, density and molecular mean free path are to be expected.

The importance of the study of rarefied gas flows becomes clear when recent trends in the aerospace community are considered. In recent years, one of the main technical goals focuses on the development and use of miniaturized satellites with the ability to perform spacecraft formation flying. As a result, the development of small propulsion systems based on Micro-Electro-Mechanical-Systems (MEMS) and able to provide low and precise thrust and impulse levels has become an important research topic. As pointed out in [MnHQ08], such propulsion systems are also advantageous for conventional spacecrafts that require very precise attitude control systems (e.g., earth observation for scientific purposes). The low propellant mass flow rates and the very small length dimensions required for very precise miniaturized propulsion systems result in gas rarefaction effects in the micronozzle becoming important for the flow and performance characteristics of the propulsion systems. Therefore, the effect of high Knudsen numbers in micronozzle flows and its numerical modelling have continuously gained importance. Recent studies on these topics include the continuum-based work described in [MnHQ08]. Here, numerical simulations with second-order slip boundary conditions for the flow velocity of cold and hot nitrogen micronozzle flows are performed. Furthermore, the gas flow simulations are coupled with a heat conduction model in the solid region of the nozzle and the results focus on the effects of the viscous heating on the flow characteristics and propulsion performance. The study presented in [LH10] focuses on the transient behaviour of a high-temperature flow of decomposed hydrogen peroxide (H_2O_2) through a micronozzle and uses a continuum approach with a no-slip boundary condition. The example presented in [BAB97] employs a Navier-Stokes-based approach with first-order slip boundary conditions. Here, the numerical results for the mass flow rates are compared with experimental data for different nozzle geometries. Approaches based on the kinetic modelling of micronozzle flows have also been widely used. They include the work described in [ALFG05, AFG$^+$06]. In these studies, the DSMC method is used for the modelling of a nitrogen micronozzle flow in the transition flow-regime. Furthermore, the flow results are coupled with a thermal model for the nozzle body and the transient thrust results for different values of the stagnation pressure and thermal boundary conditions at the external nozzle wall are analysed. In the study presented in [ALG+02], DSMC simulations of high-temperature gases in micronozzles are performed in order to examine the effects of the DSMC accommodation factors on the numerically determined performance characteristics of the studied nozzles. The DSMC method is also used in [GA15] for the numerical and experimental study of linear aerospikes, an alternative nozzle design for the expansion of exhaust gases for propulsion purposes. In this work, argon flows at several operating conditions and in the transition regime are expanded through the aerospike system. The resulting kinetic boundary layer, its growth over the spike surface, its effect on the flow field as well as the thrust produced by the system are analysed. Comparisons between the kinetic (DSMC) and continuum approaches include the studies in [ACG+02, AGLC00].

In this chapter, the cold-gas expansion of the noble gases xenon, krypton, argon and neon through the Laval nozzle of the INGA III thruster in the slip-flow and transition regimes is investigated. In the study, first-order slip boundary conditions are not considered. The reason behind this is that although simple, first-order slip boundary conditions are, in general, accurate in the slip-flow regime and not in the transition regime also investigated in this work. Furthermore, second-order boundary conditions, although theoretically accurate in the slip and transition regimes, also rely on coefficients which often require calculation/calibration from experimental data and are likely geometry-dependent [LREB04]. In addition, slip-models become very complex when the transition and Knudsen regimes are considered [DSD09]. Therefore, experimental tests, DSMC simulations and simulations based on the compressible form of the Navier-Stokes equations with no-slip boundary conditions are performed and analysed in this chapter. From the comparison between the continuum-based calculations with the experimental data, a relation describing the deviation of the pressure drop along the nozzle as a function of the Knudsen number is obtained. In this work, it is shown that the experimental pressure results deviate strongly from the no-slip assumption for gases with small collision cross sections. The corresponding deviation is mathematically best described by a second degree polynomial. For gases with large collision cross sections, the deviation as a function of Kn is closer to a linear function. From the analysis of the developed function, it is possible to correct the pressure results for the studied gases, both in the slip-flow and transition regimes, by using only four gas-independent coefficients. The structure of the chapter is as follows: the experimental and numerical setups are described in Sections 4.2 and 4.3. The experimental results, the Navier-Stokes and the DSMC results are presented in Sections 4.4.1, 4.4.2 and 4.4.3 respectively. The relation describing the deviation of the pressure drop along the nozzle as a function of the Knudsen number is described in Sections 4.4.5 and 4.4.6. Section 4.4.7 is dedicated to the impact of the low Reynolds numbers and the associated viscous boundary layer on the flow fields and thrust performance. The chapter concludes with a brief summary of the cold-gas investigation in Section 4.5.

4.2 Experimental setup

4.2.1 Vacuum and measurement systems

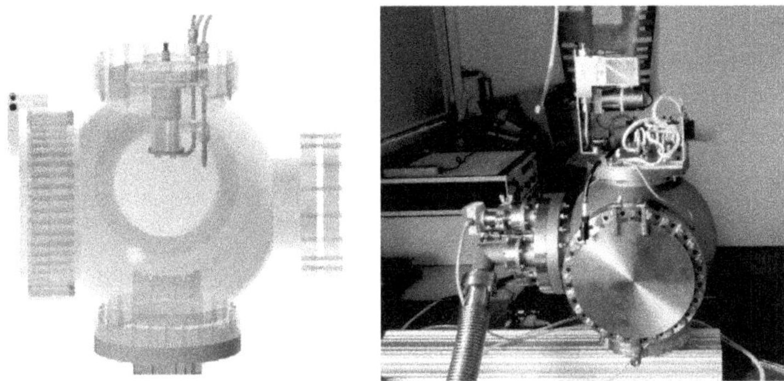

Figure 4.1: Experimental setup.

The experimental setup used for the cold-gas study is shown in Fig. 4.1. The INGA III arcjet thruster is placed inside a spherical vacuum chamber with a total volume of approx. 10 litre. Vacuum conditions are achieved using a rotary vane pump of type Trivac D 40 B from the company Oerlikon Leybold (Cologne, Germany). Steady pressure values below 1 Pa can be reached in the vacuum chamber during operation without any gas flows. The pressure in the vacuum chamber increases when propellant gas is fed into the chamber through the nozzle and reaches a maximum value of 19 Pa for neon at its highest studied mass flow rate $\dot{m} = 1.784$ mg/s. The pressure in the vacuum chamber is measured with a linear convection vacuum gauge APGX-H from the company Edwards. The studied noble gases xenon, krypton, argon and neon are led into the arcjet thruster from standard pressure bottles and through a pressure regulator of type FMD 562-16 BCF from GCE (Malmö, Sweden). The pressure regulator adjusts the gas pressure upstream of the mass flow controller to an absolute value of 0.8 bar, as recommended by the manufacturer of the controller. The mass flow controller is of type EL-Flow manufactured by Bronkhorst High-Tech B.V. (Ruurlo, The Netherlands) and is calibrated for operation with argon at a maximum volumetric flow rate of 2 ml/s at standard conditions (0 °C and 1.013 bar) and for a pressure range between 0.8 and 1.05 bar upstream of the controller. A volumetric flow rate of 2 ml/s for argon at standard conditions equals a maximum possible mass flow rate of $\dot{m} = 3.568$ mg/s for argon. The maximum mass flow rate through the controller, however, varies depending on the used gas. As a result of its lower density, the maximum achievable mass flow rate for neon through the mass flow controller is approx. 50% of the argon value, i.e., 1.784 mg/s. On the other hand, mass flow rates higher than 3.568 mg/s are possible with the current setup for krypton and xenon. However, in order to facilitate the comparison between gases, values higher than 3.568 mg/s are not considered in this study. A pressure sensor manufactured by Sensortechnics GmbH (Puchheim, Germany) is used to measure the pressure difference between the ionisation chamber of the arcjet thruster (inlet of the Laval nozzle) and the vacuum chamber. The distance between the nozzle exit and the wall of the vacuum chamber is 230 mm, which corresponds to approximately 7 times the nozzle length. Therefore, any influence of the chamber wall on the flow fields in the nozzle and at its exit is considered to be negligible.

The described experimental setup can also be used for the study of arcjet thrusters in hot-gas operation mode. As such, it disposes of two power supplies for the controlled generation and maintenance of an electric arc. Since this chapter focuses on the cold-gas expansion of gases through the Laval nozzle, the power supplies and additional arcjet related appliances are not further described.

4.2.2 Arcjet thruster and Laval nozzle

The arcjet thruster used for the experimental series is shown in Fig. 4.2 (a) and consists mainly of a tungsten cathode, a Laval nozzle which also serves as anode during hot-gas operation of the thruster, an insulator made of sintered boron nitride and several sensor and gas supply lines. Since this chapter focuses on cold-gas transonic flows, only the nozzle geometry, shown in Fig. 4.2 (b), is described in detail. Apart from the eccentrically mounted gas supply line, the nozzle can be considered as rotationally symmetrical. The inlet diameter is 12 mm. After approx. 6 mm its value decreases towards the nozzle throat, which in turn has a diameter of 1.8 mm. The nozzle throat length is 2 mm. The diameter of the diffuser increases linearly and reaches a value of 16.36 mm at the nozzle exit. The tungsten cathode of the arcjet thruster has a diameter of 2.4 mm. It is located in the convergent section of the nozzle (see Fig. 4.2 (a)) and remained at the same position during the cold-gas experimental series. The distance between the tip of the cathode and the start of the throat section is 2.3 mm.

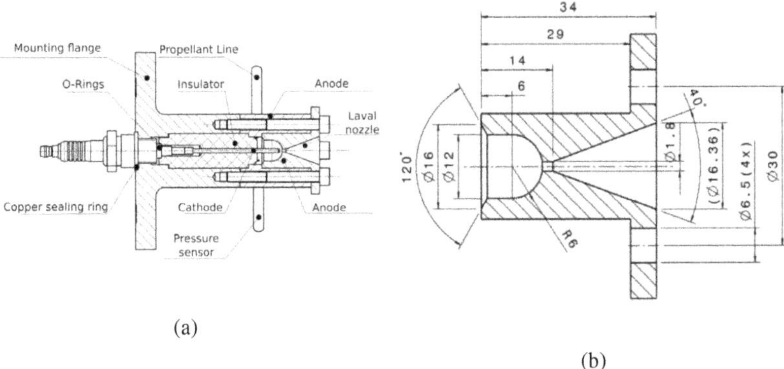

Figure 4.2: Schematic representation of the arcjet propulsion system (a) and the Laval nozzle (b) with length dimensions in mm.

4.2.3 Experimental series

Table 4.1 shows the investigated mass flow rates for the experimental series in mg/s and as percent of the maximum considered mass flow rate (3.568 mg/s). As a result of technical restrictions regarding the mass flow controller capabilities, the maximum mass flow rate for neon is limited to 1.784 mg/s (s. Section 4.2.1). Furthermore, the minimum mass flow rate for xenon is limited by the minimum valve opening position of the controller. This results in a minimum achievable value of 0.268 mg/s for xenon instead of 0.178 mg/s, the minimum for the remaining gases.

4.3 Numerical setup

4.3.1 Solved equations and numerical solver

In addition to the experimental series, continuum-based simulations of the cold-gas expansion through the described Laval nozzle (s. Fig. 4.2 (b)) are performed for every case

Mass flow rate					
[mg/s]	[%]	Xenon	Krypton	Argon	Neon
0.178	5.0		✓	✓	✓
0.268	7.5	✓			
0.357	10	✓	✓	✓	✓
0.535	15	✓	✓	✓	✓
0.714	20	✓	✓	✓	✓
0.892	25				✓
1.070	30	✓	✓	✓	✓
1.249	35				✓
1.427	40	✓	✓	✓	✓
1.605	45				✓
1.784	50	✓	✓	✓	✓
2.141	60	✓	✓	✓	
2.497	70	✓	✓	✓	
2.854	80	✓	✓	✓	
3.211	90	✓	✓	✓	
3.568	100	✓	✓	✓	

Table 4.1: Studied mass flow rates in experimental test series.

defined in Table 4.1. The simulations are performed using the open source CFD software package *OpenFOAM®*. The flow through the nozzle is mathematically modelled by the transient, compressible form of the continuity equation (Eq. 2.14):

$$\frac{\partial \rho}{\partial t} + \frac{\partial \rho u_j}{\partial x_j} = 0$$

the Navier-Stokes equations for conservation of momentum (Eq. 2.43):

$$\frac{\partial \rho u_i}{\partial t} + \frac{\partial \rho u_i u_j}{\partial x_j} = -\frac{\partial p}{\partial x_i} + \frac{\partial}{\partial x_j}\left[\mu\left(\frac{\partial u_i}{\partial x_j} + \frac{\partial u_j}{\partial x_i} - \frac{2}{3}\delta_{ij}\frac{\partial u_k}{\partial x_k}\right)\right]$$

where gravity has been neglected and the conservation equation for the internal energy e (Eq. 2.63):

$$\rho\left(\frac{De}{Dt}\right) = \nabla\cdot(\kappa\nabla T) + \tau_{ij}\frac{\partial u_i}{\partial x_j}$$

In addition, it is assumed that the ideal gas law (Eq. 2.74) holds for all the studied gases:

$$pV = nR_u T$$

where R_u represents the universal gas constant. Therefore, the gas expansion through the nozzle can be modelled by an equation system consisting of the above four equations and the four variables u, ρ, T and p. The equation system is solved using the numerical *OpenFOAM®* solver *rhoCentralFOAM*. In its source code, the solver is described as a density-based compressible flow solver based on the central-upwind schemes proposed by Alexander Kurganov and Eitan Tadmor in 2000 [KT00]. A coupled-implicit smooth solver with Gauss-Seidel as smoother is used. Furthermore, the gradient and divergence terms are discretized using the standard Gaussian scheme with linear interpolation. For the laplacian terms, the Gaussian scheme with linear interpolation for the diffusion coefficients and explicit non-orthogonal correction for the surface normal gradients is used. The time derivatives are discretized using the standard implicit Euler scheme. The simulations are performed on the super computer system of "The North-German Supercomputing Alliance" (HLRN).

4.3.2 Numerical mesh and boundary conditions

In order to simplify the numerical mesh and to reduce the number of cells, a rotationally symmetrical approach is chosen. This allows for the nozzle to be represented by a wedge geometry with an opening angle of 5°. This approach ignores the eccentrically mounted gas supply line and assumes a uniform gas entry through the inlet surface. Comparison with full three-dimensional simulations does not show any considerable deviation between both approaches. A large outlet-volume representing the vacuum chamber is added at the nozzle exit. The complete mesh comprises 11,000 cells. Grid independence has been corroborated through comparison of the pressure results in the used mesh with a much finer one (approx. 47,000 cells). A xenon flow with $\dot{m} = 3.568$ mg/s is used as reference case for the grid-independence check and the comparison is conducted as described in [Li08]. The numerical deviation for the pressure results $\delta(p)$ is determined as follows:

$$\delta(p) = \frac{\sum_{n=1}^{N} |p_n^{coarse} - p_n^{fine}|}{\sum_{n=1}^{N} |p_n^{fine}|} \tag{4.1}$$

with N denoting the total number of cells in the fine mesh and p_n^{coarse} representing the pressure results obtained with the coarse mesh (11,000 cells) and mapped onto the fine mesh. From Eq. 4.1 a value for $\delta(p)$ of 0.006 (0.6 %) is obtained. It can therefore be concluded, that the results with the employed mesh are accurate and remain essentially unchanged after an increase of the mesh resolution.

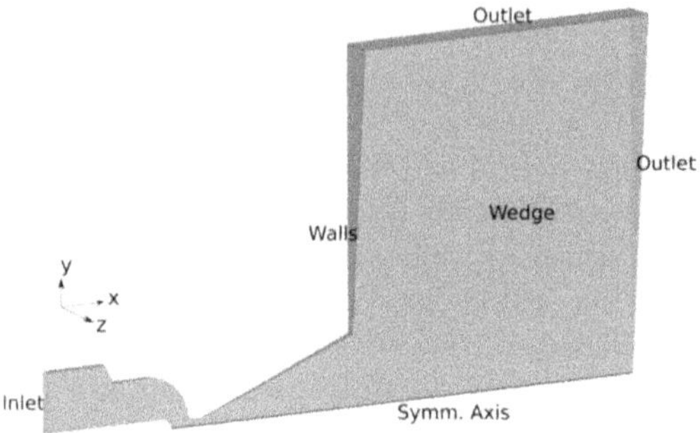

Figure 4.3: Boundary surfaces. Lime: inlet; red: outlet; olive: walls; blue: symmetry axis; gray: wedge. Outlet-volume not true-to-scale.

The physical boundary surfaces of the numerical mesh can be classified as inlet surface, wall surfaces and outlet surfaces. Non-physical boundaries of the mesh are the symmetry axis (which does not require any special boundary condition in *OpenFOAM*®) and the wedge surfaces (with the special boundary condition "*symmetryPlane*"). It should be noted that "*symmetryPlane*" was preferred to the special boundary condition "*wedge*" because for the latter, non-physical mass loss across the wedge surfaces was observed. Comparison of this setup with reference 3D-computations does not show any appreciable deviation. The location of the boundaries in the numerical mesh is shown in Fig. 4.3. The associated boundary conditions using the example of xenon with 100% mass flow rate are listed in Table 4.2. The abbreviation z.G stands here for the boundary condition "*zeroGradient*" which assumes that the face normal gradient of the corresponding field is zero. The abbreviation s.P. stands for "*symmetryPlane*". Regarding the boundary condition for the velocity, the mass flow rate at the inlet is defined according to the values in Table 4.1 with a 100% mass flow rate equalling 3.568 mg/s. Furthermore, the pressure value at the outlet of the computational grid, i.e., in the vacuum chamber depends on the mass flow rate and the studied gas. The values for each of the cases are obtained experimentally and used in the simulations in order to ensure comparability between the numerical and the experimental results. The complete list of pressure values measured in the vacuum chamber is shown in Table 4.4 in Section 4.4.1. The values are applied to the simulations through a "*waveTransmissive*" boundary condition for the outlet surface perpendicular to the nozzle axis and a "*totalPressure*" boundary condition for the outlet surface parallel to the nozzle axis.

Table 4.3 shows the thermophysical properties of the studied gases used for the simulations. The values for the specific heat capacity at constant pressure c_p listed in Table 4.3

Boundary	Velocity	Pressure	Temperature
Inlet	$\dot{m}$=3.568 mg/s	z.G.	293.15 K
Outlet	z.G.	3.2 Pa	z.G.
Walls	$u = \{0, 0, 0\}$ m/s	z.G.	z.G.
Wedge	s.P.	s.P.	s.P.
Rot. axis	-	-	-

Table 4.2: Boundary conditions for xenon and $\dot{m} = 3.568$ mg/s.

are obtained from the VDI Heat Atlas [VDI13] and are independent of temperature. The dynamic viscosity of the investigated gases, in contrast, is influenced by temperature. The relation is given by Sutherland's law [Sut93]:

$$\mu = \mu_{ref} \left(\frac{T}{T_{ref}} \right)^{3/2} \frac{T_{ref} + T_s}{T + T_s} \tag{4.2}$$

where T_{ref} stands for a reference temperature, μ_{ref} for the gas viscosity at the reference temperature and T_s, for the Sutherland temperature. In *OpenFOAM*®, Eq. 4.2 is implemented as follows:

$$\mu = \frac{A_s T^{3/2}}{T + T_s} = \frac{A_s \sqrt{T}}{1 + T_s/T} \tag{4.3}$$

with the Sutherland coefficient A_s defined as:

$$A_s = \mu_{ref} \frac{T_{ref} + T_s}{T_{ref}^{3/2}} \tag{4.4}$$

In the present work, the values A_s and T_s for the investigated gases were derived from the experimental viscosities at low temperatures reported in [CS68] and [CS69]. To this end, values for A_s and T_s fitting Eq. 4.3 were numerically determined. The obtained coefficients for Sutherland's law are listed in Table 4.3. The Reynolds numbers for the studied cases range between *Re*=8 for neon with $\dot{m} = 0.178$ mg/s and *Re*=256 for xenon with $\dot{m} = 3.568$ mg/s. Therefore, the simulations are performed with a laminar solver.

4.3.3 Numerical setup for DSMC simulations

In order to validate the experimental results and to evaluate the results of the Navier-Stokes based computations, simulations with the Direct Simulation Monte Carlo method (DSMC) are also performed. This approach is widely regarded as an accurate

Gas	c_p [J/kgK]	A_s [Pa·s/$\sqrt{K}$]	T_s [K]	M [kg/kmol]
Xenon	158	2.3833×10^{-6}	231.4456	131.29
Krypton	248	2.3903×10^{-6}	185.9452	83.80
Argon	520	1.9487×10^{-6}	144.8247	39.95
Neon	1,030	2.1332×10^{-6}	47.0223	20.18

Table 4.3: Thermophysical properties.

computational method for the modelling of flows in rarefied conditions, specially in the transition and free molecular flow regimes. Because of the high computational demands, only one gas, argon, is evaluated. The simulations are performed with a modified version of the *OpenFOAM®* solver *dsmcFOAM*, which is based on the code developed by Prof. Graeme Bird [Bir94] and can be employed for the simulation of three-dimensional, transient flows. The modified version of the *dsmcFOAM* solver enables the accurate definition of a particle flow rate across the inlet surface of the studied nozzle, and therefore, of the mass flow rates listed in Table 4.1 for argon. This guarantees the comparability between the experimental, the Navier-Stokes and the DSMC results. The variable hard sphere approach (VHS) (s. Section 3.2.2) is used to model the collisions between argon particles. The viscosity index $\omega = 0.93$ is used to approximate the viscosity-temperature dependency for atomic argon. This value is calculated as $\omega = ln(\mu_1/\mu_2)/ln(T_1/T_2)$ from the data presented in [KKM+84] and is valid in the temperature interval 50 K $< T <$ 293 K. It should be noted, that the widely used value $\omega = 0.81$ for argon presented in [Bir94] is based on the data in [CC70] and is valid in the temperature interval 20 °C$< T <$ 100 °C. This interval, however, is not representative of the temperature values reached in the studied cold-gas thruster (s. Section. 4.4.2). The dependency between the viscosity index and the temperature is discussed in detail in [SS13]. Moreover, the viscosity coefficient is defined as 2.108×10^{-5} Nsm^{-2}, which corresponds to the reference viscosity for $T = 273$ K. The resulting reference diameter calculated with Eq. 4.62 from [Bir94] is 4.43×10^{-10} m. For the interaction between particles and walls, two models are employed: the Maxwell model, which assumes a diffuse reflection of the particles after collision with a surface and the specular reflection model, in which the tangential speed of the particles relative to the wall surfaces remains unchanged after a collision. The same numerical mesh as for the Navier-Stokes simulations is used. The particles are initialized at the nozzle inlet at ambient temperature $T = 293.15$ K. The particle flux $\dot{n}_x/A_x$ injected through the outlet of the computational grid is calculated as follows:

$$\frac{\dot{n}_x}{A_x} = \frac{1}{4}n\bar{c} = n\sqrt{\frac{k_B T}{2\pi m}} = \frac{p}{\sqrt{2\pi m k_B T}} \tag{4.5}$$

with m representing the atomic mass of argon, T the ambient temperature and p the pressure values in the vacuum chamber obtained experimentally and listed in Table 4.4.

4.4 Results and discussion

In the following sections, the results of the experimental and numerical studies for the cold-gas operation of the INGA III thruster are presented. In this discussion, the terms "nozzle inlet" and "ionisation chamber" are interchangeable and denote a location in the convergent nozzle section in which the cross section average flow velocity reaches its minimum and the pressure, its maximum, i.e., far upstream of the nozzle throat.

4.4.1 Experimental results

During the experimental series, the pressure in the vacuum chamber p_{vac} and the pressure difference between vacuum and ionisation chamber Δp are measured after steady state is reached. The results for the pressure in the vacuum chamber are listed in Table 4.4 for all the studied cases. The values are used as pressure boundary conditions for the mesh outlet surfaces in the numerical simulations as described in Sections 4.3.2 and 4.3.3. As can be seen from Table 4.4, a lower molar mass leads to higher pressure values in the vacuum chamber for any given mass flow rate. This observation can be explained by the existing anti-proportionality between pressure and molar mass in the ideal gas law and a possible performance drop of the vacuum pump when operating with very light gases.

The pressure in the ionisation chamber of the arcjet thruster p_{ch} corresponds to the sum of the measured pressure in the vacuum chamber p_{vac} and the measured pressure difference between vacuum and ionisation chamber Δp. The resulting experimental absolute pressure in the ionisation chamber p_{ch} is shown in Fig. 4.4 for all studied gases and mass flow rates. Comparison of the different gases at any given mass flow rate shows that p_{ch} decreases with higher values of molar mass. The pressure values for neon (the lightest studied gas) are considerably higher than the values for the heavier argon, a trend that continues with increasing molar masses and results in the lowest inlet pressure values being achieved with xenon, the heaviest studied gas. This behaviour can

$\dot{m}$ [mg/s]	Xenon	Krypton	Argon	Neon
0.178	-	0.9	1.6	3.1
0.268	0.76	-	-	-
0.357	0.89	1.3	2.6	5.4
0.535	1.1	1.7	3.5	7.3
0.714	1.2	2.0	4.3	9.3
0.892	-	-	-	11.0
1.070	1.6	2.6	5.8	12.0
1.249	-	-	-	14.0
1.427	1.9	3.2	7.1	15.0
1.605	-	-	-	17.0
1.784	2.2	3.6	8.3	19.0
2.141	2.4	4.0	9.6	-
2.497	2.7	4.5	10.0	-
2.854	2.9	4.9	11.0	-
3.211	3.1	5.3	12.0	-
3.568	3.3	5.7	13.0	-

Table 4.4: Pressure in the vacuum chamber p_{vac} in Pa.

be explained through a gas-kinetic analysis: In order for a given mass flow rate to be achieved, a higher number of neon atoms must enter the nozzle compared with a heavier gas such as xenon. The resulting higher number densities lead to higher intermolecular and wall collision rates, which in turn result in higher momentum losses and higher pressure gradients through the nozzle. The result is a higher pressure at the nozzle inlet (i.e., in the ionisation chamber) for lighter gases. As expected, higher mass flow rates for any gas result in a higher pressure drop through the nozzle and higher absolute pressure values p_{ch} in the ionisation chamber.

For xenon with $\dot{m} = 0.268$ mg/s a pressure ratio between the ionisation and vacuum chamber of approx. 53 is reached. An increase of the mass flow rate leads to higher pressure ratios and a maximum for p_{ch}/p_{vac} of approx. 112 is observed for xenon with $\dot{m} = 3.568$ mg/s. For krypton, the pressure ratio ranges between 45 for $\dot{m} = 0.178$ mg/s and 82 for $\dot{m} = 3.568$ mg/s. For argon, values between 34 and 54 are reached. Because of the strong increase of p_{vac} as a function of the mass flow rate observed for neon, the ratio p_{ch}/p_{vac} does not show a clear tendency and a fairly constant value of approx. 30 is reached at all studied mass flow rates. As already mentioned, this is explained by the relation between pump performance and injected gas and the relatively strong increase of p_{vac} for lighter gases as shown in Table 4.4.

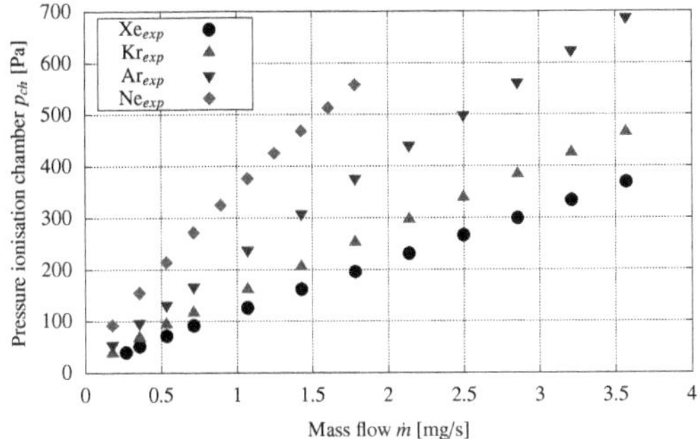

Figure 4.4: Experimentally measured pressure in the ionisation chamber p_{ch}.

4.4.2 Navier-Stokes simulations

The numerical simulations are performed using the open source CFD software package *OpenFOAM®* and the transient, compressible solver *rhoCentralFOAM*. Steady state conditions are reached after a simulated time of 80 ms for all studied cases. Fig. 4.5 shows the streamlines for the simulation with xenon and a mass flow rate of 3.568 mg/s. The fluid is accelerated from the inlet in the left hand side of Fig. 4.5 and through the nozzle throat. As described by the working principle of the Laval nozzle (s. Section 2.7), the flow velocity continues to increase downstream of the throat and through the diffuser. However, numerical results show that the maximum velocity is reached halfway through the diffuser and not at its exit. This behaviour is discussed in detail later in this chapter. Nonetheless, the still high flow velocity downstream of the nozzle exit leads to a pressure drop in its trail zone. The resulting pressure gradient induces a gas flow from outside of the numerical grid through the horizontal boundary of the outlet-volume (top of Fig. 4.5). This flow is first accelerated towards the nozzle axis and then redirected towards the outlet boundary and out of the numerical grid on the right hand side of Fig. 4.5.

Figure 4.6 shows the Mach number, velocity, temperature and pressure fields for the simulation with xenon and $\dot{m} = 3.568$ mg/s after steady state is reached. A maximum

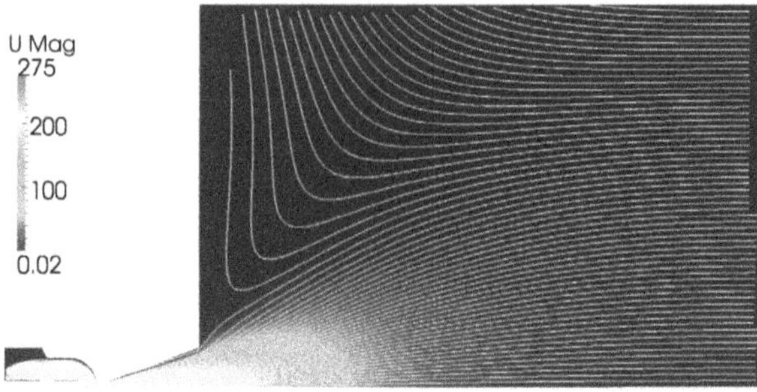

Figure 4.5: Streamlines for the simulation with xenon, $\dot{m} = 3.568$ mg/s, Re=256 and $t = 80$ ms.

Mach number of 3.25 is reached halfway through the diffuser. From this point, the Mach number decreases through the diffuser. The maximum velocity (275 m/s) and minimum temperature (67.3 K) are both reached at the same location where the Mach number reaches its maximum, which agrees with the definition of the Mach number for an ideal gas:

$$Ma = \frac{u}{a} = \frac{u}{\sqrt{\gamma \frac{p}{\rho}}} = \frac{u}{\sqrt{\frac{\gamma R_u T}{M}}} \tag{4.6}$$

Here, u represents the velocity, a the speed of sound, γ the isentropic exponent, $R_u = 8,314$ J/kmolK the universal gas constant and M, the molar mass as listed in Table 4.3. After the maximum values of Ma and u are reached, the fluid temperature in the diffuser starts to increase. The resulting higher thermal energy of the flow is compensated by a decrease of its kinetic energy, which leads to lower velocities and Mach numbers towards the nozzle exit. Despite this, supersonic conditions are still present at the nozzle exit. This behaviour has been observed in several studies [MnHQ08, Rae71, BAB97, ACG+02, AGLC00, GA15]. For the Navier-Stokes numerical results with no-slip boundary condition depicted in Fig. 4.6, this phenomenon is a consequence of the high ratio of surface area to volume in the modelled micronozzle, which leads to wall shear and viscous dissipation effects dominating the fluid behaviour. The high influence of the viscous losses on the flow fields shown in Fig. 4.6 is also evident from the very low Reynolds numbers of the studied cases as depicted in Fig. 4.7. The highest value obtained in the present study, $Re = 256$, is reached with xenon at $\dot{m} = 3.568$ mg/s and the lowest, $Re = 8$, with neon and $\dot{m} = 0.178$ mg/s. The shock-less decrease of Ma and u in Fig. 4.6 results therefore from the relatively high thickness of the viscous boundary layer in comparison with the nozzle dimensions. On the other hand, for the actual rarefied nozzle flow, this behaviour is more accurately explained not in terms of the viscous but of the *Knudsen boundary layer*. As described by [GA15], the diffusive nature of the collisions between gas atoms and walls leads to a loss of momentum in the flow direction and to the appearance of the Knudsen boundary layer. The conversion of kinetic energy into thermal energy in the Knudsen layer and its continuous growth along the divergent nozzle wall leads to a shock-less decrease of Ma and u.

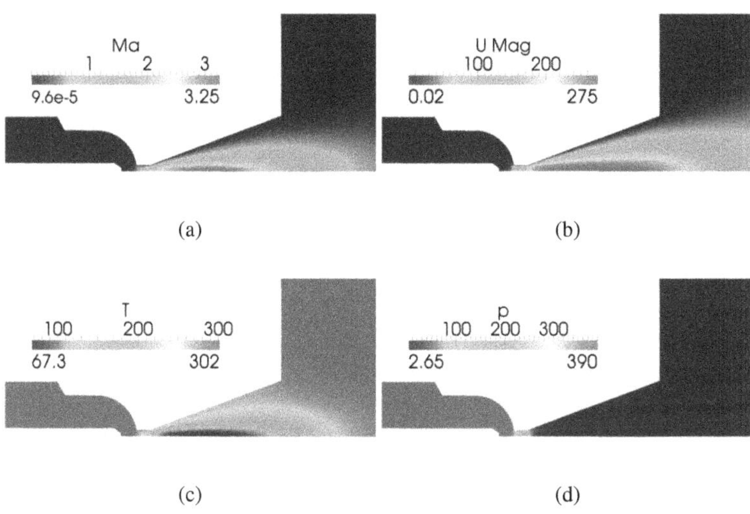

Figure 4.6: Flow fields for the simulation with xenon, $\dot{m} = 3.568$ mg/s, *Re*=256. (a) Mach number; (b) velocity [m/s]; (c) temperature [K]; (d) pressure [Pa].

Figure 4.8 shows the normalized velocity profiles in the diffuser, with xenon as propellant and $\dot{m} = 3.568$ mg/s, *Re*=256. Here, X represents the position along the x-axis measured from the start of the diffuser and normalized by the total diffuser length. Y represents the relative radial distance from the nozzle axis, with $Y = 0$ describing the nozzle axis and $Y = 1$, the diffuser wall. As can be seen, the thickness of the boundary layer increases towards the nozzle exit, therefore reducing the exit velocity and the thrust delivered by the micronozzle. Thrust performance is further discussed in Section 4.4.7. As described in Section 4.3.2, the pressure at the outlet of the numerical mesh

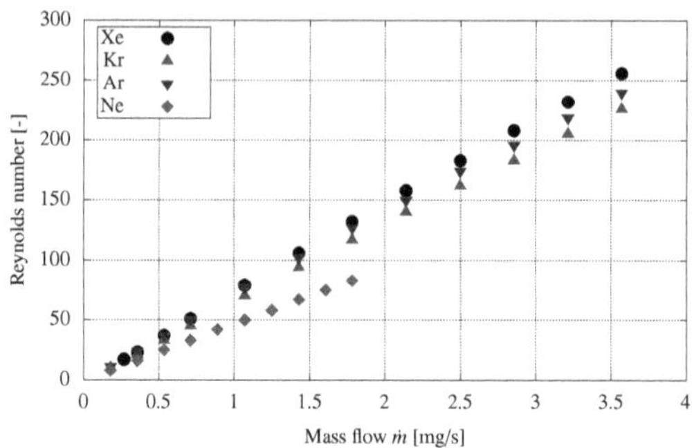

Figure 4.7: Numerically determined Reynolds numbers.

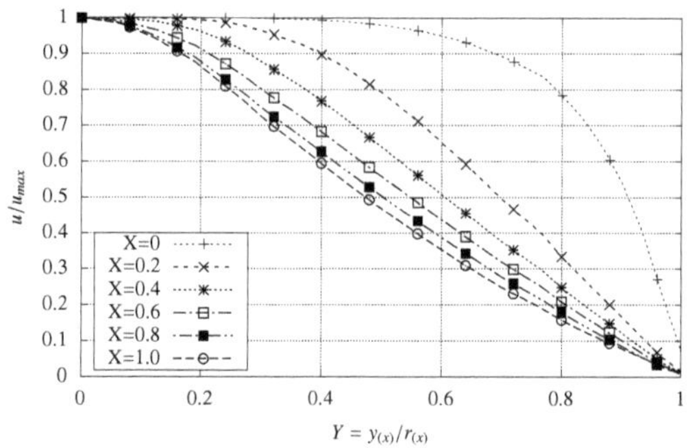

Figure 4.8: Numerical results for the normalized velocity profiles at different locations in the diffuser for xenon and $\dot{m} = 3.568$ mg/s, *Re*=256. The velocity is normalized by its maximum value (reached at the nozzle axis).

is defined according to the experimentally determined pressure in the vacuum chamber. For xenon and a mass flow rate of 3.568 mg/s, *Re*=256, the pressure in the vacuum chamber reaches 3.3 Pa (s. Table 4.4) and the pressure in the ionisation chamber p_{ch}, 390 Pa (s. Fig. 4.6 (d)). The Navier-Stokes results for p_{ch} are influenced by the specified outlet pressure, the mass flow rate and the thermophysical properties of the propellant gas. They are compared with the DSMC and the experimental results in the following section.

4.4.3 DSMC results

Figures 4.9 and 4.10 show the experimental, the Navier-Stokes and the DSMC results for the absolute pressure in the ionisation chamber for argon as a function of the investigated mass flow rates and experimental Knudsen numbers respectively. Here, Ar_{sim} represents the Navier-Stokes results, DSMC_{diff} the DSMC results with the diffuse (Maxwell) wall interaction model and DSMC_{spec}, the results obtained with the specular model. The estimation of Kn is described in detail in Section 4.4.5. As can be seen from Figs. 4.9 and 4.10, the experimental pressure results lie between the DSMC results with the diffusive and the specular model at all studied mass flow rates.

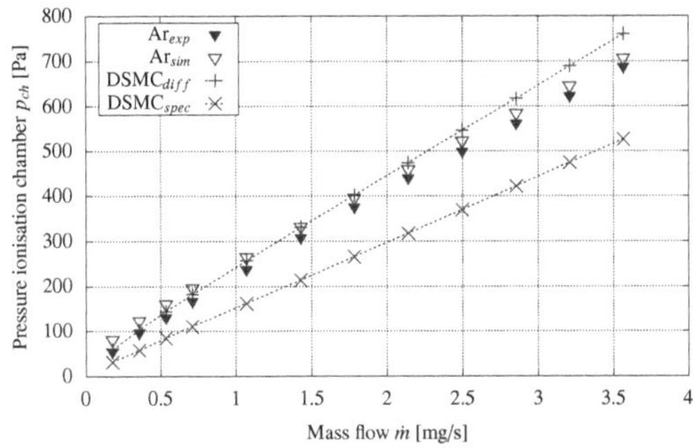

Figure 4.9: Experimental and numerical pressure in the ionisation chamber as a function of mass flow rate for argon.

In order to better compare the numerical approaches employed, the percentage deviations for the pressure results are calculated as follows:

$$Deviation = \frac{p_{ch}^{sim} - p_{ch}^{exp}}{p_{ch}^{exp}} \times 100\% \tag{4.7}$$

with p_{ch}^{exp} representing the experimental and p_{ch}^{sim} the numerical results. The least accurate results are obtained with the specular model, whose deviations range between 23 % and 41 %. This is easily explained by the lack of tangential momentum transfer between particles and nozzle walls leading to numerical pressure losses through the nozzle far below the experimental values. As a consequence, this approach considerably underpredicts the pressure in the ionisation chamber p_{ch}. On the other hand,

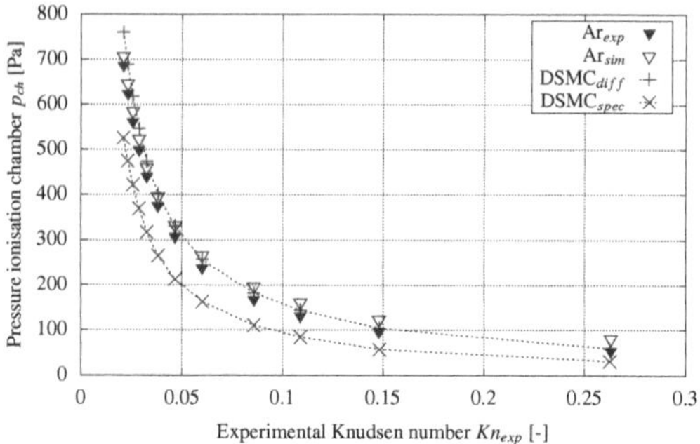

Figure 4.10: Experimental and numerical pressure in the ionisation chamber as a function of Knudsen number for argon.

the DSMC simulations with the widely used Maxwell model slightly overpredict the experimental pressure at all mass flow rates. For low mass flow rates, the pressure deviations obtained with the Maxwell model are low (on average around 8%). This is also evident from the absolute pressure values depicted in Figs. 4.9 and 4.10, which show that the Maxwell model produces particularly accurate results at very low mass flow rates and high values of Kn. Thus, it can be concluded that the experimental and DSMC pressure values are in good agreement in the range $Kn \gtrsim 0.1$. However, it is also evident from Figs. 4.9 and 4.10 that the DSMC results with the Maxwell model drift away from the experimental values with increasing mass flow rates and decreasing Knudsen numbers. The highest deviation is reached at the highest investigated mass flow rate $\dot{m} = 3.568$ mg/s, which corresponds to the lowest Knudsen number $Kn = 0.02$ (slip-flow regime). Interestingly, the Navier-Stokes results are here far more accurate, although a no-slip boundary condition was employed. In summary, the DSMC method with the specular model for the particle-wall interaction significantly underpredicts the pressure in the ionisation chamber and produces the least accurate results at all studied mass flow rates and Kn numbers. Furthermore, the DSMC method with the Maxwell model predicts the ionisation chamber pressure more accurately than the Navier-Stokes simulations for mass flow rates under 1.427 mg/s and $Kn \gtrsim 0.05$. For higher values of $\dot{m}$ and $Kn \lesssim 0.05$ (but still higher than the continuum limit $Kn = 0.01$), the opposite is the case. One possible explanation can be obtained from the analysis of the temporal progression of the pressure in the ionisation chamber p_{ch} shown in Fig. 4.11 for different values of the numerical weight W employed in the DSMC computations with the Maxwell model. As can be seen, for the lowest mass flow rate value $\dot{m} = 0.178$ mg/s, $Kn = 0.26$, the number of argon atoms represented by a DSMC particle W has, in the

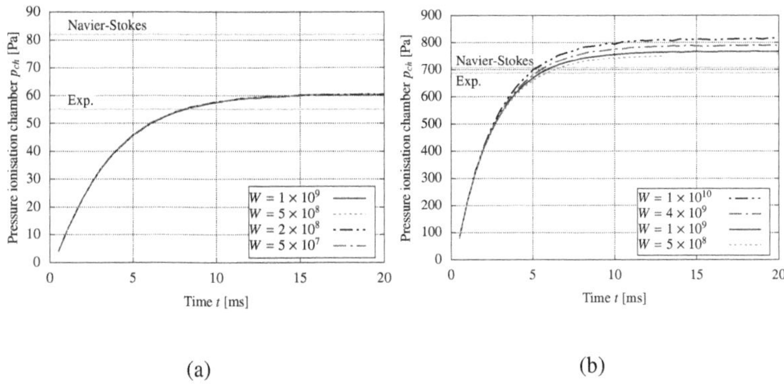

Figure 4.11: Pressure in the ionisation chamber p_{ch} vs. simulated time for argon with $\dot{m} = 0.178$ mg/s, $Kn = 0.26$ (a) and $\dot{m} = 3.568$ mg/s, $Kn = 0.02$ (b). The experimental and Navier-Stokes results are shown as green, horizontal lines.

investigated range, no appreciable influence on the stationary pressure result. Moreover, the DSMC pressure result is here far more accurate than the Navier-Stokes value. For the highest studied mass flow rate $\dot{m} = 3.568$ mg/s, $Kn = 0.02$, the opposite is the case. Here, a reduction of the value of W and the associated increase of DSMC particles in the computational domain have a considerable effect on the stationary pressure result. Lower values of W lead, however, to a steep increase in the computational time required to reach steady state, which becomes a challenge when numerous complex simulations are to be performed. Hence, the results depicted in Fig. 4.11(b) explain the overprediction of the pressure results by the DSMC method and the Maxwell model compared to the Navier-Stokes values in Fig. 4.10 for high mass flow rates and $Kn \lesssim 0.05$. Furthermore, Fig. 4.11(b) suggests that it might be possible to fine-tune the DSMC simulations in order to improve the pressure results in the range $Kn \lesssim 0.05$. However, because of the high computational cost involved, this exercise is non-trivial and beyond the scope of this work. On the other hand, the experimental and DSMC pressure results with the Maxwell model are in good agreement in the range $Kn \gtrsim 0.1$.

A comparison of the velocity fields and profiles obtained from the Navier-Stokes and DSMC simulations with the Maxwell model is shown in Figs. 4.12 and 4.13 for argon as propellant, $\dot{m} = 0.178$ mg/s, $Kn = 0.26$. As can be seen in Fig. 4.12, the no-slip boundary condition results in a strong flow deceleration. With the DSMC model, the maximal velocity reached in the diffuser is 259 /ms. In contrast, the no-slip boundary condition leads to a value of only 188 m/s. The associated overprediction of the vis-

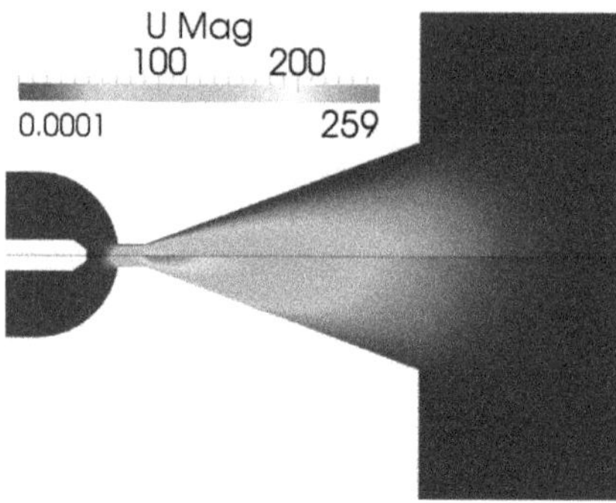

Figure 4.12: Velocity fields [m/s] for argon, $\dot{m} = 0.178$ mg/s. Top: Navicr-Stokes; Bottom: DSMC

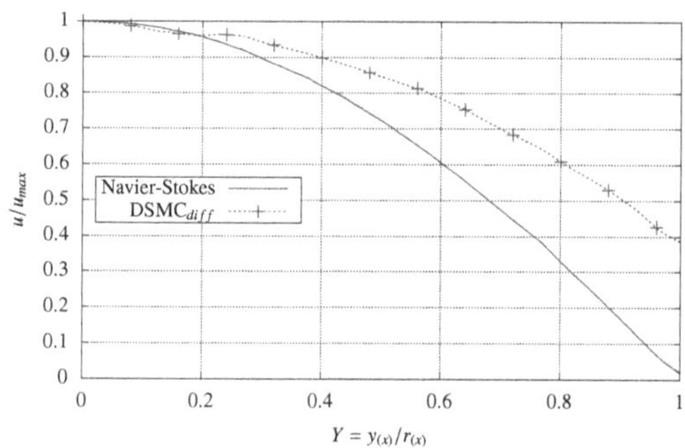

Figure 4.13: Normalized velocity profiles for argon, $\dot{m} = 0.178$ mg/s.

cous losses of kinetic energy in the throat and diffuser areas with the no-slip boundary condition is compensated by a higher pressure gradient through the nozzle. For high values of Kn, this leads to higher results for the inlet pressure with the Navier-Stokes approach compared to the experimental and the DSMC results as shown in Fig. 4.10. The strong flow deceleration induced by the no-slip boundary condition is evident from the velocity profiles shown in Fig. 4.13. Here, the profiles are determined at 50% of the diffuser length. In the figure, Y represents the relative radial distance from the nozzle axis, with $Y = 0$ describing the nozzle axis and $Y = 1$, the diffuser wall. As can be seen, with the DSMC model the wall slip velocity reaches 40% of the maximal velocity at the nozzle axis, a large difference compared to the no-slip boundary condition which greatly affects the flow fields and performance characteristics of the nozzle. In the next sections, the experimental and Navier-Stokes pressure results are analysed in detail and

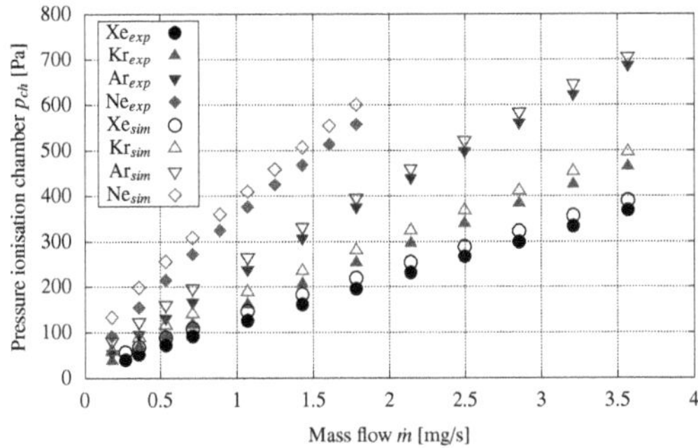

Figure 4.14: Experimental and numerical pressure in the ionisation chamber as a function of mass flow rate.

a function describing the deviation between the numerical and the experimental values as a function of the Knudsen number is developed.

4.4.4 Comparison between Navier-Stokes and experimental results

Figure 4.14 shows both the experimental and the Navier-Stokes results for the absolute pressure in the ionisation chamber of the arcjet thruster and Fig. 4.15, the percentage deviation for all the studied gases. As can be seen, the highest deviation values are obtained at the lowest studied mass flow rate $\dot{m} = 0.178\ mg/s$. Here, the deviations lie between 45% and 54%, with krypton exhibiting the highest value. The deviations decrease strongly with higher values of $\dot{m}$, which can be explained as follows. Very low mass flow rates result in low pressure and density values and hence, in high molecular mean free paths in the ionisation chamber. Furthermore, the geometrical dimensions of the studied arcjet thruster are small. The Knudsen numbers, defined as the ratio of the molecular mean free path to the characteristic length of the geometry, are high. As shown in the next section, the estimated Knudsen numbers in the nozzle inlet indicate that the continuum limit is exceeded for all the studied cases. Specifically, the studied gas flows are either in the slip-flow or in the transition regime. The momentum transfer between atoms and walls in these regimes is not as high as in the continuum regime, which has an important effect on the pressure losses through the nozzle. As a result, the numerical pressure drop calculated from Navier-Stokes simulations with a standard no-slip boundary condition is, for the low mass flow range, considerably higher than the experimental result. This explains both the strong deviations for low values of $\dot{m}$ as well as their decrease with higher mass flow rates as visible in Fig. 4.15.

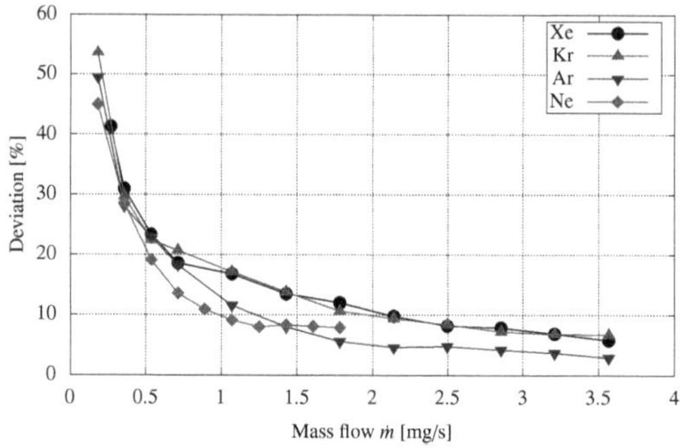

Figure 4.15: Deviations between experimental and numerical results for the pressure in the ionisation chamber.

From the comparison between the continuum-based results with no-slip boundary conditions and the experimental data, a relation for the deviation of the pressure drop along the nozzle as a function of the Knudsen number is obtained in Sections 4.4.5 and 4.4.6. The function corresponds to a second degree polynomial and can subsequently be used to correct the numerical pressure of the ionisation chamber of the arcjet thruster for the studied gases at all mass flow rates and Knudsen numbers. The main advantage of the proposed function is that it produces accurate pressure results for the studied gases while avoiding the case distinction between continuum, slip-flow and transition regimes, therefore eliminating the need for additional first- or second-order slip boundary conditions for some of the studied cases. It should be noted that such an approach is only practical if a rigorous numerical distinction between the flow regimes is not required for the considered problem. The development of the Knudsen function and the corrected pressure results are presented in detail in the following sections.

4.4.5 Knudsen-dependent correcting function for the dimensionless pressure drop

As described in the previous section, the high relative deviations between the Navier-Stokes and the experimental results for lower mass flow rates can be explained by the Knudsen numbers in the nozzle flow. In Sections 4.4.5 and 4.4.6, the continuum-based and experimental results are compared in detail and a function describing the deviation of the pressure drop along the nozzle as a function of the Knudsen number is developed.

The dimensionless Knudsen number Kn can be calculated as follows:

$$Kn = \frac{\lambda}{L} = \frac{k_B T}{\sqrt{2}\pi\sigma^2 p L} \tag{4.8}$$

where λ represents the mean free path of the molecules, L the characteristic length of the geometry, σ the molecular diameter and k_B, the Boltzmann constant. In this study, L is defined as the nozzle throat diameter $L = 1.8$ mm. Thus, Eq. 4.8 can be used to calculate the values of Kn as a function of the gas pressure and temperature. Since the studied gases are led into the nozzle at ambient temperature, the temperature at the nozzle inlet, i.e., in the ionisation chamber is 293.15 K. This value also corresponds to the inlet temperature boundary condition for the simulations. With both numerical and experimental pressure values available, Eq. 4.8 can be used to calculate both a numerical Kn_{sim} and a experimental Kn_{exp} value of the local Knudsen number in the ionisation chamber. It should be noted that the Knudsen numbers presented in this chapter are always estimated at the ionisation chamber. The results for Kn_{exp} as a function of $\dot{m}$ for all studied gases are shown in Fig. 4.16.

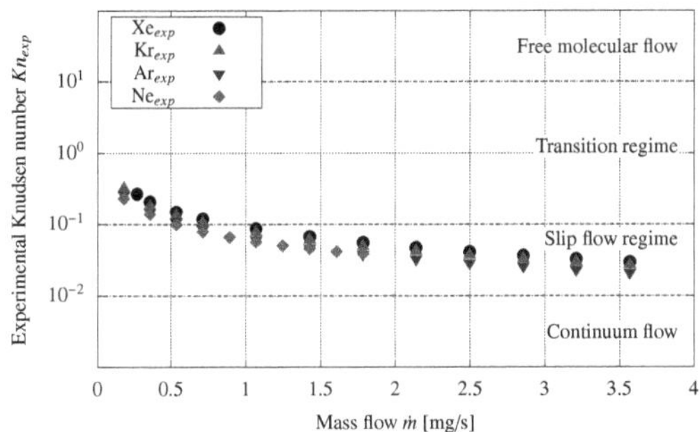

Figure 4.16: Experimental values of the Knudsen number in the ionisation chamber as a function of mass flow rate.

As can be seen from Fig. 4.16, the experimentally determined Knudsen numbers in the ionisation chamber of the nozzle are higher than the continuum flow limit $Kn = 1 \times 10^{-2}$ for all studied cases. The majority of the flows are in the slip-flow regime with the exception of the cases in the lower mass flow ranges $\dot{m} \lesssim 0.714$ mg/s where the transition regime is reached. Considering that the Knudsen numbers in Fig. 4.16 are calculated in the ionisation chamber and the fact that at this location, pressure p and number density

n reach their maximum, the presented values for Kn_{exp} are conservative estimations. Because of the inverse proportionality between Kn and n and the fact that pressure and density decrease through the nozzle towards the vacuum chamber, the local Knudsen numbers in the flow are expected to increase even further through the diffuser and towards the nozzle exit. As a result, the mean Knudsen numbers of the studied cold-gas flows are expected to lie further up on Fig. 4.16, either well inside the transition regime or even close to the free molecular flow zone. This reinforces the hypothesis that the standard no-slip boundary condition used for the Navier-Stokes simulations is the main reason for the deviations between experimental and numerical results, especially for lower mass flow rates where higher Knudsen numbers are reached. In order to further explore this assumption, the relation between the deviation and Knudsen number is illustrated in Fig. 4.17 for xenon as propellant. Here, a positive correlation between the deviation and the experimental Knudsen number in the ionisation chamber of the nozzle can be clearly observed. Furthermore, the same trend is established for all the studied gases. For high values of Kn, this correlation can be fairly well approximated by a linear function as evident from Fig. 4.17. Therefore, the deviation between numerical and experimental results for the pressure in the ionisation chamber can be explained by the Knudsen numbers in the nozzle and the discrepancy between the no-slip boundary condition for the simulations and the actual slip-flow and transition regime conditions established from the experimental pressure results.

In order to avoid a numerical case distinction between continuum, slip-flow and transition regimes and to eliminate the need for the definition of additional slip boundary conditions for some of the studied cases, a correcting function is developed in the following. The correcting function takes advantage of the correlation between the Knudsen numbers and the deviations shown in Fig. 4.17. The main goal of the approach is to accurately predict the pressure in the ionisation chamber of the studied nozzle by applying the developed Knudsen function to the numerical results with a no-slip boundary condition. Furthermore, the proposed approach must provide good results regardless of the mass flow rate, Knudsen number and studied gas.

In order to develop the Knudsen function, a dimensionless pressure drop S based on the dimensionless number presented in [MTJW03] is introduced:

$$S = \frac{\Delta p w b^3}{Q_v \mu L} \tag{4.9}$$

where Δp represents the pressure drop in the nozzle, w=b=D_t the nozzle throat diameter, Q_v the volumetric flow rate, μ the dynamic viscosity and L, the nozzle length. By expressing Q_v as a function of the mass flow rate $\dot{m}$ and the fluid density ρ and through

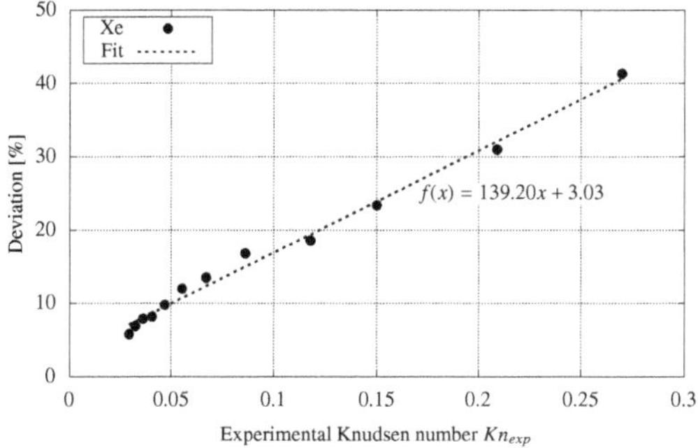

Figure 4.17: Pressure deviation as a function of experimental Knudsen number for xenon.

combination with the ideal gas law, one obtains:

$$S = \frac{\Delta p D_t^4}{\frac{\dot{m}}{\rho}\mu L} = \frac{\Delta p D_t^4}{\dot{m}\mu L}\frac{p}{\frac{R_u}{M}T} \tag{4.10}$$

in which p represents the pressure, R_u the universal gas constant and T, the temperature. The above expression can be rewritten as:

$$S = \Delta p \frac{p D_t^4}{\dot{m}\mu L R T} \tag{4.11}$$

where R is the specific gas constant. The values p, μ and T are local quantities and change along the nozzle. Since p and T in the ionisation chamber are known for both the experimental series and simulations, the final expression for the dimensionless pressure drop is defined as:

$$S = \Delta p \frac{p_{ch} D_t^4}{\dot{m}\mu_{ch} L R T_{ch}} \tag{4.12}$$

with the dynamic viscosity μ calculated from Sutherland's law (Eq. 4.3). Since two different values for the pressure in the ionisation chamber are available (p_{ch}^{exp} and p_{ch}^{sim}), two different values S_{exp} and S_{sim} can be calculated for each studied gas and mass flow rate. The same applies for the Knudsen numbers Kn_{exp} and Kn_{sim}. It should be noted that the main goal of the correcting function is the prediction of the actual pressure drop through the nozzle from simulations with the no-slip boundary condition, even when accurate experimental data is unavailable. Therefore, the proposed Knudsen function is developed based on the numerical rather than on the experimental Knudsen numbers.

To this end, Kn_{sim} is calculated from Eq. 4.8 for each of the studied cases taking advantage of the numerical Navier-Stokes results for the pressure in the ionisation chamber p_{ch}^{sim}. Fig. 4.18 shows the inverse of the dimensionless pressure drop $1/S$ for both the experimental series and simulations as a function of the respective values of Kn_{sim} and with xenon as propellant gas.

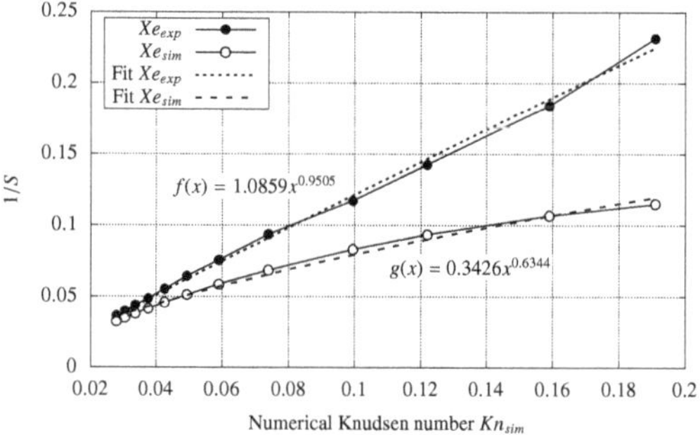

Figure 4.18: Inverse of the dimensionless pressure drop vs. numerical Knudsen number in the ionisation chamber for xenon.

Figure 4.18 highlights the discrepancy between the Navier-Stokes and the experimental results. For high Knudsen numbers, the values of $1/S_{sim}$ are significantly lower than $1/S_{exp}$. The numerical dimensionless pressure drop S_{sim}, therefore, considerably exceeds the experimental result S_{exp} for high values of Kn, reflecting the same trend already established for the absolute pressure values. As can be seen in Fig. 4.18, the results for $1/S$ can be well approximated by a power function $f(x) = ax^b$, with the coefficients a and b for the numerical results considerably smaller than those of their experimental counterparts.

The correcting Knudsen function is developed as follows: the corrected dimensionless pressure drop S_{corr} is defined as a function of the numerical dimensionless pressure drop with no-slip boundary condition S_{sim} and an arbitrary unknown Knudsen function $f(Kn)$:

$$S_{corr} = S_{corr}(S_{sim}, f(Kn)) \tag{4.13}$$

In addition, the Knudsen function should not influence the numerical results when Kn

approaches zero (continuum regime). Mathematically:

$$\lim_{Kn \to 0} S_{corr}(S_{sim}, f(Kn)) = S_{sim} \tag{4.14}$$

Furthermore, increasing values of Kn should lead to a reduction of the numerical dimensionless pressure drop, which can be written as:

$$\lim_{Kn \to \infty} S_{corr}(S_{sim}, f(Kn)) = 0 \tag{4.15}$$

The requirements defined in Eqs. 4.13, 4.14 and 4.15 can be fulfilled by a simple expression of the form:

$$S_{corr} = \frac{S_{sim}}{1 + f(Kn)} \tag{4.16}$$

The exact form of the Knudsen function can now be determined by taking advantage of the experimental results for the pressure drop. Replacing S_{corr} with S_{exp} in Eq. 4.16 and solving for $f(Kn)$ leads to following expression for the Knudsen function:

$$f(Kn) = \frac{S_{sim}}{S_{exp}} - 1 \tag{4.17}$$

The function $S_{sim}/S_{exp} - 1$ for xenon is shown in Fig. 4.19. The resulting relation is mathematically best described by an expression of the form:

$$f(Kn) = A_2 Kn^2 + A_1 Kn + A_0 \tag{4.18}$$

with A_2, A_1 and A_0 as the coefficients of the polynomial function. Since the proposed correcting function only deals with pressure deviations related to high Knudsen numbers, the independent term of the polynomial A_0 must equal zero. A non-zero value for A_0 would otherwise artificially correct deviations, whose root might not have a direct relation to the flow regime. The Knudsen function for xenon obtained from Fig. 4.19 can therefore be written as follows:

$$f(Kn) = A_2 Kn^2 + A_1 Kn \tag{4.19}$$

with $A_1 = 4.204$ and $A_2 = 4.118$. The same approach is used in order to determine the coefficients A_1 and A_2 for the remaining studied gases. The resulting coefficients are listed in Table 4.5 and shown in Fig. 4.20 as a function of molar mass and squared atomic diameter of the propellant, the latter being proportional to the collision cross section of the gas.

Based on Fig. 4.20, a trend for the coefficients of the Knudsen function as a function of the molar mass and the collision cross section of the gas can be clearly established.

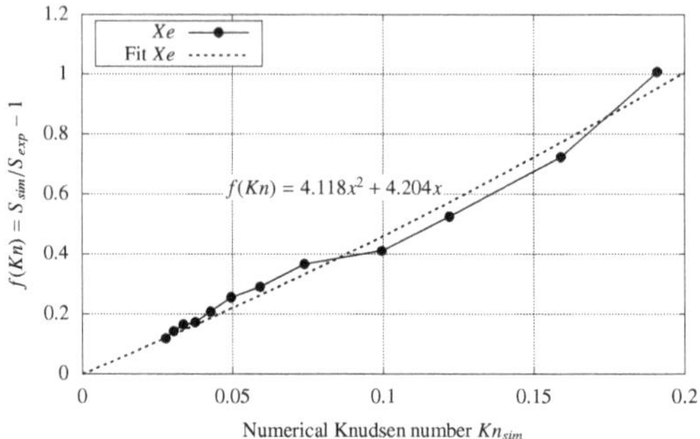

Figure 4.19: Definition of the correcting function $f(Kn)$ for xenon.

Gas	A_2	A_1
Xenon	4.118	4.204
Krypton	9.477	4.278
Argon	21.658	3.359
Neon	27.356	2.807

Table 4.5: Dimensionless Knudsen coefficients.

The highest value for the coefficient of the quadratic term $A_2 = 27.356$ is reached for neon, the lightest studied gas. The value A_2 decreases with higher values of the molar mass reaching its minimum, $A_2 = 4.118$, with xenon, the heaviest studied gas. For the coefficient of the linear term A_1, the opposite trend can be observed. The lowest value $A_1 = 2.807$ is reached with neon and the highest, $A_1 \approx 4.2$, with krypton and xenon. This implies that the second-order coefficient of the correcting function $f(Kn)$ plays a decisive role for light gases with small collision cross sections such as neon and argon. For heavier gases, the importance of the second-order coefficient gradually diminishes and the first-order coefficient, on the other hand, becomes more influential. However, the second-order coefficient A_2 for xenon is clearly still large enough to affect the results and cannot be neglected. It can therefore be concluded that the correcting function, and thereby the deviation in the pressure drop through the nozzle as a function of the Knudsen number, is closer to a linear function for gases with large collision cross sections. For gases with smaller collision cross sections, the experimental pressure results deviate more strongly from the no-slip assumption with increasing Knudsen numbers and the second degree coefficient in the Knudsen function becomes more important.

Figure 4.21 shows the obtained Knudsen functions $f(Kn)$ for the studied gases in the

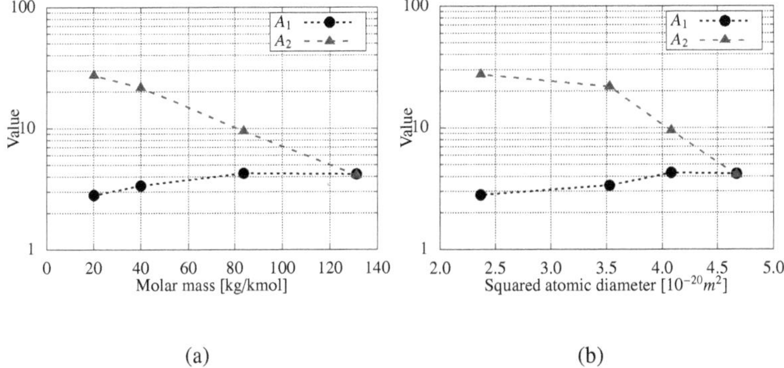

Figure 4.20: Coefficients A_1 and A_2 for the Knudsen function $f(Kn)$ as a function of molar mass and squared atomic diameter of propellant gas.

range $0 < Kn < 0.3$. For Knudsen numbers in the continuum regime ($Kn < 0.01$) the function approaches zero. For relatively low values of Kn in the range $0.01 < Kn < 0.05$, the second-order term Kn^2 becomes much smaller than the first-order term and the correcting functions can be approximated by linear functions with A_1 as their slope. In this range, the value of $f(Kn)$ for xenon and krypton is slightly higher than for argon and neon. Approaching the boundary between slip-flow and transition regime the second-order coefficients start to dominate the form of $f(Kn)$.

For $Kn > 0.1$ the relation of the correcting function values for the studied gases can be expressed as:

$$f_{Ne}(Kn) > f_{Ar}(Kn) > f_{Kr}(Kn) > f_{Xe}(Kn) \tag{4.20}$$

which also corresponds to the relation between the second-order gas coefficients:

$$A_{2,Ne} > A_{2,Ar} > A_{2,Kr} > A_{2,Xe} \tag{4.21}$$

Since the highest values of $f(Kn)$ are reached for neon, it can be concluded that neon deviates the most from the no-slip assumption. In general, the results suggest that lighter gases have a higher tendency to exhibit slip-behaviour. This is also supported by the low values of $f(Kn)$ for xenon, the heaviest studied gas. The Knudsen functions for krypton and argon lie, as expected, between those of neon and xenon.

The Knudsen function in Eq. 4.19 can alternatively be written as a combination of two linear functions as follows:

$$f(Kn) = \frac{Kn}{C_1}\left(1 + C_2 Kn\right) \tag{4.22}$$

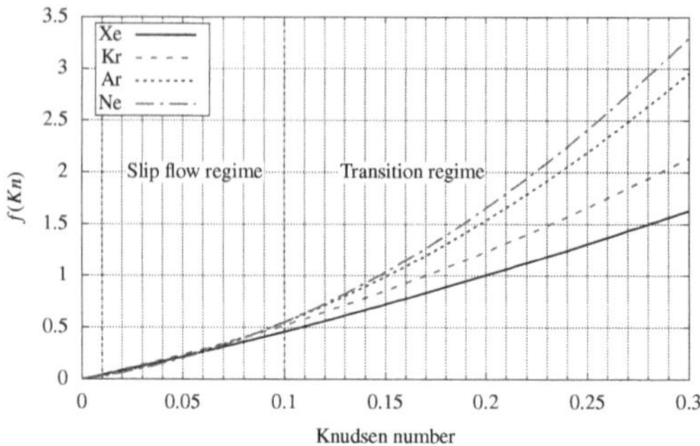

Figure 4.21: Comparison of the Knudsen functions for the studied gases.

where the coefficients of the Knudsen function are defined as:

$$C_1 = \frac{1}{A_1} \text{ and } C_2 = \frac{A_2}{A_1} \tag{4.23}$$

The resulting values for the coefficients C_1 and C_2 are shown in Table 4.6. In the modified form of the Knudsen function in Eq. 4.22, the C_2 coefficients are a combination of the values A_1 and A_2 in Table 4.5. This modification is useful for the discussion of the relation between the C coefficients and the molar mass of the propellant gas in Section 4.4.6.

Gas	C_2	C_1
Xenon	0.980	0.238
Krypton	2.215	0.234
Argon	6.448	0.298
Neon	9.746	0.356

Table 4.6: Dimensionless Knudsen coefficients.

The numerical, dimensionless pressure drop S can now be corrected using either the function with A coefficients (Eq. 4.19) or with C coefficients (Eq. 4.22). Combination of the expression in Eq. 4.16 with the obtained Knudsen function with C coefficients produces:

$$S_{corr} = \frac{S_{sim}}{1 + \frac{Kn_{sim}}{C_1}\left(1 + C_2 Kn_{sim}\right)} \tag{4.24}$$

where Kn has been defined as the numerical Knudsen number Kn_{sim} (the reason for the use of Kn_{sim} for the pressure correction is described in page 135). The experimental, numerical and corrected values for the dimensionless pressure drop are shown in Fig. 4.22 for xenon. As can be seen, the proposed approach corrects the high pressure deviations for high values of the Knudsen number. Furthermore, the numerical results for relatively low Knudsen numbers are, in comparison, only slightly influenced by the Knudsen function.

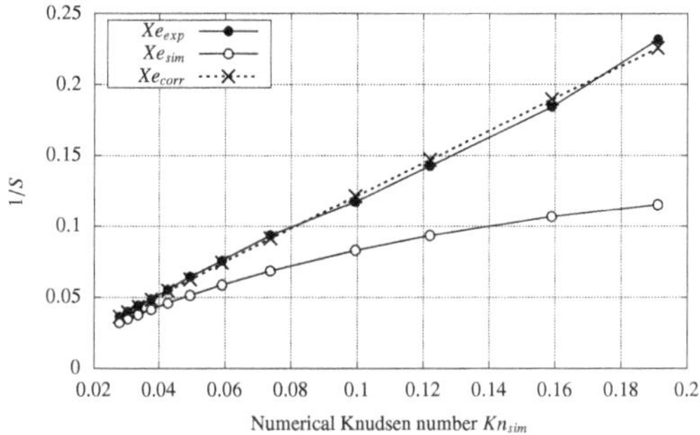

Figure 4.22: Inverse of the experimental, numerical and corrected dimensionless pressure drop vs. Knudsen number in the ionisation chamber for xenon.

The corrected absolute pressure in the ionisation chamber of the nozzle can now be calculated based on S_{corr}. To this end, Δp in the definition of the dimensionless pressure drop S (Eq. 4.12) is replaced by the difference between the pressure in the ionisation and the vacuum chamber $\Delta p = p_{ch} - p_{vac}$. Furthermore the pressure p in Eq. 4.12 is replaced by p_{ch}. Subsequent transformation of the equation produces a second degree polynomial with p_{ch}^{corr} as a function of S_{corr}. The results for the experimental and the corrected absolute pressure are shown in Fig. 4.23 for all studied gases as a function of the mass flow rate. As can be seen, the results from the proposed approach show very good agreement with the experimental results for studied gases at all mass flow rates. Fig. 4.24, shows the strongly reduced deviations after correction and the original relative

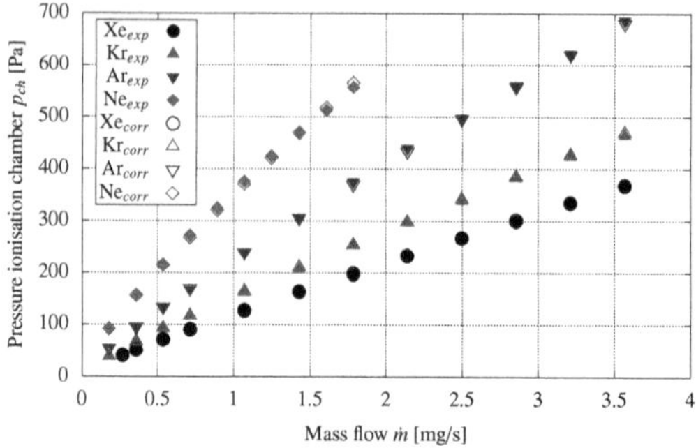

Figure 4.23: Experimental and corrected pressure in the ionisation chamber for the studied gases.

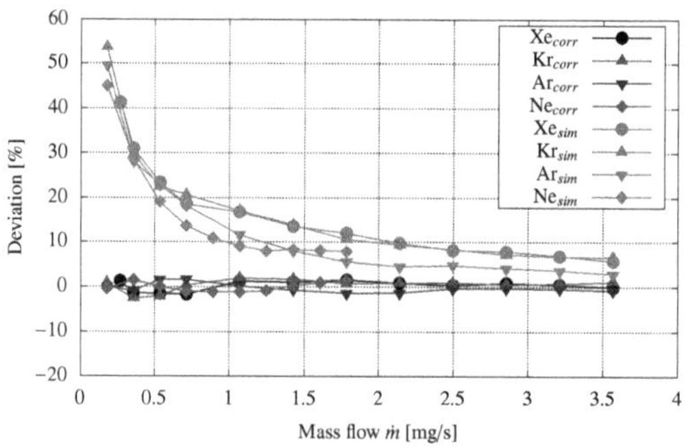

Figure 4.24: Deviations between experimental and corrected results for the pressure in the ionisation chamber. Original deviations are shown in gray.

deviations from the simulation with standard no-slip boundary conditions. While the original deviations exceeded 50%, the deviations after correction of the pressure results with the proposed approach are not higher than 3%.

4.4.6 Molar mass dependency of the Knudsen function coefficients

The values of the C_1 coefficients listed in Table 4.6 are shown in Fig. 4.25 as a function of the molar mass of the gas. As can be seen, C_1 can be well approximated by a power function of the form:

$$C_1 = \left(\frac{M}{M_1^*}\right)^{-\beta_1} \tag{4.25}$$

with $M_1^* = 0.278$ kg/kmol, $\beta_1 = 0.243$ and M, the molar mass of the propellant gas. The results for the coefficient C_2 are shown in Fig. 4.26. The values of C_2 can be fairly well approximated by a power function of the same form as Eq. 4.25 with the parameters $M_2^* = 236.492$ kg/kmol and $\beta_2 = 0.939$. However, in contrast to the results for C_1, slight deviations between the function and the C_2 values can be observed. Nonetheless, the good agreement between the C values in Table 4.6 and the proposed power functions shown in Figs. 4.25 and 4.26 suggests that it is possible to correct the continuum-based numerical pressure results for the four studied gases using only four gas-independent parameters: M_1^*, β_1, M_2^* and β_2. Note that the coefficients of determination of the proposed power functions are $R^2 = 0.953$ for C_1 and $R^2 = 0.949$ for C_2, with power functions producing the best overall results. In contrast, an exponential function produces good results for C_2 ($R^2 = 0.998$) but a poor fit for C_1 ($R^2 = 0.793$), while a polynomial function of the form $C_i(M) = a_i M^2 + b_i M$ completely fails to fit the data.

In order to correct the continuum-based numerical pressure results using the four gas-independent parameters M_1, β_1, M_2 and β_2, the values C_1 and C_2 in the Knudsen function (Eq. 4.22) are replaced by the respective power functions shown in Figs. 4.25 and 4.26. This leads to:

$$f^*(Kn_{sim}) = \left(\frac{M}{M_1^*}\right)^{\beta_1} Kn_{sim}\left[1 + \left(\frac{M}{M_2^*}\right)^{-\beta_2} Kn_{sim}\right] \tag{4.26}$$

where $f^*(Kn)$ represents the Knudsen function with gas-independent coefficients, M the molar mass of the propellant gas and M_1^*, M_2^*, β_1 and β_2 the constant parameters listed in Table 4.7 and valid for all the studied gases. The numerical, dimensionless pressure drop S can subsequently be corrected according to the definition in Eq. 4.16:

$$S_{corr}^* = \frac{S_{sim}}{1 + f^*(Kn_{sim})} \tag{4.27}$$

in which S_{corr}^* represents the corrected dimensionless pressure drop based on the gas-independent coefficients. In the same way as for S_{corr} in Section 4.4.5, S_{corr}^* can be used to calculate the corrected absolute pressure in the ionisation chamber of the studied nozzle. The results for the absolute pressure in the ionisation chamber corrected with the gas-independent Knudsen model $f^*(Kn)$ are shown in Fig. 4.27 as dashed lines for each gas. The corrected pressure results based on the gas-dependent C coefficients and the experimental results are also shown in the figure.

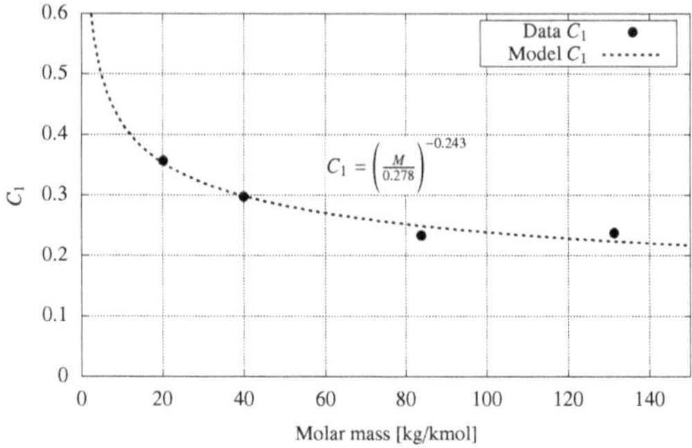

Figure 4.25: Coefficients C_1 for the Knudsen function $f(Kn)$ vs. molar mass of propellant gas.

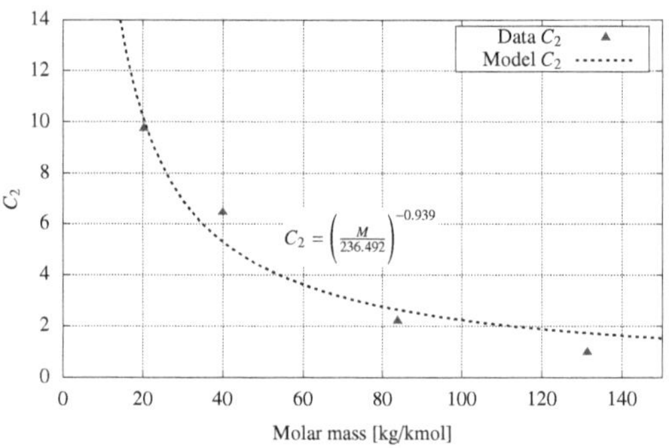

Figure 4.26: Coefficients C_2 for the Knudsen function $f(Kn)$ vs. molar mass of propellant gas.

As can be seen from Fig. 4.27, the approach for the Knudsen function with gas-independent coefficients in Eq. 4.26 produces very good results for the absolute pressure in the ionisation chamber. The results are similar to the values obtained with the function $f(Kn)$ with gas-dependent coefficients C_1 and C_2 and both approaches show very good agreement with the experimental results for the studied gases and nozzle geometry.

By applying the correcting function $f(Kn)$ with the gas-dependent coefficients C_1 and C_2, the percentage deviations between numerical and experimental pressure values are reduced from over 50% for low mass flow rates to under 3%. On the other hand, the approach with the Knudsen function $f^*(Kn)$ with gas-independent coefficients M_1^*, M_2^*, β_1, and β_2 produces corrected pressure values whose deviations from the experimental

M_1^* [kg/kmol]	M_2^* [kg/kmol]	β_1	β_2
0.278	236.492	0.243	0.939

Table 4.7: Gas-independent coefficients for the Knudsen function.

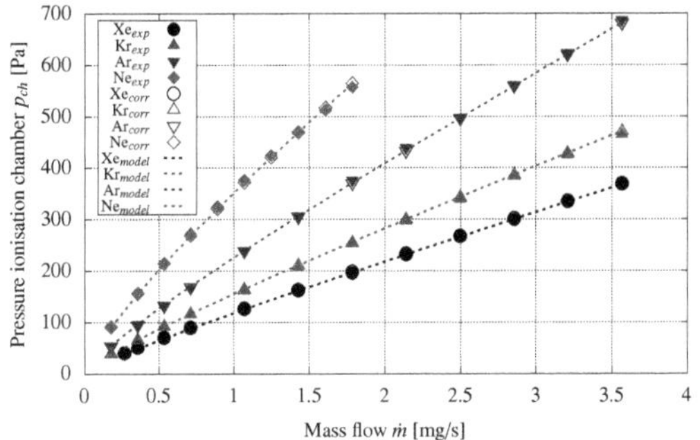

Figure 4.27: Experimental and corrected inlet pressure for the studied gases. The dashed lines represent the results from the model $f^*(Kn)$ with gas-independent coefficients.

data do not exceed 5% for the studied nozzle. The main advantage of the latter approach is the fact, that instead of eight, only four coefficients plus the molar mass of the gas are needed to fully define the correcting Knudsen function. The Knudsen function $f^*(Kn)$ in Eq. 4.26 together with the coefficients in Table 4.7 can therefore be used to correct the numerical pressure deviations resulting from the no-slip boundary condition being applied to slip- and transition flows in the studied micronozzle. The coupling with the Knudsen number guarantees that the pressure correction is only performed for flows with high Knudsen numbers while not influencing cases with low values of Kn and low pressure deviations.

4.4.7 Thrust and specific impulse

The thrust F_t produced by the studied thruster in cold-gas operation can be estimated from the numerical results as follows:

$$F_t = \iint_S \rho u(u \cdot n) \mathrm{d}S + \iint_S (p_{exit} - p_\infty) \mathrm{d}S \tag{4.28}$$

where u represents the velocity vector, n the normal vector to the imaginary exit surface of the nozzle, p_{exit} the pressure at the nozzle exit and $p_{\infty} = p_{vac}$, the ambient pressure in the vacuum chamber. In the simulations, the force resulting from the difference between the ambient pressure and the pressure at the nozzle exit during steady state operation is small in comparison with the momentum thrust and can be neglected. Furthermore, since the nozzle axis and the x-axis of the coordinate system in the simulations area aligned, the resulting thrust vector has only the component along the x-axis. Eq. 4.28 can therefore be written as follows:

$$F_{tx} = \iint_S \rho u_x (u \cdot n) \mathrm{d}S \tag{4.29}$$

The specific impulse of the cold-gas thruster can be obtained from the following equation:

$$I_{sp} = \frac{F_{tx}}{\dot{m} g} \tag{4.30}$$

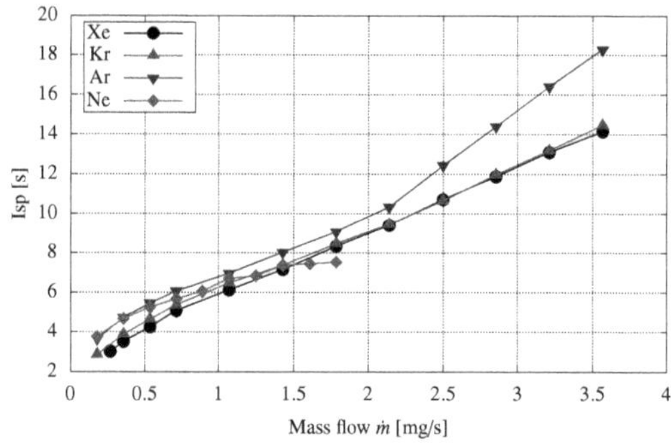

Figure 4.28: Numerically determined I_{sp} of the cold-gas thruster for the studied gases as a function of the mass flow rate.

It should be noted that the thrust values are calculated based on the velocity profiles at the nozzle exit obtained through Navier-Stokes simulations with no-slip boundary conditions. Unlike the pressure, the numerical velocity results are not subjected to a Knudsen-dependent correction in this work. At this point it is assumed that the mean density at the nozzle exit plane is dominated by the ambient pressure in the vacuum chamber p_{vac}. Since the experimental and numerical ambient pressure values are equal (measured values are used as boundary conditions, s. Table 4.4), the difference between the numerical and the actual mean density at the nozzle exit is expected to be small. Because of this fact and taking mass conservation into account, the numerical mean exit velocities can be considered as a good approximation of the actual values. The total thrust values calculated from the simulations constitute therefore an accurate prediction of the thrust produced by the cold-gas thruster in the experimental setup, despite the expected local deviations in the velocity and density profiles associated with the slip-flow and transition regimes.

The results for the specific impulse of the studied cold-gas thruster are shown in Fig. 4.28 as a function of the mass flow rate for all the studied gases. As can be seen, the highest I_{sp} values for a given mass flow rate are reached with argon as propellant with a maximum specific impulse of approx. 18 s for $\dot{m} = 3.568$ mg/s, $Re = 240$. There is little difference between the results for krypton and xenon, with a maximum I_{sp} of approx. 14 s with $Re = 226$ for krypton and $Re = 256$ for xenon. The neon curve follows closely the argon results at low values of $\dot{m}$. Then it drifts away and intersects the curves for krypton and xenon. This behaviour can be explained by both the numerical pressure boundary condition at the outlet and the ratio of ionisation chamber to vacuum chamber pressure p_{ch}/p_{vac}. As described in Section 4.3.2, the pressure boundary condition at the outlet is defined based on the experimental values measured in the vacuum chamber. With this approach, the pressure conditions in the outlet-volume of the simulations and in the vacuum chamber in the experimental setup are similar. This guarantees the comparability of the numerical and the experimental results. However, the pump performance across the different gases and mass flow rates is not homogeneous, as can be inferred from the experimental values for the pressure in the vacuum chamber listed in Table 4.4 and the resulting values of p_{ch}/p_{vac}. For xenon, pressure ratios between 53 for $\dot{m} = 0.268$ mg/s and 112 for $\dot{m} = 3.568$ mg/s are established. For krypton, the pressure ratio ranges between 45 for $\dot{m} = 0.178$ mg/s and 82 for $\dot{m} = 3.568$ mg/s. With argon values between 34 for $\dot{m} = 0.178$ mg/s and 54 for $\dot{m} = 3.568$ mg/s are reached. An increase of $\dot{m}$ for xenon, krypton and argon leads therefore to higher pressure ratios. In contrast, an almost constant value $p_{ch}/p_{vac} \approx 30$ is observed for neon. This results from the strong increase of p_{vac} with mass flow rate. The pressure in the vacuum chamber p_{vac} climbs from 3 Pa to 19 Pa in the range 0.268 mg/s $\leq \dot{m} \leq$ 1.784 mg/s for neon (s. Table 4.4). In comparison, the values of p_{vac} for krypton lie between 0.85 Pa and 3.6 Pa in the same mass flow range. The higher pressure values at the nozzle outlet for neon have, as expected, a negative effect on the flow acceleration through the nozzle. The resulting reduction of the produced thrust and specific impulse explains the unusual behaviour of neon in Fig. 4.28. While the outlet pressure also increases with $\dot{m}$ for the heavier gases, the negative effect on thrust is weaker because p_{vac} does not rise as strongly and a continuous increase of the ratio p_{ch}/p_{vac} with $\dot{m}$ is still observed. The numerical results shown in Fig. 4.28, including the trend for neon, are to be expected if the produced thrust is experimentally measured in the studied setup.

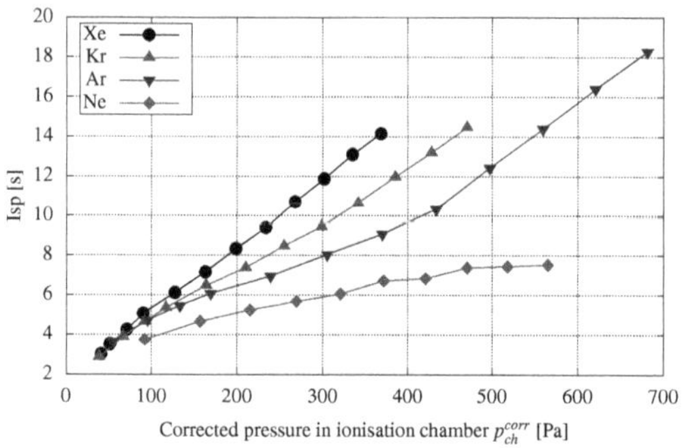

Figure 4.29: Numerically determined I_{sp} of the cold-gas thruster for the studied gases as a function of the corrected pressure in the ionisation chamber.

Figure 4.29 shows the specific impulse as a function of the corrected pressure in the ionisation chamber. Since the deviations between the corrected and the experimental pressure are small (s. Fig. 4.24), it is possible to use the results in Fig. 4.29 in order to predict the thrust produced by the cold-gas thruster as a function of the pressure in the ionisation chamber measured experimentally. As can be seen from Fig. 4.29, at any given pressure in the ionisation chamber, the highest I_{sp} is reached with xenon and its value decreases for lighter propellant gases. This can be explained by the fact that the pressure drop between ionisation and vacuum chamber at any given mass flow rate is lower for the heavier gases. In other words, the heavier the gas, the higher the mass flow rate necessary to reach a given pressure value in the ionisation chamber. The higher mass flow rate leads to higher exit velocities and therefore to the higher I_{sp} values for the heavier gases at a given pressure value in Fig. 4.29. The maximum value of both I_{sp} and p_{ch}^{corr} in Fig. 4.29 is reached with argon. However, as described in Section 4.2.1, the maximum investigated mass flow rate for neon was only 50% of the value for the remaining gases. With $\dot{m} = 3.568$ mg/s for neon and homogeneous pressure conditions in the vacuum chamber across all mass flow rates and gases (closer to actual space operating conditions), the maximum value of both I_{sp} and p_{ch}^{corr} is expected to be reached with neon instead of argon. The thrust produced by the cold-gas thruster is shown in Fig. 4.30 for all the studied gases and mass flow rates. As for the I_{sp} in Fig. 4.29, Fig. 4.30 can be used in order to predict the thrust produced by the cold-gas thruster in the experimental setup. The highest value of the studied cases is reached with argon. For the highest mass flow rate $\dot{m} = 3.568$ mg/s, $Re = 240$, a thrust of 640 μN or 0.64 mN is achieved. With a mass flow rate for neon of 3.568 mg/s and homogeneous pressure conditions at the outlet of the nozzle, the maximum value of F_t is expected to be reached with neon instead of argon.

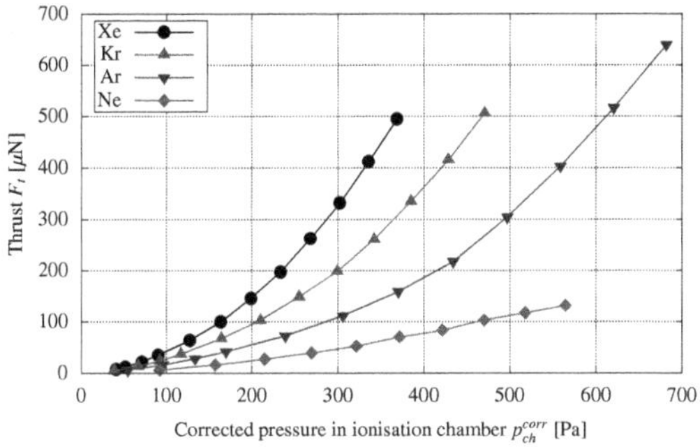

Figure 4.30: Thrust of the cold-gas thruster for the studied gases as a function of the corrected pressure in the ionisation chamber.

Finally, we compare the thrust results for each of the studied cases with the theoretical thrust values obtained from the quasi-1D inviscid theory. The theoretical maximum thrust can be calculated as follows:

$$F_{tx,max} = [(\rho_{exit} \times u_{x\,exit}^2) + (p_{exit} - p_\infty)]A_{exit} \tag{4.31}$$

As in Eq. 4.28, the pressure term $(p_{exit} - p_\infty)A_{exit}$ in Eq. 4.31 is small compared to the momentum thrust and is neglected. The theoretical isentropic flow velocity at the nozzle exit is calculated from the definition of the Mach number as $u_{exit} = Ma_{exit} \times a_{exit}$. The Mach number at the nozzle exit is obtained from the area-Mach number relation in Eq. 2.93:

$$\left(\frac{A}{A^*}\right)^2 = \frac{1}{Ma^2}\left[\frac{2}{\gamma+1}\left(1+\frac{\gamma-1}{2}Ma^2\right)\right]^{(\gamma+1)/(\gamma-1)} \tag{4.32}$$

by replacing A with A_{exit} and A^* with the area of the nozzle throat. Furthermore, the speed of sound at the nozzle exit can be calculated from the flow exit temperature as follows:

$$a_{exit} = \sqrt{\frac{\gamma R_u T_{exit}}{M}} \tag{4.33}$$

and the exit temperature is obtained by replacing Ma with Ma_{exit} in the isentropic relation in Eq. 2.90:

$$\frac{T_0}{T} = 1 + \frac{\gamma - 1}{2} Ma^2 \tag{4.34}$$

where $T_0 = 293.15$ K corresponds to the inlet temperature of the flow. In Eqs. 4.32 to 4.34 γ represents the heat capacity ratio, $R_u = 8,314$ J/kmolK the universal gas constant and M, the molar mass as listed in Table 4.3. Finally, the density at the nozzle exit can be calculated based on the mass flow rate of the studied cases (constant along the nozzle) and the theoretical isentropic flow exit velocity:

$$\rho_{exit} = \frac{\dot{m}}{u_{exit} A_{exit}} \tag{4.35}$$

Since the focus lies on the ideal, theoretical maximum thrust from the quasi-1D inviscid theory, it is important that the values used for its calculation are not influenced by rarefaction or viscosity related effects. By using the approach described in Eqs. 4.32 to 4.35 for the calculation of the isentropic u_{exit} and ρ_{exit}, one takes advantage of two boundary conditions T_0 and $\dot{m}$ which are not influenced by the consideration of the slip-flow or transition regimes. T_0 (*stagnation temperature*, in this case assumed to be equal to the ambient temperature) has the same value across all gases and mass flow rates. Furthermore, the mass flow rate used as numerical boundary condition is regulated in the experimental setup by a mass flow controller at the nozzle inlet, which guarantees accurate mass flow values for all the studied cases, even with gas rarefaction effects occurring downstream. For the comparison between the numerically obtained and the theoretical thrust values, the thrust efficiency η is defined as follows (cp. [LH10]):

$$\eta = \frac{F_{tx}}{F_{tx,max}} (\times 100\%) \tag{4.36}$$

The results for the thrust efficiency are shown in Fig. 4.31 as a function of the mass flow rate and the Reynolds number. As can be seen, the thrust efficiency for all the studied gases increases with higher values of $\dot{m}$ and Re. Higher mass flow rates lead to higher fluid velocities and hence, to higher Reynolds numbers. In the example shown in Figs. 4.6 and 4.8, the viscous boundary layer was shown to induce a shock-less decrease of Ma and u and an increase of the flow temperature. This is, in turn, associated with a conversion of kinetic energy into thermal energy. Since thermal energy does not

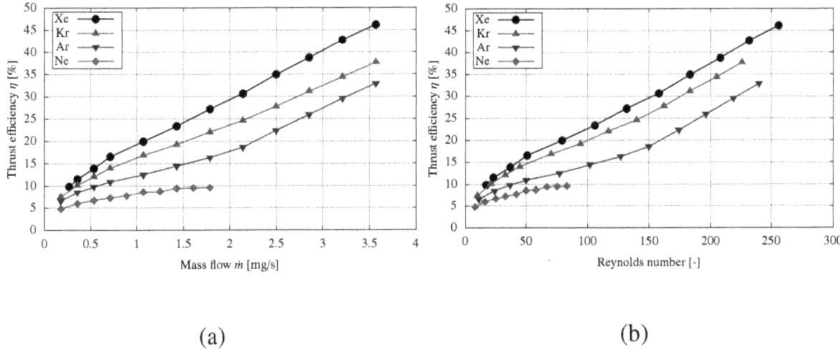

Figure 4.31: Thrust efficiency for the studied gases as a function of the mass flow rate (a) and the Reynolds number (b).

contribute to the thrust produced by the device, this process leads to a net reduction of thrust efficiency. With higher values of *Re*, the influence of the viscous boundary layer, the main responsible for the loss of kinetic energy in the flow, becomes smaller. The consequence is a net improvement of thrust efficiency as visible in Fig. 4.31. One additional factor playing an important role can be identified from the comparison of the velocity profiles for xenon at different values of the mass flow rate in Fig. 4.32. Here, Y represents the relative radial distance from the nozzle axis, with $Y = 0$ describing the nozzle axis and $Y = 1$, the diffuser wall. Higher mass flow rates result in higher pressure values in the ionisation chamber. As visible in Fig. 4.32, this leads to a slight decrease of the relative thickness of the boundary layer and hence, to a reduction of viscous losses and an increase of exit velocity and thrust efficiency. The reduction of the boundary layer thickness as a result of higher stagnation pressures has also been reported in [ALFG05, AFG$^+$06], albeit for the thermal boundary layer. This phenomenon can also be qualitatively observed in the velocity fields for xenon at different mass flow rates depicted in Fig. 4.33. The low velocity blue region near the diffuser wall becomes thinner with increasing values of $\dot{m}$ and *Re*. In addition, the position in the diffuser where the velocity starts to decrease, moves towards the nozzle exit with higher mass flow rates.

It is therefore clear, that the viscous boundary layer has a considerable effect on the flow velocity and thrust values of the investigated micronozzle. Furthermore, it has already been established that the thickness of the boundary layer in the diffuser increases towards the nozzle exit (s. Fig. 4.8). In order to study the effect of the increasing boundary layer thickness on nozzle performance, the thrust as a function of the diffuser length is calculated for xenon with $\dot{m} = 3.568$ mg/s. To this end, additional simulations

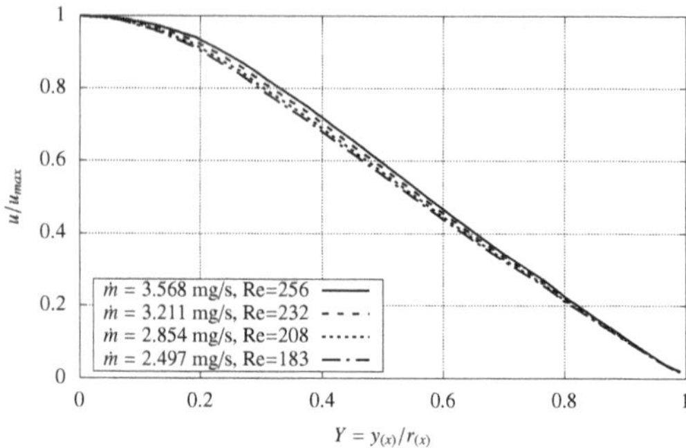

Figure 4.32: Velocity profiles for xenon at various mass flow rates. The velocity is normalized by its maximum value (reached at the nozzle axis). The profiles are determined at half the diffuser's length.

for nozzle geometries with shorter diffusers are performed and the produced thrust is calculated at the respective nozzle exit. The results depicted in Fig. 4.34 show that the optimum thrust is reached for $D_L \approx 0.15$, i.e., with a diffuser length of 15% of the current design value (20 mm). For longer diffusers, viscous effects become more dominant and a higher fraction of flow kinetic energy is converted into thermal energy. For the specific case with xenon and $\dot{m} = 3.568$ mg/s, a reduction of the diffuser length to its optimal value would lead to a thrust increase of approx. 55% from 495 μ N to almost 770 μ N. Furthermore, the specific impulse would climb from 14 s to 22 s, while the Reynolds number reached in the divergent section of the nozzle remains constant (Re =256). These results illustrate the importance of the precise characterization of the viscous boundary layer and the viscous losses for thrust optimization in propulsion systems using micronozzles.

In order to make a conclusive comparison of thrust and specific impulse between the studied gases, the pressure conditions in the outlet region of the nozzle should be the same for all cases. While this might be the case for actual propulsion systems in space, constant pressure conditions across all gases and mass flow rates cannot be reached with the available experimental setup. Nonetheless, the main goal of the cold-gas study is the comparison between numerical and experimental results and the subsequent prediction of inlet pressure and thrust for the existing experimental setup. In this context, the results discussed in this section remain valid for the prediction and optimization of the thrust and specific impulse produced by the INGA III experimental setup during cold-gas operation.

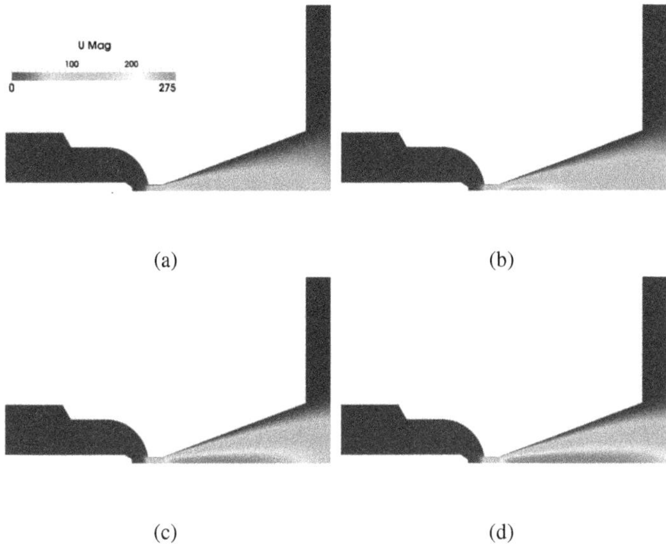

(a) (b) (c) (d)

Figure 4.33: Velocity field for xenon simulations at different mass flow rates. (a) $\dot{m}$ =0.714 mg/s, *Re*=51; (b) 1.427 mg/s, *Re*=106; (c) 2.854 mg/s, *Re*=208; (d) 3.568 mg/s, *Re* =256.

4.5 Summary

The cold-gas expansion of the noble gases xenon, krypton, argon and neon through a millimetre-scale Laval nozzle at mass flow rates ranging between 0.178 mg/s and 3.568 mg/s and with Knudsen numbers in the range $0.02 < Kn < 0.33$ has been studied experimentally and numerically. The experimentally measured pressure in the ionisation chamber is compared with numerical results from Navier-Stokes based simulations with a no-slip boundary condition and with DSMC results.

The DSMC and the experimental results show good agreement in the range $Kn > 0.1$, while the Navier-Stokes results describe the experimental data more accurately than the DSMC method for $Kn < 0.05$. For low mass flow rates and high Kn numbers, the pressure values in the ionisation chamber determined through the continuum-based simulations are considerably higher than the experimental results and relative deviations of over 50% are established. These are the result of the discrepancy between the no-slip boundary condition used for the Navier-Stokes simulations and the high experimental Knudsen numbers which lie in the slip-flow and transition regimes.

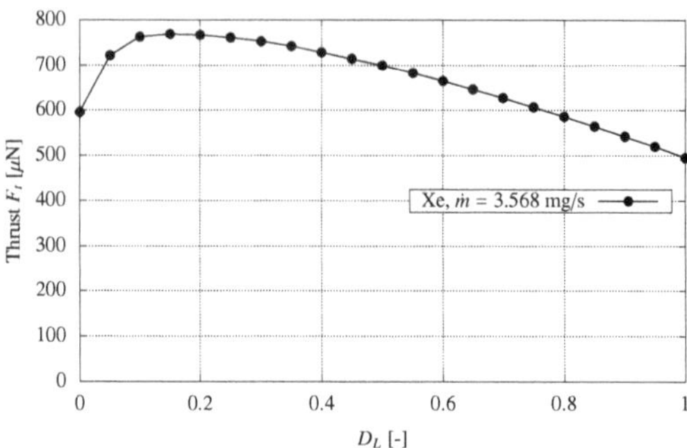

Figure 4.34: Numerically determined thrust for xenon with $\dot{m} = 3.568$ mg/s, $Re = 256$ as a function of the normalized diffuser length with $D_L = 1$ representing the design diffuser length (20 mm).

Based on the comparison between the experimental and numerical results, a relation describing the deviation of the pressure drop through the nozzle as a function of Kn with gas-dependent and gas-independent accommodation coefficients is obtained. The Knudsen-function shows that for gases with small collision cross sections, the experimental pressure results strongly deviate from the no-slip assumption in a relation mathematically best described through a second degree polynomial. For gases with large collision cross sections, the deviation as a function of Kn is closer to a linear function. Based on the developed function, the pressure results for the studied gases, both in the slip-flow and transition regimes, are corrected. As already the case for standard slip-flow boundary conditions, the coefficients of the presented Knudsen function are geometry-dependent and are determined based on experimental results. The main advantage is that once the coefficients are calibrated, the described approach allows the prediction of the pressure drop and the absolute pressure values at the inlet of the studied nozzle geometry based on numerical results from usually computationally inexpensive Navier-Stokes based solvers with no-slip boundary conditions.

The corrected pressure results show very good agreement with the experimental values. The approach with gas-dependent coefficients leads to a reduction of the relative pressure deviations from over 50% to under 3% across all gases and mass flow rates. The approach with only four gas-independent coefficients also provides very good results, with the relative deviations below 5% for all gases and mass flow rates. The flexibility of the proposed approach is useful for the numerical modelling of experimental setups operating in a large range of mass flow rates, where several flow regimes might appear. Through the proposed Knudsen functions, it is possible to avoid a numerical case distinction between continuum, slip-flow and transition regimes, thus eliminating the need for the definition of additional slip boundary conditions for some of the studied cases, provided that a rigorous numerical distinction between continuum, slip-flow and transition regimes is not essential for the problem at hand. Since the coefficients are determined from dimensionless numbers, the results are expected to be qualitatively valid for other comparable Laval nozzle geometries. In this regard, the effect of the diffuser's opening angle and of the nozzle throat diameter can be the subject of future complementary studies. The transferability of the obtained coefficients to other propellants such as diatomic gases as well as the widely used hydrazine or ammonia gases in arcjet thrusters might also be explored in the future.

The thrust and specific impulse delivered by the cold-gas thruster are estimated based on the numerical results. The maximum thrust value reached for all gases and mass flow rates is 640 μN and the maximum specific impulse is 18 s, both achieved with argon at a mass flow rate of 3.568 mg/s, $Re = 240$. Furthermore, an increase of the thickness of the viscous boundary layer through the diffuser of the micronozzle is observed. This results in a shock-less decrease of the Mach number and the flow velocity, which penalizes thrust efficiency. The negative effect of the viscous boundary layer on thrust efficiency can be lowered using higher mass flow rates (higher Re numbers) and through a reduction of the diffuser length. The results show that an increase in thrust performance of approximately 55% can be achieved for xenon with $\dot{m} = 3.568$ mg/s, $Re = 256$ through an 85% reduction of the diffuser length. This result illustrates the importance of the characterization of the viscous losses for thrust optimization in propulsion systems operating with micronozzles.

Chapter 5

Development of a Kinetic Plasma Model for Electric Propulsion Systems

In this chapter, a numerical model for the simulation of plasma phenomena inside electric propulsion systems like the INGA III arcjet thruster during hot-gas operation is developed. The model is based on a kinetic description of the ionised gas and takes advantage of the Particle-In-Cell (PIC) (s. Section 3.3), the Monte-Carlo-Collision (MCC) (s. Section 3.2) and the Finite-Volume-Method (FVM) (s. Section 3.1.2) approaches. Sections 5.1 and 5.2 are dedicated to the working principles of electric propulsion systems for spacecraft as well as to the state of the art concerning the numerical modelling of arcjet, ion and hall-effect thrusters. Based on this information, a global numerical concept is designed consisting of a coupled fluid-submodel and and a kinetic PIC-MCC submodel. In the frame of the present work, the kinetic PIC-MCC submodel is developed. The obtained kinetic solver, called *dsmcPlasmaFoam*, includes a PIC algorithm, short-range Coulomb collisions between charge carriers, electron-neutral interactions including collisional ionisation as well as additional numerical features and is described in detail in Section 5.3.

5.1 Electric propulsion systems for spacecraft

In recent years, the space industry has seen a considerable increase in the use of electric propulsion systems for orbit and attitude control of satellites complementing classic chemical and cold-gas devices. This trend results mainly from two key characteristics of electric propulsion systems. On the one hand, the use of electric devices enables the generation of very low and precise thrust values, a feature which is key for the attitude control of miniaturized satellites (e.g., CubeSats) as well as for the envisioned coordinated formation flight of satellite constellations (s. [KSBY00, SBM01]). On the other hand, the specific impulse,

$$I_{sp} = \frac{F}{\dot{m} \cdot g} \tag{5.1}$$

achieved by electric systems is considerably higher than the one produced by their cold-gas and chemical counterparts. In Eq. 5.1, F corresponds to the magnitude of the produced thrust, $\dot{m}$ to the mass flow of expelled propellant and g, to the earth gravity. The advantage associated with higher I_{sp} values can be easily identified using Tsiolkovsky's rocket equation:

$$\begin{aligned} \Delta v &= v_e \ln \frac{m_0}{m_f} \\ \Delta v &= g I_{sp} \ln \frac{m_0}{m_f} \end{aligned} \tag{5.2}$$

where Δv corresponds to the change of velocity of the system, v_e to the effective exhaust velocity of propellant, m_0 to the initial total mass of the system and m_f, to the final total mass ($m_0/m_f > 1$). From Eq. 5.2 it is evident that for a given Δv, a higher I_{sp} will lead to a decrease of the ratio m_0/m_f and hence, to a reduction of the propellant mass necessary to achieve the desired increase in velocity. Therefore, the higher I_{sp} values associated with electric propulsion systems reflect an inherent superior efficiency compared to cold-gas and chemical systems. The application regions of some common electric propulsion devices are shown schematically in Fig. 5.1. Typical performance values for thrust F and specific impulse I_{sp} of different systems are listed in Table 5.1.

The described advantages have promoted, in recent years, the gradual replacement of chemical and cold-gas systems by electric propulsion devices. As of 2005, 115 spacecraft with hydrazine resistojets, 36 with hydrazine arcjet thrusters, 25 with xenon ion engines, 14 with hall thrusters and one vehicle with pulsed plasma thrusters were in service [Boy05]. Because of their very high I_{sp} values, the development of ion and Hall thrusters for commercial applications has been intensified in recent years [GdA15]. Moreover, the scientific community and the space industry continue their work on pulsed plasma thrusters, resistojets and arcjet thrusters [GdA13].

Despite the increasing use of electric propulsion systems, one of the biggest challenges continues to be the empirical character of their development. Electric propulsion sys-

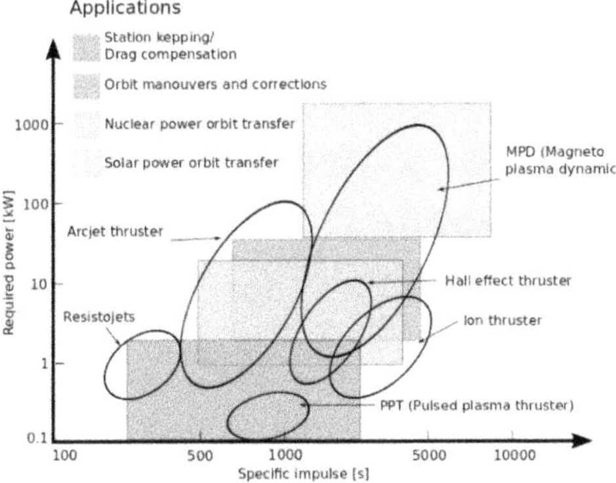

Figure 5.1: Application regions of common electric propulsion concepts as a function of the electric requirements and specific impulse [SB10].

tems have been developed for roughly 50 years using mainly empirical models and experimental data [Boy05]. The reason for this is the high complexity of the phenomena taking place in the systems and the resulting challenges associated with the development of physically consistent models for different propulsion concepts [Boy05]. However, because of the increasing relevance of electric propulsion systems in the space industry, the accurate numerical modelling of the associated plasma phenomena becomes more important every day. Furthermore, the continuously growing computational capabilities offer an additional incentive for intensified research in this field. From an economic standpoint, accurate modelling approaches can help lowering the overall development costs by reducing the amount of experimental research and testing done in earth-based gravity-bounded laboratories and by revealing optimisation opportunities. On the other hand, the development and validation of modelling approaches offer the opportunity to better understand physical plasma processes inside electric propulsion systems.

5.2 State of the art and previous works

In this section, an overview of the working principles and modelling concepts for electric propulsion systems is presented. The focus lies on the most common propulsion devices already in operation. They are resistojets, arcjet thrusters, ion thrusters and Hall thrusters.

Type	Thrust [mN]	Specific Impulse [s]
Resistojet (thermal)	200-300	200-350
Arcjet Thruster (thermal)	200-1,000	400-1,000
Ion Thruster	0.01-500	1,500-8,000
Pulsed Plasma Thruster (PPTs)	0.05-10	600-2,000
Magnetoplasma dynamic Thruster (MPD)	0.001-2,000	2,000-5,000
Hall Thrusters	0.01-2,000	1,500-2,000
Monopropellant rocket [a]	30-100,000	200-250

Table 5.1: Characteristic values of typical electric propulsion systems [SB10].

[a] Listed for comparison

5.2.1 Resistojets

In a resistojet, the propellant flow is accelerated over an electric element which in turn, increases the fluid temperature. The propellant is subsequently expanded to supersonic velocities through a Laval nozzle generating this way thrust. The higher gas temperature resulting from the interaction of the propellant fluid with the heating element leads to a higher critical temperature and speed of sound at the nozzle throat. This results in higher fluid velocities in the diffuser and at the nozzle exit and hence, in an increase of the produced thrust compared to a cold-gas device. Since the thrust increase is only the result of higher fluid temperatures, resistojets fall into the category of electrothermal devices. Compared to other electric propulsion devices, resistojets are both simple and robust. Because of the relative simplicity of the processes taking place, numerical models for resistojets are usually based on standard continuum or DSMC approaches with the fluid inlet temperature adjusted for the heating effect of the electric element. These approaches have been shown to produce accurate results compared to experimental data (cp. [Boy05]).

5.2.2 Arcjet thrusters

As in a resistojet, the propellant temperature in an arcjet thruster is increased using electric energy. To this end, an electric arc is generated in the convergent nozzle section between a cathode and the body of the Laval nozzle which acts as the anode. The pressurized propellant crosses the electric arc, where its temperature increases through *Joule heating* (*Ohmic heating*). In the process, the propellant gas is partially ionised and a plasma appears. As in a resistojet, the higher thermal energy now present in the gas is converted into kinetic energy and hence, into thrust through a Laval nozzle (cp. [Boy05]). The main components of a typical arcjet thruster are shown in Fig. 5.2.

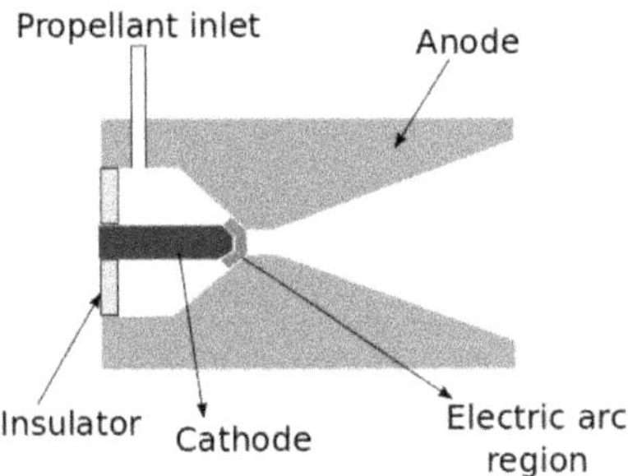

Figure 5.2: Schematic representation of an arcjet thruster.

The majority of the developed numerical models for arcjet thrusters are based on continuum approaches. One example is the one-phase and two-phase continuum models described in [BK92]. In this study, a 10 kW arcjet thruster with H_2 as propellant is simulated and the numerical results for I_{sp} are compared with experimental data. The model consists of transport equations for mass, momentum and energy. Additional terms for the momentum and energy exchange between the heavy particles and the free electrons in the ionised gas are added to the equations. Furthermore, electromagnetic relations based on the Maxwell equations are also considered.

A similar two-phase approach for a H_2 arcjet thruster is presented in [MMS96]. Here, macroscopic neutrality ($n_i = n_e$) as well as a strong coupling between ions and neutral atoms ($u_i = u_n = u$, $T_i = T_n = T_g$) is assumed. In these relations, the subscript i stands for ions, n for neutral atoms, e for electrons and g for gas. Therefore, the developed equations system consists of individual continuity equations for ions and neutral atoms as well as for the global mass density. Furthermore, one equation for the conservation of momentum for the gas as a whole and energy equations for electrons and heavy particles (ions and neutral atoms) are considered together with ionisation and dissociation rates.

In the study presented in [MKB96], a comparable continuum model using modified Navier-Stokes equations, two energy equations (electrons and heavy particles) and rates for the ionisation and dissociation processes is described. The numerical results for the analysed hydrazine arcjet thruster are in relative good agreement with experimental data. The observed deviations between the results are explained by the authors as the result of the incomplete treatment of the inelastic collisions between electrons and heavy particles.

A continuum approach for a 100 kW H_2 arcjet thruster based on three energy conservation equations (heavy particles, vibrational energy and electron energy) is presented in [AKGH$^+$98]. The model also includes individual mass continuity equations for the considered species in the plasma, a momentum conservation equation, the Maxwell equations and the Joule heating equation,

$$\frac{\mathrm{d}P}{\mathrm{d}V} = J \cdot E \tag{5.3}$$

where the LHS represents the heating power per unit volume, P the heating power, J the current density and E the electric field. The electric current is calculated from the differential form of Ohm's law as follows:

$$J = \sigma E \tag{5.4}$$

where σ stands for the electrical conductivity.

One of the weaknesses of the described continuum approaches is the often required estimation of physical parameters like electrical conductivity, anode temperature and transport coefficients, properties that can often be approximated only from empirical data. In multi-phase approaches, the largely different time scales dominating the electrons and the heavy particles behaviour constitute an additional major challenge [BK92]. The limitation regarding the estimation of physical parameters can be, in theory, circumvented by taking advantage of kinetic approaches. One example of such a model is given in [Boy97b] and [Boy97a]. In these studies, the modelling of an arcjet thruster is performed based on the DSMC method. Momentum exchange is implemented through effective collision cross sections for the considered species in a H_2 plasma. Dissociation and ionisation processes are modelled by temperature-dependent rates and Joule heating is described through a phenomenological approach. Additional electromagnetic factors like the Lorentz forces are not considered in the model. The modelling approaches for arcjet thrusters described in this section are summarized in Table 5.2 with green coloured cells indicating species implementation (continuum-based or kinetic approach). Moreover, the blue coloured cells in the final row of the table represent the modelling concept developed in the present work and whose kinetic submodel is described in detail in Section 5.3.

Model	Continuum			Kinetic			Miscellaneous
	Electrons	Ions	Neutral	Electrons	Ions	Neutral	
[BK92]	Eqs.: ρ, p, E	As heavy particles with Eqs.: ρ, p, E					G_B, Faraday
[MMS96]	Eqs.: ρ, E	As heavy particles with Eqs.:, ρ, p, E					1xOhm, 1xϕ
[MKB96]	Eqs.: E	Eqs.: 1xE for heavy particles, 1xp for all species					1xOhm, 1xJ
[AKGH+98]	Eqs.: E	As heavy particles with Eqs.: ρ, p, E, $E_{vibrational}$					1xOhm
[Boy97b]				No PIC, No Lorentz forces, Joule heating phenomenological, $u_e = u_i$			
[Boy97a]				No PIC, No Lorentz forces, Joule heating phenomenological, $u_e = u_i$			
Developed concept			As fluid; 3D; With Eqs.: ρ, p, E	PIC-MCC approach	PIC-MCC approach	Neutrals cloud as collision partner for e^- and ions	*Mapping* of neutral fluid fields onto particles cloud; Joule heating of neutrals through source term

Table 5.2: Numerical models for arcjet thrusters. ρ: Mass continuity equation; p: Conservation of momentum; E: Conservation of energy; G_B: Gauss's law for magnetic fields; Faraday: Faraday's law; Ohm: Ohm's law; ϕ: Electric potential; J: Current density; u_e =Electron velocity; u_i =Ion velocity; e^-: Electrons.

5.2.3 Ion thrusters

In an ion thruster, a propellant gas is led into a discharge chamber where it is ionised through collisions with electrons generated at a hollow cathode. The resulting ions are then focused and accelerated by a set of electrically charged grids, also known as ion optics, before being neutralised by a second hollow cathode at the exit of the thruster. The main components of an ion thruster are the discharge chamber, the ion optics and the hollow cathodes [Boy05]. They are shown schematically in Fig. 5.3.

The majority of the numerical approaches for the modelling of the discharge chamber in ion thrusters are based on kinetic methods for the involved species. One of the earliest approaches is described in [AI91]. In this two-dimensional model, the Finite-Element-Method (FEM) is employed for the computation of the magnetic field and a Monte-Carlo approach is used for the electrons generated at the hollow cathode (primary electrons). Collisions between electrons and neutral atoms as well as Lorentz forces are implemented. Ion motion is modelled with a relatively simple approach consisting of Poisson's equation (Eq. 3.122), the mass continuity equation and Newton's second law. Some of the model assumptions include constant temperatures for electrons, ions and

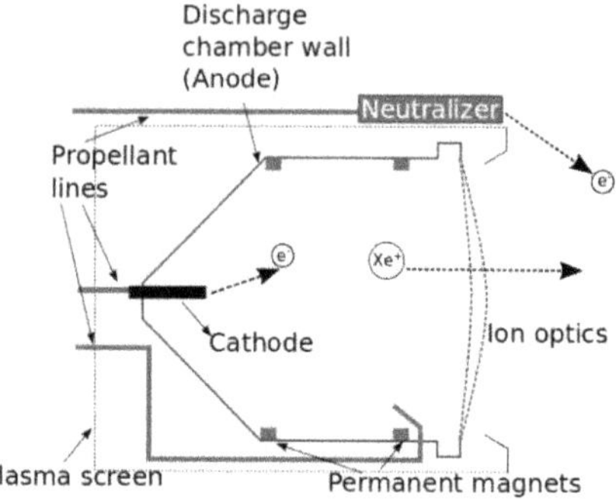

Figure 5.3: Schematic representation of an ion thruster with xenon as propellant gas [Boy05].

neutral atoms (cp. [WK04]), as well as a constant number density for the neutral atoms.

A full kinetic approach is described in [JH01]. The injected neutral xenon atoms, the xenon ions and free electrons are modelled with the DSMC method. Furthermore, a Lorentz force computation is implemented in order to include the effect of external magnetic fields and gas ionisation is modelled with a probabilistic formulation. However, the model description does not include information about the implementation of long-range particle interactions through a PIC approach. The same applies for the exact type of short-range interactions (collisions) included in the model.

One common characteristic of current numerical models for ion thrusters is the externally performed computation of the magnetic field. The field is assumed to be constant and is subsequently used for the calculation of the Lorentz force acting on the electrons. This approach is, for instance, followed by the three-dimensional electron submodel presented in [Stu05]. Here, only long-range interactions resulting from Lorentz forces are implemented, while short-range Coulomb interactions between particles are neglected. The submodel for the neutral atoms, presented in a separate work, is also based on a kinetic approach.

One additional example of a discharge chamber model is given in [MM05] and [MM06]. In this PIC-based approach, an external calculation of the stationary magnetic field is performed. In the model, primary and secondary electrons as well as ions are implemented. Furthermore, the external electric field is computed using Laplace's equation,

$$\Delta\phi = 0 \tag{5.5}$$

with constant boundary conditions for the electric potential ϕ. Note that the use of Eq. 5.5 implies $\rho = 0$, i.e., charge neutrality is assumed in each of the numerical cells.

After the calculation of the external magnetic and electric fields, the internal electric field is updated in the numerical solver using the charge carrier distribution as source term for Poisson's equation (Eq. 3.122). The updated electric field is subsequently used for the modelling of the long-range interactions. Moreover, electron-neutral and ion-neutral collisions are implemented with the Monte-Carlo-Collision-Method (MCC). Electron-electron and electron-ion Coulomb collisions are not considered, which leads to an underestimation of short-range effects between charge carriers. One further limitation is the assumption of a homogeneous number density for the neutral gas, also called *"background gas"*. Although the initial results of the study are promising, a stationary solution could not be reached.

A further development of the model in [MM05] and [MM06] is presented in [MCL$^+$10]. The approach is also based on the coupling of the PIC and MCC methods and includes the external computation of the magnetic and electric fields and the dynamic update of the internal electric field with Poisson's equation. One of the improvements in the new model is the implementation of Coulomb collisions between ions and electrons and between doubly charged xenon ions. Furthermore, the neutral *"background gas"* is now modelled with a kinetic approach. As expected, the accurate kinetic modelling of several species leads to very high computational requirements. In order to counteract this, a number of measures like scaling down of the discharge chamber geometrical dimensions as well as the use of different plasma parameters are implemented. However, stationary solutions could not be reached. In [MCL$^+$11], the same author presents additional results obtained with a similar model. Here, an artificial modified value of the electric permittivity ε is employed in order to reach a fast initial solution for the discharge chamber. In the next steps, the value of ε is gradually corrected. This approach leads to a considerable reduction of the computational time and relatively good results.

One example of a hybrid approach is given in [WK04]. In this model, primary electrons are handled kinetically with the Monte-Carlo-Method and the Lorentz equation. Ions and secondary electrons are modelled as fluid with transport equations. The neutral gas is modelled kinetically taking advantage of a so-called *optical model* in which constant neutral particle fluxes from the geometric boundaries are determined prior to the actual computation. For the neutral gas, only collisions between particles and walls are implemented (cp. [WK03]). Interactions between primary electrons and neutral atoms are described using empirical collision cross sections which in turn, depend on the specific collision type (ionisation, electronic excitation or elastic collisions). The modelling approaches for the discharge chamber of ion thrusters described in this section are summarized in Table 5.3 with green coloured cells indicating species implementation (continuum-based or kinetic approach). Moreover, the blue coloured cells in the final row of the table represent the modelling concept developed in the present work and whose kinetic submodel is described in detail in Section 5.3.

5.2.4 Hall thrusters

In a Hall thruster, a neutral gas, typically xenon or krypton, is led through a holed anode into a ring-shaped acceleration chamber. At the same time, electrons produced by an external cathode enter the chamber from the opposite (outlet) side. The associated spatial distribution of electrons produces an axial electric field inside the acceleration chamber. This field interacts with a magnetic field applied in radial direction creating an azimuthal force, which in turn acts on the electrons and captures them at a specific distance from the anode. The resulting electronic cloud is penetrated by the neutral atoms from the holed anode and in this process, ionisation of the propellant gas takes place through electron-neutral collisions. The ions created during the ionisation process are then accelerated towards the thruster outlet by the existing electric field. The ions are finally neutralized and exit the thruster at very high velocities, producing this way thrust. The main components of a Hall thruster are shown schematically in Fig. 5.4.

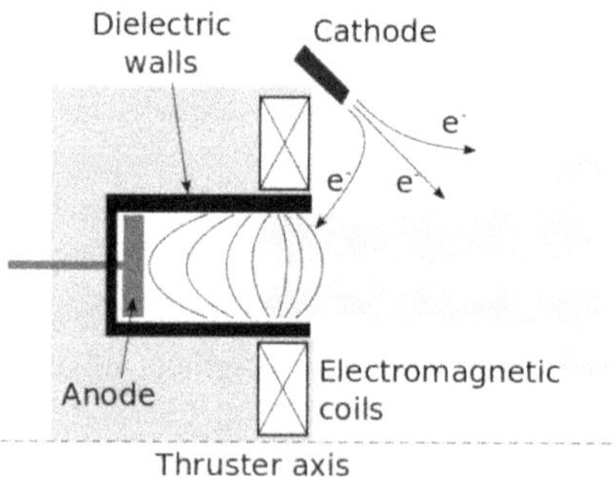

Figure 5.4: Schematic representation of a Hall thruster [Boy05].

A high number of approaches for the modelling of plasma phenomena inside Hall thrusters employ a hybrid approach for the description of the involved species. In most of these models, the electrons are handled as a fluid while the ions and neutrals are treated kinetically as particles. Examples of such approaches are given in [FMS95] and [FMSS$^+$97]. Here, xenon is used as propellant gas and the heavy particles in the system are modelled with the PIC method. The main model assumptions are quasi-neutrality of the plasma and the non-consideration of Coulomb interactions. Furthermore, the voltage between the cathode and anode of the thruster is assumed to be constant, while the current is handled as a variable.

A comparable model is presented in [HBGB02]. Here, the neutral xenon atoms are modelled as particles using a Monte-Carlo approach and taking into account wall collisions only. The neutral atoms, which disappear during the ionisation process, are estimated from an exponential weighting function as a function of time, local plasma density, local mean electron energy and an initial weighting factor. This way, the deletion of individual ionising neutral particles is circumvented. In the model, ions are handled kinetically, but collision-free, with a Monte-Carlo approach. Furthermore, ion-wall collisions are assumed to lead to neutralization. Electrons are modelled as fluid, which requires the formulation of several assumptions like quasi-neutrality in the plasma. Finally, energy losses resulting from the collisions between electrons and walls are described using a phenomenological approach.

One considerable disadvantage of the continuum-based description of electrons is the required definition of a higher electron mobility orthogonal to the magnetic field lines, a phenomenon experimentally observed in Hall thrusters. Without an appropriate increase in electron mobility, it would be impossible for large numbers of electrons from the external cathode to cross the magnetic lines on their way into the acceleration chamber which, in turn, would contradict the working principle of Hall thrusters. For this reason, approaches for the description of the higher electron mobility like the *near-wall conductivity* (s. [MEK+72]) and the *Bohm diffusion* (s. [GK08]) are often used in combination with continuum-based electron models. However, it should be noted that these approaches include empirical parameters often calibrated to obtain a good agreement between numerical and experimental results.

A comparison of the *near-wall conductivity* and *Bohm diffusion* models for the increase in electron mobility is presented in [KB04]. The results using the Bohm diffusion model show a good agreement with experimental results for thrust and specific impulse. On the other hand, the modified near-wall conductivity approach, which is related to a electron-wall collision correction, produces in this study better results for the electric potential distribution. Empirical values for the description of the near-wall conductivity and the Bohm diffusion are also implemented in the model presented in [GHB+03]. Here, xenon ions and neutral atoms are handled with a PIC-MCC approach, while electrons are described using transport equations. The focus of this study, in which quasi-neutrality is assumed, lies on the effects of the topology of the magnetic lines on the behaviour of the considered Hall thruster. One additional example of a hybrid approach is described in [PAFMS06]. Here, an improved description of the interaction between ions and the electron-repelling layer located at the side channel walls and at the anode surface of the Hall thruster is implemented through addition of numerical algorithms aiming to meet the Bohm condition (s. [HT59])

Empirical parameters for the description of the electron mobility are also employed in the hybrid model presented in [BHG+04]. The focus of this study lies on the comparison between numerical results and experimental data for the static and dynamic behaviour of a Hall thruster. The advantages of a fully kinetic approach regarding potential improvements in the understanding of the processes inside Hall thrusters are outlined in this work. Specifically, the understanding of specific aspects which are observed in experiments and require modelling using empirical parameters, like the increased electron mobility and the energy losses from electron-wall collisions, could greatly benefit from a kinetic handling approach for the electrons (cp. [BHG+04]).

One approach based on a kinetic model for the electrons is briefly described in [BHG+05] for a two-stage Hall thruster. However, since the obtained data is analysed only on a qualitative basis and the numerical model is not described in detail, the results of this study should be considered as preliminary in nature. The main challenge of a fully kinetic approach lies on the associated high computational requirements. This issue has motivated several studies with the aim to reduce these demands while attempting to conserve, at least at some acceptable level, the accuracy of the numerical results. Approaches like the reduction of the transit time of the slow neutrals through the artificial reduction of their mass combined with an increase of the vacuum permittivity (s. [SJ01]) or the scaling of the geometrical thruster dimensions (s. [TLCS05]) have disadvantages concerning the conservation of the fidelity of the plasma dynamic processes and the comparability of the obtained results. In the work described in [LWY+10], an approach consisting of an increase of the vacuum permittivity, which leads to a reduction of the required mesh resolution and number of iterations, combined with a fluid-based modelling of the neutral gas is proposed. To this end, a one-dimensional stationary mass continuity equation for the neutral gas is implemented. The results of the transport equation are used to create a cloud of neutral particles, which in turn, interacts with the kinetically modelled charge carriers. Ionisation and recombination rates are determined in the kinetic submodel and used as source terms in the continuity equation of the neutrals.

One additional example of an approach based on the kinetic handling of electrons is described in [CG14]. Here, a strong geometrical simplification is performed and only axial and azimuthal directions are considered. The neutral gas is handled using one-dimensional fluid-like equations while the electron-wall interactions are implemented using a relatively simple approach. The computational requirements are kept in check, again, through an increase of the vacuum permittivity and the associated increase in the characteristic Debye length, while the focus of the analysis lies on the transient behaviour of the plasma flow.

An example of a fully continuum approach is provided in [MKH11]. Here, the feasibility of magnetic shielded channel walls, aiming to improve the service life of Hall thrusters, is studied. In the presented two-dimensional model, both the electrons and ions are handled as a fluid. This is combined with an indirect implementation of the neutral gas, which leads to a reduction of the computational requirements. However, the employed transport equations for the electrons include the already mentioned empirical parameters associated with the electron mobility. Because of this, the potential of such an approach regarding an improved understanding of the plasma phenomena in the thruster still has some limitations. The modelling approaches for Hall thrusters described in this section are summarized in Table 5.4 with green coloured cells indicating species implementation (continuum-based or kinetic approach). Moreover, the blue coloured cells in the last row of the table represent the modelling concept developed in the present work and whose kinetic submodel is described in detail in Section 5.3.

Model	Continuum			Kinetic			Miscellaneous
	Electrons	Ions	Neutrals	Electrons	Ions	Neutrals	
[AI91]				MC; W/ Lorentz force for motion; Collisions w/ neutrals only	No MC; Ion motion w/ Poisson's Eq. + Conti Eq. + Newton's 2nd law	No MC; N_n=const.	2D; T_e=const.; T_i=const.; T_n=const.; W/o Coulomb (short-range) interactions
[JH01]				MC	MC	MC	W/ Lorentz forces; Ionisation w/ probabilistic approach; W/o long-range interactions
[WK04]	Ions and secondary electrons as quasi-neutral fluid; Eqs.: ρ, p			MC; W/ Lorentz forces; Numerous collision types included		Kinetically handled w/ optical model	Hybrid model; Collisions between primary electrons and neutrals w/ empirical collision cross sections; W/o long-range interactions
[Stu04]				MC; W/ Lorentz forces; W/o collisions; W/o long-range interactions			3D; Only electron model described; Ionisation, excitation and recombination not modelled
[Stu05]						MC w/ particle-particle and particle-wall collisions	3D; Only neutral model described; Complements approach in [Stu04]
[MM05, MM06]				PIC for primary and secondary electrons	PIC	Included for collisions w/ e^- and ions; N_n=const. "*Background gas*"	W/o Coulomb collisions (Short-range interactions underestimated); Static E w/ Laplace-Eq.; Dynamic E w/ Poisson's-Eq. and charge distribution
[MCL+10]				PIC for primary and secondary electrons	PIC for Xe^+ and Xe^{++}	MC	Further development of [MM05, MM06]; PIC-MCC w/ Coulomb collisions (Electron-Xe^+); Geometrical dimensions and plasma parameters scaled for reduction of computational requirements; Stationary solution not reached
[MCL+11]				PIC for primary and secondary electrons	PIC for Xe^+ and Xe^{++}	MC	[MCL+10] w/ modified ε_0 in order to reach partial solution, ε_0 corrected stepwise
Devel. concept			As fluid; 3D; W/ Eqs.: ρ, p, E	PIC-MCC approach	PIC-MCC approach	Neutrals cloud as collision partner for e^- and ions	*Mapping* of neutral fluid fields onto particles cloud; Joule heating of neutrals through source term

Table 5.3: Models for discharge chamber of ion thrusters. MC: Monte-Carlo approach; N_n: Number density of neutrals; T_e: Electron Temperature; T_i: Ion Temp.; T_n: Neutrals Temp.; ρ: Mass continuity Eq.; p: Conservation of momentum; e^-: Electrons; E: Electric field; PIC: Particle-In-Cell; MCC: Monte-Carlo-Collisions; ε_0: Vacuum permittivity; W/: With; W/o: Without.

Model	Continuum			Kinetic			Miscellaneous
	Electron	Ions	Neutrals	Electrons	Ions	Neutrals	
[FMS95, FMSS+97]	As fluid; Eqs.: p, n, E				PIC; W/o particle-particle collisions		2D; Rot. Sym.; Quasi-neutrality; Const. voltage
[HBGB02]	As fluid; Eqs.: p, J, E; Phenomenological Appr. for e^- mobility; Energy losses through e^--wall collisions				MC; W/o particle-particle collisions	MC; W/o particle-particle collisions	2D; Rot. Sym.; Ionised neutrals from exponential weighting function; E from e^- distribution
[GHB+03]	As fluid; Eqs.: J, E				PIC-MCC		2D; Rot. Sym.; Empirical values for description of near-wall e^- conductivity and Bohm diffusion
[KB04]	As fluid; Eqs.: J, E				PIC-MCC		2D; Rot. Sym.; Comparison of models for increase in e^- mobility
[PAFMS06]	As fluid; Eqs.: p, I, E				PIC-MCC; W/ Algorithms to fulfil Bohm's condition		2D; Rot. Sym.; Empirical values for description of near-wall e^- conductivity and Bohm diffusion
[BHG+04]	As fluid; Eqs.: p, J, E				PIC		2D; Rot. Sym.; Empirical values for description of near-wall e^- conductivity and Bohm diffusion
[BHG+05]				MC	PIC	PIC	Preliminary results; W/o detailed model description
[LWY+10]			As fluid; Eqs.: n (1D)	PIC	PIC		2D; Rot. Sym.; Neutral gas as fluid + Neutrals cloud generation; Neutrals cloud interaction w/ e^- and ions; Ionisation and recombination as source terms in fluid-eqs.; Scaling of ε_0 for reduction of computational requirements
[MKH11]	As fluid	As fluid	Indirect handling				2D, Rot. Sym.; Empirical values for description of near-wall e^- conductivity and Bohm diffusion
[CG14]			As fluid (1D)	PIC	PIC		2D; Radial direction not considered; Geometry simplified; Scaling of ε_0 for reduction of computational requirements
Devel. concept			As fluid; 3D; W/ Eqs.: ρ, p, E	PIC-MCC approach	PIC-MCC approach	Neutrals cloud as collision partner for e^- and ions	*Mapping* of neutral fluid fields onto particles cloud; Joule heating of neutrals through source term

Table 5.4: Models for Hall thrusters. PIC: Particle-In-Cell; MC: Monte-Carlo; MCC: Monte-Carlo-Collisions; p: Conservation of momentum; n: Continuity equation (number density); ρ: Mass continuity equation; E: Conservation of energy; Rot. Sym.: Rotational symmetry; J: Current density; I: Electric current; ε_0: Vacuum permittivity; e^-: Electrons; W/: With; W/o: Without.

5.3 Development of a kinetic plasma model

As described in Section 5.1, electric propulsion systems for space applications are designed mostly based on empirical models and experimental data. Two of the main reasons behind this are the high complexity of the plasma phenomena involved and the usually very high computational requirements associated with plasma numerical modelling. However, the always increasing computational power opens the possibility to explore new approaches for the modelling of such systems, which at the same time offers a chance to improve the understanding of plasma phenomena, energy exchange and transport mechanisms, not only in electric propulsion devices, but also in other plasma-related applications like plasma arc welding and electron beam processing. Some of the challenges regarding the description and numerical modelling of plasma phenomena in electric propulsion systems are:

- Determination of plasma-related parameters, like electrical conductivity, inside electric propulsion systems while avoiding the use of empirical approaches.
- Development of efficient hybrid models for technical applications in which large differences of length or time scales in the physical mechanisms exist.

The main goal of the work presented in this chapter is the development of a numerical modelling concept potentially able to accurately describe the dynamic plasma phenomena inside electric propulsion devices, like the INGA III thruster, while at the same time having acceptable computational requirements. The envisioned hybrid model concept, briefly introduced in Tables 5.2, 5.3 and 5.4, is based on a combination of a kinetic PIC-MCC submodel for plasma modelling and its coupling with a fluid submodel for heavy particles. The global numerical concept aims to use nozzle outlet pressure, propellant mass flow and electric current as boundary conditions in order to produce numerical results for, among others, voltage drop between cathode and anode, plasma temperature and ionisation chamber pressure. The modelling concept is kept as general as possible in order to facilitate its application in future studies to further electric propulsion devices like ion, Hall or HEMP thrusters. In the present work, the focus lies on the development of the kinetic PIC-MCC plasma submodel and the validation of the obtained numerical algorithm and solver. For completeness, the envisioned global modelling concept including both its kinetic as well as its fluid submodel is presented in Section 5.3.1. The development of the kinetic PIC-MCC plasma submodel and the corresponding solver *dsmcPlasmaFoam* are discussed in detail in Sections 5.3.2 through 5.3.9.

5.3.1 General modelling concept

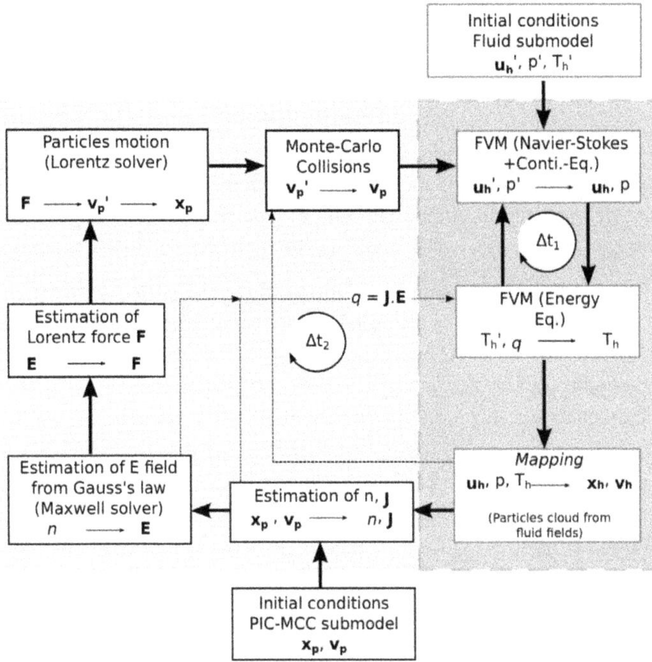

Figure 5.5: Numerical plasma modelling concept. Gray: Kinetic PIC-MCC submodel. Red: Fluid-submodel. u: Velocity field; p: Pressure; T: Temperature; $\dot{q}$: Source term (Joule Heating); x: Position (Test particle/kinetic); v: Particle velocity (kinetic); Subscripts: h: Heavies (Fluid submodel); p: Plasma (charge carriers in kinetic submodel).

The envisioned numerical plasma modelling concept for the description of an electric propulsion system is shown schematically in Fig. 5.5. The model consists of a kinetic PIC-MCC submodel for charge carriers and a continuum-based FVM fluid submodel for the neutral propellant gas. The main aspects of the hybrid model are described in the following.

- As a consequence of high frequency plasma oscillations and the low electron mass, the time step required for numerical stability and accuracy of a plasma solver is generally small, which in turn, leads to high computational demands. Considering this, the final model should allow an independent tracking of plasma phenomena in its small time scale. Therefore, the model includes a kinetic handling method for the charge carriers making up the plasma in a time scale called *micro time scale* as part of the PIC-MCC submodel (Gray in Fig. 5.5). Furthermore, a fluid-like handling of the heavy particles (neutrals) is performed in a time scale called *macro time scale* as part of the fluid submodel (Red in Fig. 5.5). The charge carrier treatment in the micro time scale allows the resolution of dynamic plasma phenomena, specially, of the electrons motion and its effects. Furthermore, the consideration of the heavy neutrals in the macro time scale enables the study of macroscopic parameters like thrust, while avoiding the high computational requirements of a fully kinetic approach.

- The PIC approach is selected as basis for the kinetic charge carriers submodel. Through the use of a PIC approach, the model is capable of handling long-range interactions between charge carriers, a mechanism of great importance for typical plasma behaviour. Based on the spatial distribution of particles and their velocities, fields for the number density n and the current density J can be computed. In the next step, the electric field E can be calculated using Gauss's law for electric fields (Eq. 3.117). The final step in a typical PIC algorithm involves the determination of the Lorentz forces F acting on the charge carriers and the particles motion in the numerical mesh.
- As discussed in Section 3.3, a typical PIC approach is usually associated with an underprediction of the short-range particle interactions. This weakness is compensated through combination of the PIC method with a Monte-Carlo approach for charge interactions. The modelling of the short-range collisions used in this work is based on the study "Theory of cumulative small-angle collisions in plasmas" published by K. Nanbu in 1997, in which a high number of small-angle binary Coulomb collisions over a given period of time is treated as a large individual collision (s. [Nan97]).
- Besides interactions between the charge carriers in a plasma, several other types of collisions are implemented in the MCC part of the kinetic submodel. They include interactions between electrons and neutrals, which in this work are modelled as either elastic or inelastic collisions. Based on the energy available prior to an inelastic collision, they can in turn lead to either electronic excitation or ionisation of a neutral atom. Finally, recombination is implemented as short-range tertiary electron-electron-ion interactions based on Thomson's recombination the-

ory [Bio82].

- The fluid submodel in Fig 5.5 deals primarily with the neutrals in the macro time scale. Here, the fluid fields are computed from the transient compressible form of the Navier-Stokes equations, the mass continuity equation and the energy equation. The submodel uses boundary conditions typical of fluid mechanics problems and in case of rarefied flow conditions, the correcting Knudsen functions developed in Chapter 4 of the present work.
- A two-way coupling between the kinetic PIC-MCC and the fluid submodels is performed as follows. In the micro time scale (kinetic submodel), the average electric field E and average current density J are determined and used for the computation of the Joule heating $\dot{q} = J \cdot E$, which is subsequently used as source term in the energy equation of the neutral gas in the fluid submodel. The coupling in the opposite direction (from fluid to kinetic submodel) is based on the approach developed by Liu *et al.* (s. [LWY$^+$10]). In this model, a one-dimensional stationary fluid equation is used for the description of the neutral gas inside a Hall thruster. From the fluid results, a cloud of neutral particles is subsequently created which in turn, interacts with the PIC modelled particles through Monte-Carlo collisions. The creation of kinetic test particles based on continuum fields is referred to as *field mapping*.
- The global envisioned model uses the electric current through the arcjet thruster as boundary condition. Hence, the resulting voltage between cathode and anode corresponds to a numerical output of the model. Furthermore, magnetic effects are not considered in the developed algorithm.

By employing a kinetic method for the charge carriers, the use of empiric parameters for the description of phenomena, like increased electron mobility, can be mostly avoided. Since the direct, kinetic particle description can generally be considered to be assumption-free, the approach described above offers the possibility to gain important and new insights regarding energy exchange and transport mechanisms for charge carriers inside electric propulsion systems.

Because of the high complexity of typical plasma phenomena and the relation between the numerical algorithms in different time scales, the modelling concept outlined above and shown schematically in Fig. 5.5 is not developed entirely in the present work. The focus of the next sections lies, therefore, on the development and validation of the PIC-MCC plasma submodel (Gray submodel in Fig. 5.5). To achieve this, the neutral particles necessary for the kinetic modelling of the ionisation process are assumed to constitute a *background gas* which serves as *stage* for the dynamic plasma behaviour. Since momentum transfer between neutral particles is part of the fluid submodel to be developed in future studies, interactions between neutrals as well as between neutrals and ions are not considered in the present work.

Considering the information above, the kinetic PIC-MCC submodel, described in detail in the following sections, covers the typical plasma behaviour of the following species:

- Primary Electrons: Produced at the cathode of the electric propulsion device and modelled with a kinetic approach.
- Neutrals: Kinetically implemented as background gas representing the neutral atoms present between the cathode and the anode of the device and subject to ionisation from collisions with fast moving electrons.
- Ions: Produced during ionisation collisions between fast moving electrons and neutrals.
- Secondary Electrons: Produced during ionisation collisions between fast moving electrons and neutrals.

Furthermore, the following mechanisms for particles creation and particles loss are implemented:

- Ionisation: As inelastic binary collision between fast moving electrons and neutrals leading to the disappearance of the neutral and the appearance of one secondary electron and one ion.
- Recombination: As tertiary collision between two electrons and one ion leading to the disappearance of one ion-electron pair and the appearance of one neutral.
- Ion neutralization at walls: Ions which make contact with cathode or anode walls are replaced by neutrals.
- Electron deletion/retention at walls: In order to numerically achieve a given electric current flow between cathode and anode, electrons are not allowed to leave the numerical grid through the cathode. Moreover, from the electrons reaching the anode, only a given amount of electrons is randomly selected and allowed to exit the numerical grid, while the rest is kept in the domain creating an electron layer at the anode. The amount of electrons allowed to leave the system is defined in a way that the electric current through the anode agrees with the electric current applied at the cathode.

The interaction mechanisms between the species implemented in the kinetic PIC-MCC submodel are listed in Table 5.5. As already discussed, the neutral gas in the PIC-MCC submodel is handled as a background gas allowing ionisation processes to be modelled. At the same time and in order to keep computational requirements manageable, neutral-neutral and neutral-ion collisions are not implemented in the current model. Note that

Particle	Interaction Partner	Interaction approach	
		Short-range	Long-range
Electron	Electron	Coulomb Collision (MCC)	E field (PIC)
	Ion	Coulomb Collision (MCC) & Recombination	E field (PIC)
	Neutral	Elastic/Inelastic Collisions (MCC)	n/a
Ion	Ion	Coulomb Collision (MCC)	E field (PIC)
	Neutral	Not implemented	n/a
Neutral	Neutral	Not implemented	n/a

Table 5.5: Interaction mechanisms implemented in the kinetic PIC-MCC submodel.

neutral-neutral collisions can be included in future versions of the submodel using the standard DSMC model developed by Bird and described in Section 3.2. In addition, the same approach can be employed for a simplified description of Ion-Neutral collisions if charge transfer processes are neglected and only momentum transfer is considered.

As can be seen in Table 5.5, long-range interactions between charge carriers are covered by the PIC algorithm described in Section 3.3. Note that these interactions are the consequence of the electric field E appearing in the plasma as a result of the spatial distribution of the charge carriers. Short-range electron-electron, electron-ion and ion-ion interactions are the product of the Coulomb collisions (attraction and repulsion) of charged particles inside the plasma. The corresponding MCC algorithm for Coulomb collisions is described in detail in Section 5.3.4. Furthermore, interactions between electrons and neutrals are assumed to be of three types: elastic, inelastic collisions leading to electronic excitation and inelastic collisions leading to ionisation. The corresponding MCC procedure for electron-neutral interactions is described in detail in Section 5.3.5. The algorithm for interactions between electrons and ions leading to recombination is discussed in Section 5.3.6, while Section 5.3.7 is dedicated to the boundary conditions implemented for the particles in the plasma. Finally, a numerical tool has been developed and implemented in the developed code in order to reduce the computational requirements by limiting the number of particles within the numerical mesh. The tool is based on *dynamic weighting* of the test particles and is described in Section 5.3.8.1.

5.3.2 Basis DSMC solver

In this section, a brief overview of the software package *OpenFOAM*® and of its DSMC solver, which serves as basis for the implementation of the kinetic PIC-MCC submodel, is presented. *OpenFOAM*® (*Open Field Operation and Manipulation*) is a free and open-source software developed by OpenCFD Ltd. The software consists of a large collection of functions and numerical schemes used for the treatment of a wide variety of physical and engineering problems, specially in the field of fluid mechanics. The available functions are combined to create solvers covering a large range of problems, from simple incompressible, laminar one-phase flows to, among others, chemical reactions, turbulent flows and heat transfer. Because of the open-source nature of the project, *OpenFOAM®* offers the possibility to modify existing solvers or to develop completely new algorithms for specific user needs. *OpenFOAM®*, which is written in C++ programming language, also includes a wide variety of additional utilities covering pre-processing and post-processing tasks. The general structure of *OpenFOAM®* is depicted in Fig. 5.6.

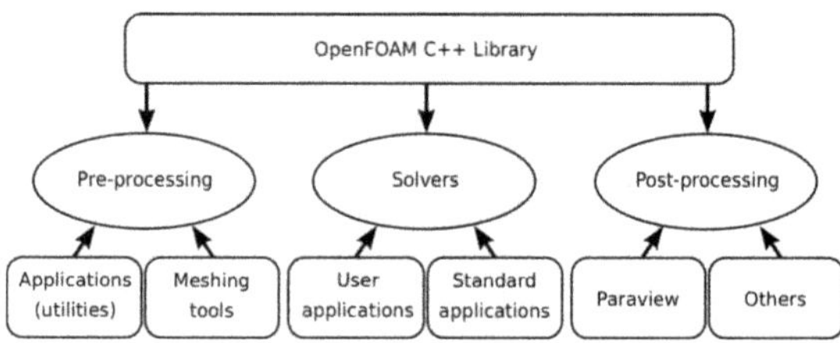

Figure 5.6: Schematic representation of the structure of the software package *OpenFOAM®* (cp. [Gre15]).

In order to use *OpenFOAM®*, the user must first create a folder containing the problem's information. This *work* folder encloses:

- The **"0" directory**, containing boundary and starting conditions for the problem. In most cases, this folder contains individual files for each of the fields in the problem, e.g., fluid velocity, pressure or temperature. If a kinetic solver is used, the "0" folder includes additional files with particle properties like their positions, velocities and species definition at the start of the simulation.
- The **constant directory**, containing at least one file named *thermophysicalProperties*. This file, also known as *dictionary*, is used to define the physical properties of the simulated fluid, like its viscosity and molar mass. In addition, the file is used to define the thermodynamic behaviour of the fluid (e.g., constant or temperature-dependent c_p, ideal or real gas behaviour). The "constant" folder includes a sub-folder named *polymesh* containing the numerical mesh information.

Depending on the used solver, additional dictionaries might also be present.

- The **system directory**, containing in most cases the following three dictionaries: The *controlDict*, used to define control parameters like start and end time of the simulation, time step and numerical triggers for the writing of solver results. The *fvSchemes* dictionary, used to set interpolation schemes and discretization approaches for the field variables treated in the solver. Finally, the *fvSolution* dictionary, which contains information regarding the linear solvers for the discretized differential equations, numerical tolerances as well as solver-specific parameters.

After the simulation has been prepared, the corresponding solver can be started. The results for the simulated fields and parameters are written during the simulation in individual folders for the corresponding simulation time. The results can be further modified using any appropriate post-processing application and can be analysed with the visualization application *ParaView*.

The *OpenFOAM®* solver *dsmcFOAM* is used as starting point for the kinetic PIC-MCC submodel developed in the present work. The solver is based on the DSMC method developed by Prof. Graeme Bird (s. [Bir91]) and described in Section 3.2. The structure of the solver is shown schematically in Fig. 5.7. As can be seen, the algorithm behind the *dsmcFOAM* solver can be divided into the following main steps.

- **dsmcInitialise**: When a DSMC simulation is started for the first time, the presence of an initial collection of particles in the numerical domain might be recommendable and sometimes even necessary. In *OpenFOAM®*, the *dsmcInitialise* application accomplishes this by using starting values of macroscopic parameters defined by the user in the DSMC specific dictionary *dsmcInitialiseDict* contained in the system directory. The required parameters are the number density of the modelled species n, the macroscopic temperature T and the macroscopic flow velocity u. The number of test particles to be initialised is determined based on n, the total volume of the numerical domain V and the particle weight W. The created particles are randomly distributed in the domain. The initial particles velocities are estimated as a random sample of the Maxwellian speed distribution function for the defined T. Specifically, each velocity component is calculated based on Eq. 2.124 as $v_i = (k_B T/m) U_i$, where U_i is a component of the unit vector U randomly sampled from the Gaussian distribution function. The macroscopic velocity u is then added to the Maxwellian velocity v and the result is used as initial particle velocity. The case of atoms with internal degrees of freedom also requires the estimation of their initial internal energy. Since the present work focuses on mono-atomic species, the reader is referred to [Bir91] and the source code of the *dsmcFOAM* solver for details.

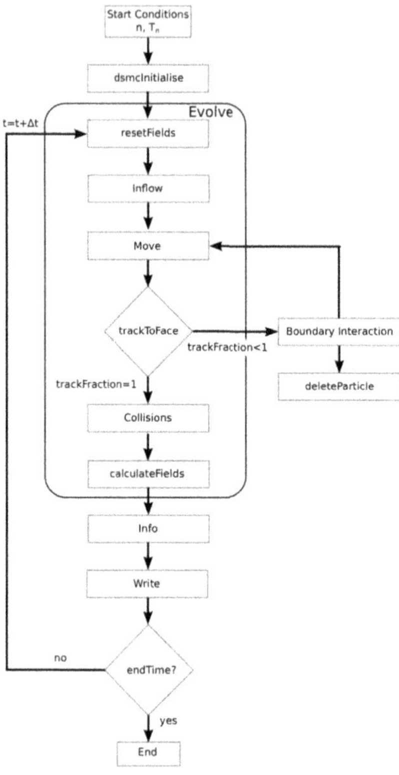

Figure 5.7: Schematic representation of the *OpenFOAM®* solver *dsmcFOAM* showing the functions included in the *evolve* method (cp. [Fas11], [AH07]).

- **resetFields**: During the course of a DSMC simulation, cell fields are calculated based on the spatial distribution and velocity of the test particles. The calculated fields include the number density n, the mass density ρ, the test particle number density n_{dsmc}, the total linear kinetic energy in the cell E_{kin}^{cell} and the net momentum in the cell p^{cell}. The calculation of the fields, which are used later for the estimation of macroscopic properties, is performed at the end of each iteration with the method *calculateFields*. Hence, the purpose of the *resetFields* method is to clear the field values at the start of each iteration.
- **inflow**: This method deals with the creation of test particles at the open boundaries of the numerical grid, i.e., at the grid's inlet and outlet surfaces. In the standard

dsmcFOAM solver provided with *OpenFOAM®* version 2.4, the user can choose between two inflow models: "*noInflow*" and "*freeStream*". As its name implies, the *noInflow* model assumes that no particles are introduced into the system, even if open boundaries are present. On the other hand, the *freeStream* model is used to define a particles flux through the boundary. To this end, the model employs the particle number densities n at the boundary surfaces set by the user in the *dsmcProperties* dictionary located in the constant directory of the work folder. Furthermore, values for the boundary temperature T and boundary fluid velocity c_0 of the macroscopic fluid are also necessary and must be defined in the files *boundaryT* and *boundaryU* of the "0" folder. The total number of particles N crossing the domain's boundary of area A in the inwards direction in the time Δt, i.e., to be introduced into the domain, is then estimated using the following expression (cp. [Bir91] Eq. 4.22):

$$N = \dot{N}A\Delta t = \frac{nc_m^{'}}{2\pi^{1/2}}[\exp(-s^2\cos^2\theta) + \pi^{1/2}s\cos\theta\{1+\mathrm{erf}(s\cos\theta)\}]A\Delta t \quad (5.6)$$

where $c_m^{'} = (2k_BT/m)^{1/2}$ stands for the magnitude of the most probable thermal velocity, $s = c_0/c_m^{'}$ is the molecular speed ratio, θ is the angle between the surface normal vector and the fluid velocity vector at the boundary and $\mathrm{erf}(x)$ is the error function. After the number of particles to be introduced is established, the new particles are randomly distributed across the boundary faces of the considered open surface. The initial particles velocities are estimated from a random sample of the Maxwellian speed distribution function for the defined T at the boundary. Note that this approach requires knowledge of both temperature and number density (i.e., pressure) at the open boundaries of the numerical domain. For the simulation of, for instance, a channel flow using a continuum-based solver, this means that pressure values at both inlet and outlet surfaces are used as boundary conditions and the mass flow rate through the channel is allowed to converge to its adequate value. Since plenty of problems require the pressure value at the system inlet to be an output of the simulation instead of a boundary condition, an additional inflow boundary model was implemented at ZARM in previous years (for details s. [Fas11]). The model, called *ZARMInOutflow*, defines the number of particles entering the system directly as:

$$N = \dot{N}A\Delta t \quad (5.7)$$

where the number of particles $\dot{N}$ entering the system per unit area and unit time is specified by the user based on the mass flow rate as follows:

$$\dot{N} = \frac{\dot{m}}{M}\frac{1}{A} \quad (5.8)$$

with $\dot{m}$ representing the mass flow rate, M the atomic mass and A, the inlet area. In addition to the direct definition of $\dot{N}$, the in-house version of the *dsmcFOAM* solver includes an additional modification regarding the handling of existing particles interacting with inlet boundaries. Since the inlet boundary is open, particles that reach this surface are immediately deleted from memory in the standard solver. In the in-house version of the *dsmcFOAM* solver, if a particle reaches the inlet surface, the boundary is treated as a wall and the particle bounces back into the numerical domain. This change is implemented in order to achieve agreement between the mass flow rate used as boundary condition and the net mass flow rate across the inlet surface in the simulation.

- **move**: The *move* and *trackToFace* methods are responsible for the particles motion in the numerical domain. The underlying procedure is described in Section 3.2.1. When a test particle is moved, the scalar $0 < trackToFace < 1$ is determined in the *trackToFace* method. This value corresponds to the fraction of the time step Δt in which the particle can move freely in the cell, i.e., without changing cells or interacting with domain boundaries. If $trackToFace < 1$ and the particle interacts with an outlet boundary, the particle is erased from the simulation. As already described, if $trackToFace < 1$ and the particle interacts with an inlet boundary or with a wall, the particle bounces off the boundary face and back into the numerical domain. This procedure is repeated until $trackToFace = 1$. At this point, the test particle has exhausted the Δt available for movement and its final position has been determined.
- **collisions**: After the motion step is performed, the collisions between the particles can be computed and the post-collision velocities, estimated. The collision method implemented in the standard *dsmcFOAM* solver corresponds to the approach described in Section 3.2.2 and is based on the no time counter method (NTC) proposed by Bird in [Bir89].
- **calculateFields**: After the post-collision velocities are computed, the fields necessary for the determination of macroscopic properties can be computed. To this end, the *calculateFields* method in the *dsmcFOAM* solver uses, for a mono-atomic gas, Eqs. 3.104, 3.105, 3.106 and 3.107 in Section 3.2.4 for the determination of the number density n, the mass density ρ, linear kinetic energy per unit volume E_{kin} and momentum per unit volume p respectively.
- **info**: In this step, useful parameters for the monitoring of the simulation, like total mass in the system, average linear momentum or average linear kinetic energy are computed.
- **write**: The final part of a typical DSMC iteration in the *dsmcFOAM* solver in-

volves the writing down of the particles positions and velocities, as well as of the field variables for the concluded iteration according to the writing parameters set by the user.

Note that the write method performs the saving of the simulation results for the iteration in which the write function is called. However, because of the statistical nature of the DSMC approach, the field results for a specific time step have little value and an averaging over a large number of time steps is highly recommended. This can be done by activating the optional function *fieldAverage* when running the solver. The average fields can subsequently be used by the additional optional function *dsmcFields* to calculate macroscopic variables like fluid velocity u, temperature T and pressure p using Eqs. 3.108, 3.109 and 3.110. As shown in Fig. 5.7, the main steps described above are part of a global function in *dsmcFOAM* called *evolve*.

Table 5.6 provides a summary of important files and variables specific to the standard *dsmcFOAM* solver and required for a typical particle simulation. The standard *dsmcFOAM* solver provided with *OpenFOAM®* version 2.4 and its in-house modification both serve as basis for the implementation of the kinetic PIC-MCC submodel described in Section 5.3.1.

For simplicity, the numerical solver corresponding to the kinetic PIC-MCC submodel developed in the present work will be henceforth referred to as ***dsmcPlasmaFoam***.

5.3.3 Implementation of PIC algorithm

The PIC component of the *dsmcPlasmaFoam* solver is based on the standard PIC algorithm depicted in Fig. 3.10 (a). The PIC method is implemented using the standard *dsmcFOAM* solver and its in-house modification as basis. The implementation of the four main steps of the PIC algorithm is described in the following.

5.3.3.1 Field equations - Maxwell solver

In order to determine the electric and magnetic forces acting on a charge carrier, the electric and magnetic fields resulting from the charge distribution and charge velocities must be determined first. The general procedure was discussed in Section 3.3.2 and is based on the Maxwell equations (Eqs. 3.117, 3.118, 3.119 and 3.120). Therefore, this element of the PIC algorithm is often referred to as "**Maxwell solver**" (cp. [Sti15]). Since the present work focuses on the application of the plasma model for the INGA III arcjet thruster and magnetic effects in this setup are expected to be negligible, the implemented PIC algorithm does not include the computation of the magnetic field B.

File	Location	Contents
boundaryT	"0"	Boundary conditions for macroscopic temperature T. Used for particle creation at open boundaries and for determination of particle kinetic energy after diffusive wall reflection (s. Section 3.2.3).
boundaryU	"0"	Boundary conditions for macroscopic fluid velocity. Used for particle creation at open boundaries and for determination of particle kinetic energy after diffusive wall reflection (s. Section 3.2.3).
dsmcProperties	constant	• Particle weight W (*nEquivalentParticles*). • Wall interaction model (specular, diffusive/Maxwellian or a blend of both). • Binary collision model (Variable Hard Sphere (s. Section 3.2.2), Variable Hard Sphere with Larsen Borgnakke internal energy redistribution). • Inflow model (none, *freeStream*, *ZARMInOutflow*). • Molecular properties (mass, internal degrees of freedom, reference viscosity μ_{ref}, viscosity index ω, reference diameter d_{ref} (s. Section 3.2.2)).
dsmcInitialiseDict	system	Number density n, Temperature T and fluid velocity u for particles collection at $t = 0$.

Table 5.6: Overview of files and parameters required for a typical simulation with the standard *dsmcFOAM* solver.

Hence, only Gauss's law for electric fields,

$$\nabla \cdot E = \frac{\rho}{\varepsilon_0}$$

is considered. Since the direct solution of Gauss's law requires boundary conditions for E, usually unavailable, Poisson's equation,

$$\nabla^2 \phi = -\frac{\rho}{\varepsilon_0} \tag{5.9}$$

is solved instead in *dsmcPlasmaFoam* with $\rho = q_e n_e + q_i n_i$ being the total charge density. Note that for the specific case of the INGA III thruster, the electric potential at the anode of the device is held by the power supply at 0 V. Furthermore, the boundary condition at the cathode is defined indirectly as an electron particle inflow boundary condition calculated from the electric current through the cathode. The specific value of the electric current is also controlled by the power supply, with values between 20 A and 30 A leading to successful ignition of the INGA III arcjet thruster. With a similar approach in the numerical model, the electric potential at the cathode, where the electrons enter the computational domain, is effectively allowed to assume any value in the simulations and the results can be used for future comparison with experimental data.

The numerical solution of Poisson's equation is performed using the already available Finite Volume discretization schemes in *OpenFOAM®*. After the electric potential field has been calculated, the corresponding electric field is computed as follows:

$$E = -\nabla\phi \tag{5.10}$$

Equations 5.9 and 5.10 are implemented in the main solver file *dsmcPlasmaFoam.C* after the *evolve* function of the solver is invoked. Hence, the electric potential ϕ and electric field E are updated each iteration after the motion and collision steps, included in the *evolve* function, are performed. The updated values are subsequently used in the next iteration for the determination of the forces acting on the charge carriers during the particles motion step. The validation of the Maxwell component of the *dsmcPlasmaFoam* solver is presented in Section 6.1.

5.3.3.2 Particle and force weighting

The particle and force weighting steps described in Section 3.3.3 constitute one of the main components of a typical PIC algorithm. They are responsible for the assignment of discrete particle quantities to grid points in the numerical domain as well as for the application of electromagnetic forces to charge carriers. In *dsmcPlasmaFoam*, the weighting steps, which involve the determination of the number density in the numerical cells based on the particles distribution as well as the estimation of the Coulomb force acting on each particle, are both implemented using the zero-order or NGP approach described in Section 3.3.3. Hence, the particle number density in a cell corresponds to the total amount of test particles contained in the cell multiplied with the particles weight and divided by the cell's volume. In a similar way, the force acting on a charge carrier in *dsmcPlasmaFoam* is calculated based on the electric field present at the centre of the cell containing the charge carrier. As shown in Section 6.3, this approach produces accurate results with acceptable computational requirements.

5.3.3.3 Particle motion - Lorentz solver

The implementation of the charge carriers motion in the solver *dsmcPlasmaFoam* is based on the *move* and *trackToFace* functions described in Section 5.3.2 for the standard *dsmcFoam* solver. In the standard solver, the particles velocities are constant during the motion's time step Δt as long as the considered test particle does not interact with wall surfaces. On the other hand, the PIC implementation requires the consideration of the Lorentz forces acting on the test particles and hence, of variable velocities. The Lorentz force is given by:

$$F = qE \tag{5.11}$$

where the magnetic force term has been neglected. Since this force is responsible for the variable particles velocity, the motion's step of the PIC algorithm is, in the following, referred to as "**Lorentz solver**" (cp. [Sti15]).

After the force acting at a given particle's position has been determined in the force weighting step, the particle is advanced in time through the computation of its motion. In contrast to the standard *dsmcFoam* solver, the motion's procedure in *dsmcPlasmaFoam* requires an update of both the particle's position and its velocity. Note that if the particle changes cells during the motion's time step Δt, the electric field and hence, the force acting on the particle might, in theory, also change. Let us consider a test particle at the time t with position x_{old}, velocity v_{old} and located in the cell i with electric field E^i_{old}. Note that E^i_{old} is estimated from the same particles spatial distribution that includes the particle's current position x_{old}. Hence, position and electric field both carry the subscript "*old*". The Lorentz force acting on the particle and estimated with the approach described in Section 5.3.3.2 is $F^i_{old} = qE^i_{old}$. Rewriting of Eqs. 3.113 and 3.114 produces the following expressions:

$$v_{new} = F^i_{old}\frac{\Delta t}{m} + v_{old} \tag{5.12}$$

$$x_{new} = v_{new}\Delta t + x_{old} \tag{5.13}$$

which are introduced in the *move* function of *dsmcPlasmaFoam* using the time-centred leapfrog algorithm described in Section 3.3.1. As shown in Fig. 3.12, the leapfrog algorithm involves the calculation of the particle's position and the force it is subjected to at the same time-level, while the particle's velocity is shifted by $\Delta t/2$. Hence, v_{old} is computed at the time $t - \Delta t/2$ and v_{new} at the time $t + \Delta t/2$. Based on this, the leapfrog algorithm combined with the *trackToFace* function in the *dsmcPlasmaFoam* solver is implemented as follows:

- At the start of the Lorentz solver step, the values of the particle's position x_{old}, the force F^i_{old} acting on the particle at the time level t as well as the particle's velocity v_{old} at the time $t - \Delta t/2$ are known.
- v_{new} is estimated using Eq. 5.12. The result corresponds to the estimated velocity at the time $t + \Delta t/2$.
- The new position x_{new} of the particle is estimated with Eq. 5.13 using the velocity at the time $t + \Delta t/2$. The estimated new position enters the *trackToFace* function where the parameter $trackFraction$, corresponding to the fraction of Δt before the particle encounters a boundary or cell face, is calculated.
- If $trackFraction = 1$, the particle does not encounter any boundary or cell faces. In this case, the estimated position x_{new} corresponds to the actual particle's position at $t + \Delta t$. Moreover, the estimated velocity v_{new} corresponds to the particle's velocity at the time level $t + \Delta t/2$. The particle's motion step is hence complete and the procedure can be applied to the next particle.
- If $trackFraction < 1$ and the particle encounters an open outlet, the particle is erased. If $trackFraction < 1$ and the particle encounters a wall, the particle's motion continues from the point where the wall collision takes place. To this end, the time is advanced by $trackFraction\Delta t$, the collision coordinates become x_{old} and the post-collision velocity is obtained from the wall temperature according to the reflection model selected (specular or diffusive). This velocity is stored as v'_{old} and is valid for the collision time level $t + trackFraction\Delta t$. In order to continue the particle's motion in a time-centred manner, this velocity needs to be shifted again in time. To this end, v'_{old} is used for a new estimation of the velocity v_{new} at the time level $t + \Delta t/2$ using the following expression:

$$v_{new} = F^i_{old}(\frac{1}{2} - trackFraction)\frac{\Delta t}{m} + v'_{old} \tag{5.14}$$

Note that if $trackFraction > 1/2$, the wall collision takes place after the time level $t + \Delta t/2$ for which v_{new} is valid. This means, that the particle's post-collision velocity v'_{old} is "pushed back in time" by Eq. 5.14 to the time level $t + \Delta t/2$ and the time-centring character of the leapfrog scheme is conserved. On the other hand, if $trackFraction < 1/2$, the particle's post-collision velocity v'_{old} is advanced to the time level $t + \Delta t/2$ by Eq. 5.14. If $trackFraction = 1/2$, the particle collides with the wall exactly at the time level $t + \Delta t/2$ and $v_{new} = v'_{old}$. The end position of the particle x_{new} is estimated again using the updated v_{new} and replacing Δt by $(1 - trackFraction)\Delta t$ in Eq. 5.13. The second estimation of x_{new} is subsequently used to determine the corresponding $trackFraction$ and the procedure is repeated until $trackFraction = 1$.

- The special case of $trackFraction < 1$ and charge carriers encountering an anode or cathode in the numerical domain is described in detail in Section 5.3.7.
- The case of $trackFraction < 1$ and the particle encountering a cell interface can be handled in two different ways. In the simplest approach, the velocity v_{new} is kept constant and the particle continues its motion without taking into account the potential different value of E in the new cell. This approach is sufficiently accurate for values of Δt which are small compared to the average transit time of the particles in the numerical cells. In the alternative approach, the effects of the electric field in the new cell are considered through a correction of the original estimation of the leapfrog velocity v_{new} using $trackFraction$. To this end, v_{new} is re-defined using a linear combination of the contributions of the electric fields in the old and new cells. Hence,

$$v_{new} = v_{old} + (trackFraction)F^{i}_{old}\frac{\Delta t}{m} + (1 - trackFraction)F^{j}_{old}\frac{\Delta t}{m} \tag{5.15}$$

  where F^{i}_{old} and F^{j}_{old} represent the force acting on the particle as a result of the electric field at the cells i (old) and j (new) respectively. The particle's motion step through the new cell is subsequently performed starting at the location where the particle changes cells x'_{old} and using the new estimation of the leapfrog velocity v_{new} from Eq. 5.15. The end position of the particle x_{new} is estimated again using the updated v_{new} and replacing Δt by $(1 - trackFraction)\Delta t$ in Eq. 5.13. The second estimation of x_{new} is subsequently used to determine the corresponding $trackFraction$ and the procedure is repeated until $trackFraction = 1$. The accuracy of both approaches is discussed in detail in Section 6.2.

After the positions of all particles at the time $t + \Delta t$ are obtained, the Lorentz solver step is finalised and the results can be used to update the electric field with the Maxwell solver. As mentioned in Section 3.3.1, the particles velocities at the start of the simulation v_0 must be shifted back in time to the level $-\Delta t/2$ in the leapfrog algorithm. This is achieved using Eq. 5.12 with $t = -\Delta t/2$ and the force at the time $t = 0$. Also note that the particles velocities must be synchronized to the same time levels as the particle's positions and electric field before writing down the numerical results for post-processing. The validation of the Lorentz component of the *dsmcPlasmaFoam* solver is discussed in detail in Section 6.2.

5.3.4 Coulomb collisions with the MCC algorithm

5.3.4.1 Model description

The treatment of Coulomb collisions inside the electric propulsion system is based on the theory of cumulative small-angle collisions developed by Kenichi Nanbu (s. [Nan97] and [Nan00]). Interactions between charged carriers inside a plasma are the result of the electrostatic forces acting on them. These forces act over a distance of the order of the Debye length and tend to produce only small-angle direction changes of the involved particles. Note that for the most part, charged particles do not actually collide but rather "interact" with each other. The approach described in [Nan97] is based on the treatment of a a large number of successive small-angle interactions as one single large-angle collision (cumulative collision). To this end, a relation between the cumulative scattering angle and the time step Δt is developed, allowing a single large-angle collision to be representative of a large number of small-angle interactions. The corresponding modelling approach is described in the following based on [Nan97] and [Nan00].

In order to determine the cumulative scattering angle after N collisions, let g_0 represent the initial velocity of an arbitrary charged particle and g_1, g_2, ..., g_k,..., g_N, the post-collision velocities after the first, second, k^{th} and N^{th} collisions with a field of fixed charged particles respectively. The cumulative deflection angle χ_N after N collisions is given by the following expression:

$$\cos \chi_N = g_0 \cdot g_N / g^2 \tag{5.16}$$

Since the collisions are assumed to be elastic, the magnitude of the velocity vector remains constant and $g = |g_0|$. In order to obtain an expression for the cumulative scattering angle χ_N that does not depend on the individual post-collision velocities, a series of coordinate system rotations is performed after each collision. The reference coordinate system (x,y,z) is defined in a way that the z-axis coincides with the initial velocity vector g_0. Let θ_1 be the polar angle of g_1 in the reference coordinate system and φ_1, the azimuthal angle. The coordinate system is rotated after the first collision in order for the z-axis to coincide again with the velocity vector g_1. The resulting coordinate system (x_1,y_1,z_1) is used as reference for the deflection angles (θ_2, φ_2) after the second collision. This procedure is repeated N times so that N pairs of angles (θ, φ) are obtained. Under the assumption that all deflections are small, θ_1, θ_2, $\theta_N << 1$, the total cumulative angle χ_n is given by:

$$\sin^2 \frac{\chi_N}{2} = \frac{1}{4} \sum_{k=1}^{N} \theta_k^2 + \frac{1}{2} \sum_{k=2}^{N} \sum_{l=1}^{k-1} \theta_k \theta_l \cos(\varphi_k - \varphi_l) \tag{5.17}$$

A detailed derivation of Eq. 5.17 can be found in the Appendix of [Nan97]. Note that the angle sets (θ, φ) are random variables. Furthermore, the individual deflection angles θ_1, θ_2... θ_N are independent of each other and can be described by the same probability law. Moreover, the polar set $\{\theta_1, \theta_2,...\}$ is independent of the azimuthal set $\{\varphi_1, \varphi_2,...\}$. Let us now examine the azimuthal angles. While the polar deflection θ depends on the impact parameter b, the azimuthal deflection φ can assume any value between 0 and 2π. The expected value $\langle\ \rangle$ of the expression $\cos(\varphi_k - \varphi_l)$ is therefore:

$$\langle \cos(\varphi_k - \varphi_l) \rangle = \frac{1}{(2\pi)^2} \int_0^{2\pi} \int_0^{2\pi} \cos(\varphi_k - \varphi_l) \mathrm{d}\varphi_k \mathrm{d}\varphi_l = 0 \tag{5.18}$$

Combination of this result with Eq. 5.17 produces the following expression for the expected value of the cumulative deflection angle:

$$\left\langle \sin^2 \frac{\chi_N}{2} \right\rangle = \frac{1}{4} \sum_{k=1}^{N} \langle \theta_k^2 \rangle \tag{5.19}$$

Since each θ_k in Eq. 5.19 obeys the same probability law, the expected value of each θ_k is the same and the sum over the individual collisions can be replaced by $N\langle\theta_k^2\rangle$. Furthermore, since the set θ_k consists of random and independent variables, their expected value $\langle\theta_k\rangle$ can be interpreted not only as the set of deflection angles for one individual particle over a large number of collision events, but also as the ensemble average over a large collection of test particles. According to this, the first deflection angle θ_1 obtained for a large collection of test particles can be used to determine the value of the RHS of Eq. 5.19. This leads to,

$$\left\langle \sin^2 \frac{\chi_N}{2} \right\rangle = \frac{1}{4} \langle \theta_1^2 \rangle N \tag{5.20}$$

Note that the RHS of Eq. 5.20 approaches infinity if N becomes too large. On the other hand, the LHS is clearly always smaller than one. Hence, Eq. 5.20 requires some kind of additional relaxation.

If the number of collision events N approaches infinity, the scattering is expected to be isotropic. In this case, the probability density function of the cumulative deflection angle χ_N after N events is $(\sin \chi_N)/2$. As shown in Fig. 5.8, the expected value of χ_N is $\pi/2$ and the LHS of Eq. 5.20 becomes:

$$\left\langle \sin^2 \frac{\chi_N}{2} \right\rangle = \sin^2 (\frac{\chi_N}{2}) = \sin^2 (\frac{\pi}{4}) = \frac{1}{2} \tag{5.21}$$

In order to further develop Eq. 5.20, let us consider the Rutherford scattering problem. If a particle with charge q_α moves with velocity g through a cloud of fixed charge

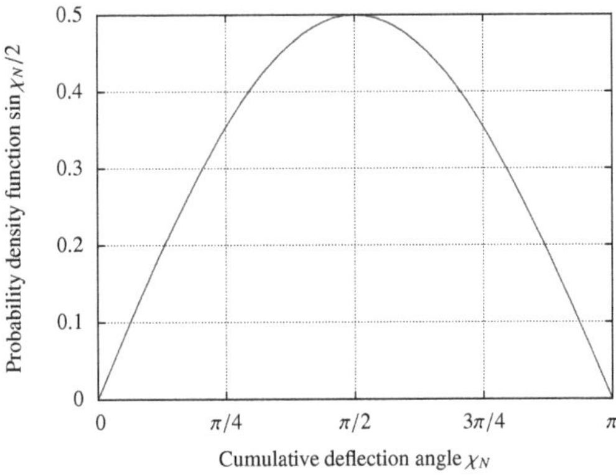

Figure 5.8: Probability distribution function for isotropic scattering.

carriers, it will interact with a particle of charge q_β and as a result, be deflected by the angle θ. This angle is, in turn, a function of the impact parameter b, which describes the minimum distance between the fixed particle and the trajectory of the moving charge as shown in Fig. 5.9. The deflection angle of the kth small-angle collision is given by,

$$\tan\frac{\theta_k}{2} = \frac{|q_\alpha q_\beta|}{4\pi\varepsilon_0 \mu g^2 b} \tag{5.22}$$

where μ represents the mass of the moving particle. As described in Section 2.8.1.2, charge carriers inside a plasma can interact with each other over a distance equal to the Debye length. Hence, the impact parameter b ranges between 0 (head-on collision) and λ_D (grazing interaction).

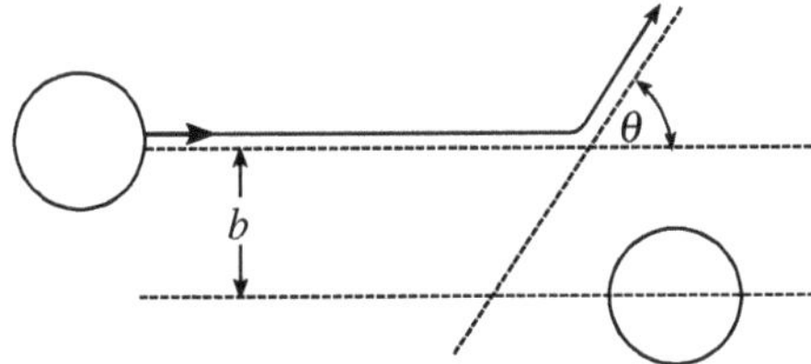

Figure 5.9: Rutherford scattering. $b =$ Impact parameter; $\theta =$ Scattering angle.

In order to evaluate the behaviour of Eq. 5.20 towards the isotropic limit ($N \to \infty$) and to develop a probability density function for χ_N, Nanbu performs numerical simulations with an ensemble consisting of 200,000 test particles that interact with each other based on Eq. 5.22 (s. [Nan97]). For the simulations, the impact parameter $0 < b < \lambda_D$ is randomized using the function $b = \lambda_D \sqrt{U}$ where $0 < U < 1$ represents a uniformly distributed random number. The chosen random function leads to the majority of the interactions in the test simulations taking place over relatively long distances, which in turn is equivalent to small-angle scattering. Eq. 5.22 can be rewritten as follows:

$$\theta_k = 2 \arctan\left(\frac{\theta_{min}}{2\sqrt{U}}\right) \tag{5.23}$$

where,

$$\theta_{min} = \frac{b_0}{\lambda_D} \tag{5.24}$$

represents the minimal deflection angle and

$$b_0 = \frac{|q_\alpha q_\beta|}{2\pi\varepsilon_0 \mu g^2} \tag{5.25}$$

a constant value. Hence, the only remaining free parameter is θ_{min}, which in turn is linked to λ_D. Small values of θ_{min} imply high values of λ_D and therefore, an increase in the number of binary particle interactions in the test simulations. Although a typical value of θ_{min} is of the order 2.6×10^{-3} deg, the values 0.5, 1 and 2 deg were used by Nanbu for the test simulations in order to keep the computational requirements low. Finally, the azimuthal deflection angle can take any value. Hence, $\varphi = 2\pi U$ with $0 < U < 1$ being another uniformly distributed random number. With the described conditions, test simulations are performed and the obtained sets $(\theta_k,\ \varphi_k)$ can be used to calculate the cumulative scattering angle χ_N using Eq. 5.17. In order to obtain the probability density function of χ_N, the expected value of θ_1^2 is computed based on Eq. 5.23 as follows:

$$\langle \theta_1^2 \rangle = 8 \int_0^1 \left[\arctan\left(\frac{\theta_{min}}{2\eta}\right)\right]^2 \eta \, \mathrm{d}\eta \tag{5.26}$$

with $\eta = \sqrt{U}$. Eq. 5.26 produces $\langle \theta_1^2 \rangle = 3.051 \times 10^{-3}$ radian for $\theta_{min} = 1$ deg. Note that the value $\theta_{min} = 1$ deg remains constant in the study in [Nan97]. Finally, let us define a function s based on the value $\langle \theta_1^2 \rangle$ and the number of collisions considered N:

$$s = \frac{1}{2} \langle \theta_1^2 \rangle N \tag{5.27}$$

The results obtained for $\langle \sin^2(\chi_N/2) \rangle$ as a function of s with the 200,000 test particles ensemble can be used to develop a numerical fit leading to the following relaxed form of Eq. 5.20:

$$\left\langle \sin^2 \frac{\chi_N}{2} \right\rangle = \frac{1}{2}(1 - e^{-s}) \tag{5.28}$$

The function in Eq. 5.28 is plotted in Fig 5.10. Note that the function accurately predicts the value of $\langle \sin^2(\chi_N/2) \rangle$ as 1/2 for $s \to \infty$ ($N \to \infty$) in Eq. 5.21.

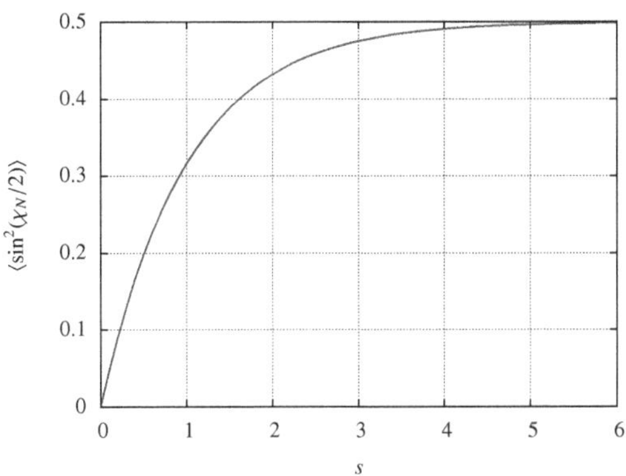

Figure 5.10: Relaxation function vs. parameter s (cp. [Nan97]).

Based on the expected value of χ_N in Eq. 5.28, the corresponding probability distribution function can now be developed. To this end, let us define $f(\chi_N)\mathrm{d}\Omega$ as a function describing the probability that the post-collision velocity g_N falls into the differential solid angle $\mathrm{d}\Omega = 2\pi \sin\chi_N \mathrm{d}\chi_N$. The probability that χ_N falls between 0 and π clearly equals unity and is given by,

$$\int_0^{\pi} f(\chi_N) 2\pi \sin\chi_N \mathrm{d}\chi_N = 1 \tag{5.29}$$

The probability density function of χ_N is therefore $2\pi f(\chi_N)\sin\chi_N$ and only $f(\chi_N)$ is to be found. As discussed above and shown in Fig. 5.8, the probability density function for isotropic scattering ($N \to \infty$) equals $(\sin\chi_N)/2$ and in this case, $f(\chi_N) = 1/4\pi$. In order to find a general $f(\chi_N)$, it is possible to take advantage of the result $\langle \sin^2(\chi_N/2) \rangle$ in Eq. 5.28 applicable to all values of N. Hence, the condition,

$$2\pi \int_0^{\pi} f(\chi_N) \sin^2 \frac{\chi_N}{2} \sin\chi_N \mathrm{d}\chi_N = \frac{1}{2}(1 - e^{-s}) \tag{5.30}$$

must be satisfied by the unknown function $f(\chi_N)$. In order to determine $f(\chi_N)$, the results of the test simulations are categorized according to fixed intervals of cumulative deflection angles $\Delta\chi_N = 5$ deg. In the next step, random event samples of different sizes in the range $100 < N < 3,000$ are defined and the mean value of $f(\chi_N)$ in the ith interval is computed as follows:

$$f_i = \frac{M_i}{M\Omega_i} \tag{5.31}$$

where M_i represents the number of deflections falling in the interval:

$$\Omega_i = 2\pi(\cos\chi_{i-1} - \cos\chi_i) \tag{5.32}$$

With this procedure, the function $f(\chi_N)$ vs. χ_n can be obtained for different values of N. The results of this analysis show a strong linear relation between $\ln[f(\chi_N)]$ and $\cos\chi_N$ for all values of N. Based on these results, Nanbu proposes the following expression for the probability density function:

$$f(\chi_N) = B\exp(A\cos\chi_N) \tag{5.33}$$

where A and B are constants. Combination of Eq. 5.33 with the condition in Eq. 5.29 enables the elimination of B, yielding:

$$f(\chi_N) = \frac{A}{4\pi\sinh A}\exp(A\cos\chi_N) \tag{5.34}$$

The constant A can be determined by taking advantage of the condition in Eq. 5.30. Combination of Eq. 5.34 with Eq. 5.30 leads to:

$$\coth A - A^{-1} = e^{-s} \tag{5.35}$$

With this approach, the parameter s can be obtained from Eq. 5.27 once the number of collision events is known. The value s can subsequently be used to compute the constant A from Eq. 5.35 and with this value, the function $f(\chi_N)$ and the probability density function:

$$F(\chi_N) = 2\pi f(\chi_N) sin\chi_N \tag{5.36}$$

Although Eq. 5.35 can be solved directly in numerical solvers employing an appropriate numerical method, the use of look-up tables containing values of the parameter A as a function of s is recommended in [Nan97]. For reference, the values of A as a function of s in the range $0.01 < s < 4$ are listed in Table 5.7. For values of s outside of the range documented in the look-up table, simplified equations may be used. For $s < 0.01$,

s	A	s	A
0.01	100.5	0.3	3.845
0.02	50.50	0.4	2.987
0.03	33.84	0.5	2.448
0.04	25.50	0.6	2.067
0.05	20.50	0.7	1.779
0.06	17.17	0.8	1.551
0.07	14.79	0.9	1.363
0.08	13.01	1	1.207
0.09	11.62	2	0.4105
0.1	10.51	3	0.1496
0.2	5.516	4	0.05496

Table 5.7: Values of the parameter A as a function of s (cp. [Nan97]).

A can be computed as $A = 1/s$ with an error of only 0.5% compared to the exact value obtained from Eq. 5.35. On the other hand, for $s > 3$, the value of A can be set to $A = 3e^{-s}$. The error here is only 0.15%.

The main advantage of the cumulative scattering angle approach is that the probability distribution function associated with the deflection angles is a function of the number of collision events taking place in the plasma in a given period of time Δt (i.e., of the parameter s). In order to examine this feature, the probability distribution function $F(\chi_N) = 2\pi f(\chi_N) sin\chi_N$ is shown in Fig. 5.11 for different values of s. As evident from Fig. 5.11, small values of s, which are associated with a low number of collision events, result in a very narrow distribution function with a high concentration of probable cumulative angles near 0 deg. This agrees with the expected behaviour of a particle experiencing barely any collisions ($\langle\chi_N\rangle \approx 0$). On the other hand, high values of s, which correspond to a high number of collision events, lead to the isotropic scattering distribution function shown in Figs. 5.8 and 5.11 with the expected value $\langle\chi_N\rangle = \pi/2$.

As formulated in Eq. 2.127 for a test particle moving through a gas of fixed particles, the total number of collisions N can be expressed as follows:

$$N = n_\beta g \pi b_{max}^2 \Delta t \tag{5.37}$$

where n_β stands for the number density of the fixed particles and b_{max}, for the maximum value of the impact parameter (i.e., λ_D). Furthermore, the value $\langle\theta_1^2\rangle$ can be analytically

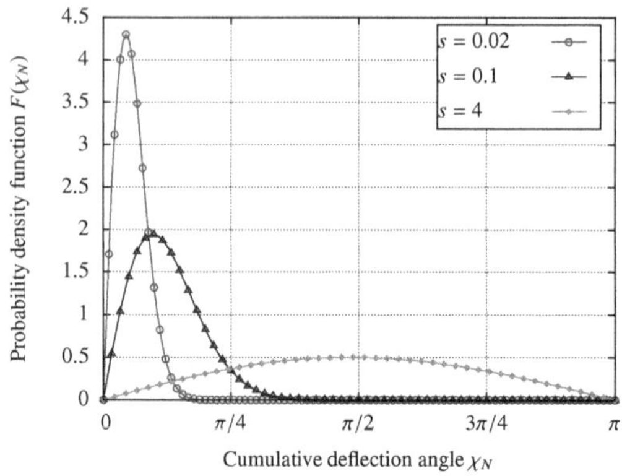

Figure 5.11: Probability density functions $F(\chi_N)$ for the cumulative scattering angle as a function of the deflection angle χ_N for different values of the parameter s.

approximated as follows (cp. [Nan97]):

$$\langle \theta_1^2 \rangle = 2 \left(\frac{b_0}{b_{max}} \right)^2 \ln \Lambda \tag{5.38}$$

with $\Lambda = \lambda_D / b_0$. Substitution of Eqs. 5.37 and 5.38 into 5.27 yields:

$$s = n_\beta g \pi b_0^2 (\ln \Lambda) \Delta t \tag{5.39}$$

which can be used in numerical applications instead of Eq. 5.27 for the determination of s as a function of the time step Δt. Because of the dynamic nature of the distribution function for the cumulative scattering angle (s. Fig. 5.11), the value Δt can be chosen arbitrarily in Nanbu's model. After the value of the parameter s has been determined, a random value for the scattering angle $0 < \chi_N < \pi$, which obeys the corresponding probability density function shown in Eq. 5.36, can be obtained using the following expression:

$$\cos \chi_N = \frac{1}{A} \ln(e^{-A} + 2U \sinh A) \tag{5.40}$$

with $0 < U < 1$ representing a random number. For very small values of s, A becomes very large which can lead to an exponential overflow in Eq. 5.40. To prevent this, for very small values of s, Eq. 5.40 may be replaced by:

$$\cos \chi_N = 1 + s \ln U \tag{5.41}$$

Furthermore, for $s > 6$ isotropic scattering is expected and Eq. 5.40 can be replaced by the simple expression:

$$\cos \chi_N = 2U - 1 \tag{5.42}$$

Since the field particles are assumed to be fixed, the colliding test particle retains its pre-collision velocity magnitude g. Hence, the post-collision velocity can be easily obtained from g, the cumulative polar deflection angle χ_N and the azimuthal deflection angle $\varphi = 2\pi U$.

The approach described above is valid for the case of a test particle moving through a fixed cloud of particles. In order to obtain the post-collision velocities for interactions between a test particle and a moving particles field, some additional considerations are necessary. Let us focus on the test particle α with initial velocity $v_{\alpha 1}$. The test particle collides with one of the moving field particles β with initial velocity $v_{\beta 1}$. The relative pre-collision velocity is therefore $g_1 = v_{\alpha 1} - v_{\beta 1}$. The test particle is scattered by the angle θ_1 and the relative post-collision velocity becomes $g_1' = v_{\alpha 1'} - v_{\beta 1'}$. Since the collisions are assumed to be elastic, $g_1' = g_1$. The test particle then collides a second time with a different field particle. Since the field particles have different velocities, the relative pre-collision velocity of the second interaction g_2 is not equal to g_1'. In summary, $g_1' = g_1$, $g_2' = g_2$, $g_k' = g_k$ but $g_1' \neq g_2$ and in general, $g_k' \neq g_{k+1}$. Since the magnitude of the relative velocity g appears in the definition of the deflection angle in Eq. 5.22, the scattering angle θ_k is, for a moving field of particles, not only a function of the impact parameter b but also of the individual relative pre-collision velocity g_k.

Let us examine the effect of this new variable on the developed method. Eq. 5.20, which is the corner stone of the approach for a fixed particles field, is derived from Eq. 5.19 using the relation:

$$\sum_{k=1}^{N} \langle \theta_k^2 \rangle = N \langle \theta_1^2 \rangle \tag{5.43}$$

which is based on the fact that the deflection angle depends only on the impact parameter b and the expected value $\langle \theta_k^2 \rangle$ is constant, i.e., $\langle \theta_1^2 \rangle = \langle \theta_2^2 \rangle = \langle \theta_k^2 \rangle$. In contrast, if the test particle interacts with a moving particles field, the change of relative pre-collision velocity between collisions will have an effect on the expected value of the deflection angle and therefore, $\langle \theta_1^2 \rangle \neq \langle \theta_2^2 \rangle \neq \langle \theta_k^2 \rangle$. In this case, the expected value of θ_k^2 averaged over the impact parameter b can be formulated in a similar way to Eq. 5.38 as follows:

$$\langle \theta_k^2 \rangle = 2 \left(\frac{b_0}{b_{max}} \right)^2 \ln \Lambda \tag{5.44}$$

with,

$$b_0 = \frac{|q_\alpha q_\beta|}{2\pi\varepsilon_0 \mu_{\alpha\beta} g_k^2} \tag{5.45}$$

and

$$\mu_{\alpha\beta} = \frac{m_\alpha m_\beta}{m_\alpha + m_\beta} \tag{5.46}$$

where m_α and m_β represent the mass of the colliding particles and g_k, the magnitude of the pre-collision velocity of the kth collision. The presence of g_k in Eq. 5.45 makes the impact of the variability of g_k on $\langle\theta_k^2\rangle$ evident.

Because of the variable nature of $\langle\theta_k^2\rangle$ for a moving particles field, $\langle\theta_1^2\rangle$ in Eq. 5.20 should not be interpreted as the expected value of the squared deflection of the first collision but as the average $\langle\theta^2\rangle$ of $\langle\theta_1^2\rangle$, $\langle\theta_2^2\rangle$, ..., $\langle\theta_N^2\rangle$ over N collisions. This way, Eq. 5.20 does not require any modifications. Furthermore, the s parameter defined in Eq. 5.27 becomes, for a moving particles field:

$$s = \frac{1}{2}\langle\theta^2\rangle N \tag{5.47}$$

Therefore, s is, strictly speaking, a function of $\langle\theta^2\rangle$ instead of $\langle\theta_1^2\rangle$. In theory, this would require the computation of the individual values $\langle\theta_1^2\rangle$, $\langle\theta_2^2\rangle$, ..., $\langle\theta_N^2\rangle$ which are themselves a function of the relative pre-collision velocities. However, since the approach aims for a cumulative description of the particles interactions, the consideration of individual g_k's would greatly diminish the applicability of the method. To avoid this, it is assumed that $\langle\theta^2\rangle \approx \langle\theta_1^2\rangle$ and hence, Eq. 5.27 retains its applicability. The s parameter can therefore computed analogue to Eq. 5.39 as follows:

$$s = n_\beta g \pi b_0^2 (\ln\Lambda)\Delta t = \frac{\ln\Lambda}{4\pi}\left(\frac{q_\alpha q_\beta}{\varepsilon_0 \mu_{\alpha\beta}}\right)^2 \frac{n_\beta}{g^3}\Delta t \tag{5.48}$$

From a numerical point of view, the assumption $\langle\theta^2\rangle \approx \langle\theta_1^2\rangle$ is introduced when the parameter s for the cumulative collision in the time Δt is calculated from Eq. 5.48 based on the "initial" relative velocity $g = g_1$. If the particle ensemble used with this method is large enough, the assumption has no negative effects because overall, a large number of collisions employing a large sample of relative velocities is computed and $\langle\theta_1^2\rangle$ approaches $\langle\theta^2\rangle$.

Because of the definition of b_0 in Eq. 5.45, the term Λ, calculated for a fixed particles field as λ_D/b_0, is also a function of the individual pre-collision relative velocities. In order for the cumulative scattering angle approach to be applied for a moving particles field, Λ is redefined as $\lambda_D/\langle b_0\rangle$ with,

$$\langle b_0\rangle = \frac{|q_\alpha q_\beta|}{2\pi\varepsilon_0\mu_{\alpha\beta}\langle g^2\rangle} \tag{5.49}$$

and

$$\langle g^2\rangle = \frac{3k_BT}{\mu_{\alpha\beta}} \tag{5.50}$$

which assumes a plasma in equilibrium.

In order to determine the post-collision velocity of the test particle α in a moving field of particles β after the time Δt, the following procedure is applied:

- A random sample of the field particle velocity v_β is obtained from the field velocity distribution function.
- The relative pre-collision velocity $g = v_\alpha - v_\beta$ is computed and used to obtain s from Eq. 5.48.
- The parameter A is subsequently determined using either Eq. 5.35 or look-up tables.
- A random sample of the cumulative deflection angle χ is obtained from Eq. 5.40. The cumulative polar deflection angle is given by $\varphi = 2\pi U$.

After the cumulative deflection angles are obtained, the post-collision velocity v'_α for the test particle α is given by,

$$v'_\alpha = v_\alpha - \frac{m_\beta}{m_\alpha + m_\beta}[g(1-\cos\chi) + h\sin\chi] \tag{5.51}$$

where the components of the vector h are defined as,

$$\begin{aligned} h_x &= g_\perp \cos\varphi \\ h_y &= -(g_y g_x \cos\varphi + g g_z \sin\varphi)/g_\perp \\ h_z &= -(g_z g_x \cos\varphi - g g_y \sin\varphi)/g_\perp \end{aligned} \tag{5.52}$$

with $g_\perp = (g_y^2 + g_z^2)^{1/2}$. The procedure above is applied for each of the test particles and the simulation is subsequently advanced to the next time step. These steps are to be performed until stationary macroscopic quantities are obtained (cp. [Nan97]).

The procedure above considers a test particle moving through a moving particles field. The field affects the test particle velocity vector but the field itself is not affected. If the studied plasma consists of different species and the simulated particles affect each other, a slight modification of the procedure above can be employed. Instead of focusing on the test particle and its interaction with the particles field, the collisions can be computed in a pair-wise manner through the multi-component particle assembly. Let us assume a two-component plasma consisting of the species α and β. If the particles have the same numerical weight W, the number densities of the particles in a control volume are given by,

$$n_\alpha = \frac{WN_\alpha}{V_c} \tag{5.53}$$

$$n_\beta = \frac{WN_\beta}{V_c} \tag{5.54}$$

where N_α and N_β represent the number of α and β particles in the cell and V_c, the cell's volume. Clearly, a particle α collides with another particle α with the probability n_α/n while collisions with the β species will take place with the probability n_β/n. The number of particles α that collides with particles β in the considered cell is therefore given by,

$$N'_\alpha = N_\alpha \frac{n_\beta}{n} \tag{5.55}$$

In an analogue way, the number of particles α that collide with other particles α is:

$$N''_\alpha = N_\alpha \frac{n_\alpha}{n} \tag{5.56}$$

The number of particles β colliding with either species can be determined in a similar way. Note that $N'_\alpha + N''_\alpha = N_\alpha$. The same applies for the species β. This implies that each particle in the cell will collide only once in the considered time interval. Furthermore, the particles will collide in each iteration and the concept of collision frequency does not apply. Note that the effect of the collision frequency is replaced by the cumulative nature of the scattering angle, i.e, small values of Δt lead to small angle deflections while large values of Δt result in high cumulative scattering angles.

In the case of a multicomponent plasma, the following procedure can therefore be implemented (cp. [Nan00]):

- The number of collisions between the particles α and β $N'_\alpha = N'_\beta$ is calculated and the corresponding collision partners are randomly selected in the cell.
- The individual cumulative collision for each of the N'_α pairs:

  $$\{(\alpha_i, \beta_i) : i = 1, 2, ..., N'_\alpha\}$$

  is computed.

- For the calculation of the collision, the parameter s is obtained analogue to Eq. 5.48 as follows:

$$s = \frac{\ln \Lambda_{\alpha\beta}}{4\pi} \left(\frac{q_\alpha q_\beta}{\varepsilon_0 \mu_{\alpha\beta}} \right)^2 \frac{n}{g^3} \Delta t \tag{5.57}$$

with $g = |v_{\alpha i} - v_{\beta i}|$ and

$$\Lambda_{\alpha\beta} = \frac{4\pi\varepsilon_0 \mu_{\alpha\beta} \langle g^2_{\alpha\beta} \rangle \lambda_D}{|q_\alpha q_\beta|} \tag{5.58}$$

where the expected value of the squared relative velocity is approximated by:

$$\langle g^2_{\alpha\beta} \rangle = \frac{3k_B T_\alpha}{m_\alpha} + \frac{3k_B T_\beta}{m_\beta} + (\langle v_\alpha \rangle - \langle v_\beta \rangle)^2 \tag{5.59}$$

with $\langle v_\alpha \rangle$ and $\langle v_\beta \rangle$ representing the macroscopic fluid velocities and the temperatures are obtained from the average kinetic energy of each of the colliding species.

- After the parameter s has been obtained, A can be either calculated from Eq. 5.35 or interpolated from prepared look-up tables.

- A random sample of the cumulative deflection angle χ is obtained from Eq. 5.40. The polar deflection is given by $\varphi = 2\pi U$.

- The post-collision velocities are computed from the deflection angles for the particles α and β using the following expressions:

$$v'_{\alpha i} = v_{\alpha i} - \frac{m_\beta}{m_\alpha + m_\beta} [g(1 - \cos\chi) + h \sin\chi] \tag{5.60}$$

$$v'_{\beta i} = v_{\beta i} + \frac{m_\alpha}{m_\alpha + m_\beta} [g(1 - \cos\chi) + h \sin\chi] \tag{5.61}$$

where the components of the vector h are determined using Eq. 5.52 and:

$$g_\perp = (g_y^2 + g_z^2)^{1/2} \tag{5.62}$$

- After the collisions between the particles α and β are computed, the same procedure is applied for the collisions between particles of the species α. To this end, the number of particles N''_α is determined, random particle pairs are formed and the same procedure as for the collisions between α and β particles is applied.

- Finally, the same steps are performed for the collisions between the particles β. At the end of the procedure, each of the plasma particles collides once with one particle, either of the same species or of a different one.

Note that if N''_α or N''_β are odd numbers, one particle would remain without collision partner. To compensate this, one particle is made to collide twice in the corresponding time step. Since a second collision would advance the corresponding particle in time by a total of $2\Delta t$, a slight modification of the collision time step for the particles system must be introduced. The reader is referred to [Nan00] for details on this procedure.

5.3.4.2 Implementation in OpenFOAM

The Coulomb collisions model described in Section 5.3.4.1 is implemented in the solver *dsmcPlasmaFoam* in the *collisions* function of the *evolve* method. The considered particles are ions, primary electrons (originated at the cathode) and secondary electrons (created in ionisation events). The particles properties, including their charge, are defined in the *dsmcProperties* dictionary. Furthermore, the new dictionary *plasmaProperties* located in the *constant* directory contains an extended version of the look-up Table 5.7 providing values of the parameter A as a function of the parameter s.

In *dsmcPlasmaFoam*, the Coulomb collisions procedure is applied in a cell-wise manner. Once the *collisions* function is called by the *evolve* method, the first step in the algorithm involves the determination of the number of charge carriers contained in the cell. Based on the individual particles velocities, the macroscopic (drift) velocity v in the cell is estimated for both the electrons and the ions species. The number of collision events is subsequently determined for electron-electron, electron-ion and ion-ion collisions using Eqs. 5.55 and 5.56. Note that the modifications for the case of an odd number of particles described in [Nan00] are also implemented in *dsmcPlasmaFoam*. Furthermore, the considered cell is divided into eight subcells. After the number of collision events has been determined, collision partners are randomly selected while prioritizing interactions between particles in the same subcell. When the two particles involved in a collision are chosen, the following procedure is applied:

- The expected value of the squared relative velocity is computed using Eq. 5.59.
- The reduced mass $\mu_{\alpha\beta}$ is computed.
- The Debye length λ_D is estimated from Eq. 2.97 with T equalling the temperature obtained from the average kinetic energy of the electrons in the cell and the Coulomb logarithm $\ln\Lambda_{\alpha\beta}$ is calculated with Eq. 5.58.
- The s parameter is subsequently obtained from Eq. 5.57 and the result is used to find the parameter A from the prepared look-up table.
- In the final step, the cumulative deflection angle is obtained from Eq. 5.40, the post-collision velocities are computed with Eqs. 5.60 and 5.61 and the procedure

is repeated for the next collision.

Note that in *dsmcPlasmaFoam*, the particles weights for electrons and ions are equal. The validation of the implemented model for Coulomb collisions is presented in Section 6.4.

5.3.5 Electron-neutral collisions

5.3.5.1 Model description

The implementation of the interactions between electrons and neutral atoms involves the consideration of a variety of collision mechanisms. For argon, for example, the possible outcomes of a collision event include elastic collisions, ionisation events and up to 25 cases of electronic excitation [Nan94]. The kind of collision that takes place during an interaction between an electron and a neutral atom is a function of the collision cross sections associated with each collision type as well as of the energy of the colliding electron. It should be noted that collision cross sections for interactions between electrons and neutral atoms are not always well documented in the literature. Regarding the implementation in the solver *dsmcPlasmaFoam*, the consideration of individual excitation states has little benefit on the expected results of the numerical approach. Hence, the present work considers three possible electron-neutral collision outcomes: elastic collision, ionisation and a "composite" excitation event which summarizes all possible electronic excitation states. Because of the amount of data available, the implemented approach for electron-neutral collisions is developed and validated for argon.

The implementation of the collisions between electrons and neutral atoms in the solver *dsmcPlasmaFoam* is based on the approach described in [Nan00] as well as on the *no time counter method* (NTC) described in Section 3.2.2.

As mentioned in Section 3.2, the collision and motion steps in a typical DSMC algorithm can be handled separately if the used time step Δt is smaller than the characteristic time interval between collisions. The probability P of a collision between two DSMC particles in the time Δt was defined in Eq. 3.79 as:

$$P = W\sigma_T c_r \Delta t / V \tag{5.63}$$

Furthermore, the NTC method also requires the definition of a maximum collision probability as introduced in Eq. 3.80:

$$P_{max} = W(\sigma_T c_r)_{max} \Delta t / V \tag{5.64}$$

Moreover, the number of potential collision candidates is estimated with the NTC approach using Eq. 3.81:

$$Candidates = \frac{1}{2}N(N-1)W(\sigma_T c_r)_{max}\frac{\Delta t}{V} \tag{5.65}$$

After the number of collision candidates has been estimated using Eq. 5.65, particles pairs are randomly selected and a collision takes place with the probability defined in Eq. 3.82:

$$P_{collision} = \frac{\sigma_T c_r}{(\sigma_T c_r)_{max}} \tag{5.66}$$

In theory, the expressions above can be applied for a multi-component gas mixture without modifications. However, if the difference between the molecular masses of the modelled species is large or if the use of different collision algorithms for specific species combinations is desired, Eqs. 5.64 to 5.66 should be slightly modified so that the NTC procedure can be applied to a specific particles combination. Since in *dsmcPlasmaFoam* a particular collisional approach is to be applied for interactions between electron and neutral atoms, the latter case applies. In a multi-component gas mixture, the number of collision candidates for interactions between the species P and Q is given by the following expression (cp. [Bir94]):

$$Candidates = \frac{1}{2}N_p N_q W\{(\sigma_T c_r)_{max}\}_{pq}\frac{\Delta t}{V} \tag{5.67}$$

where N_p and N_q stand respectively for the number of particles of the species P and Q in the control volume, σ_T for the total collision cross section, c_r for the magnitude of the relative pre-collision velocity and $\{(\sigma_T c_r)_{max}\}_{pq}$ represents the estimated maximum value of $\sigma_t c_r$ for the species combination P,Q. The value $\{(\sigma_T c_r)_{max}\}_{pq}$ is estimated for each cell at the beginning of the simulation and updated if the estimated value is exceeded during the numerical run. It should be noted that *dsmcPlasmaFoam* uses different weight factors W for neutral atoms and electrons. The reason for this being the expected large difference of number densities between the species. If the same weight factor is used for both species, the low value of W required for the accurate handling of the electrons would lead to extremely high numbers of neutral DSMC particles and hence, to prohibitively high computational requirements. On the other hand, a high value of W would keep computational requirements manageable but would also hinder an appropriate resolution of the electrons in the numerical domain. Considering this and based on [STM01], W in Eq. 5.67 is defined as the maximum weight W_{max} of the species P and Q. After the number of potential electron-neutral collision candidates has been determined, electron-neutral pairs are randomly selected and a collision takes place with the probability:

$$P_{collision} = \frac{\sigma_T c_r}{\{(\sigma_T c_r)_{max}\}_{pq}} \tag{5.68}$$

Eqs. 5.67 and 5.68, which correspond to the application of the NTC method for species P and Q in a multi-component gas mixture, are hence implemented in the solver *dsmcPlasmaFoam* for the definition of the collisions between electrons and neutral atoms. Note that a similar approach has been described in [EBWL95].

Now let us focus on the total collision cross section σ_T appearing in Eqs. 5.67 and 5.68. σ_T corresponds to the sum of the individual collision cross sections associated with each of the considered collisional events. Hence,

$$\sigma_T = \sum_{k=1}^{K} \sigma_k \tag{5.69}$$

and in a similar way,

$$P_{collision} = \sum_{k=1}^{K} P_k \tag{5.70}$$

where the label k indicates the collision type. In the solver *dsmcPlasmaFoam*, $k = 1$ corresponds to elastic collisions, $k = 2$ to the "composite" excitation collisions summarizing all possible electronic excitation states and $k = 3$ describes an ionisation event. Furthermore, the collision cross sections σ_k are a function of the electron energy $\varepsilon = mv^2/2$. The collision cross sections for the elastic and ionisation collisions between electrons and neutral argon atoms implemented in the solver take advantage of the set originally presented in [KH92] and made available online by Nanbu on [Hay]. Moreover, the cross sections for the composite excitation events are obtained from [SGJ90] and [Egg75]. The cross sections for the three electron-neutral collisional events considered in the present work are depicted in Fig. 5.12 as a function of the electron energy.

Now let us assume that a collision between an electron and a neutral argon atom takes place. As evident from Fig. 5.12, if the electron energy is below the excitation threshold (11.55 eV), the only possible outcome is an elastic interaction. If the electron energy lies between 11.55 eV and 15.76 eV, either an elastic or an excitation event can take place. If the electron energy is above the ionisation threshold (15.76 eV), elastic, excitation and ionisation events are all possible outcomes. The approach implemented in *dsmcPlasmaFoam* to determine whether a collision takes place as well as the collision

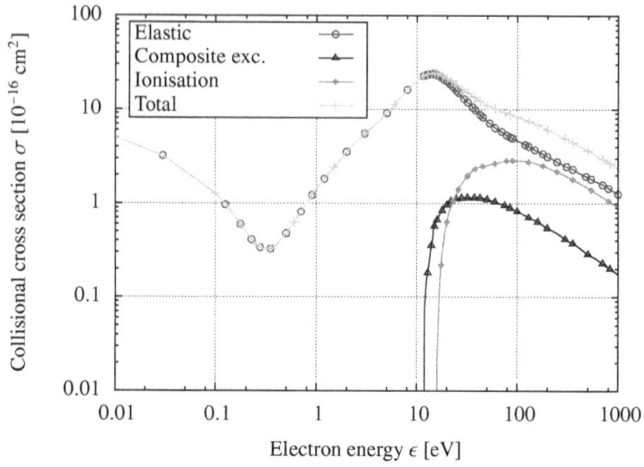

Figure 5.12: Integral cross sections σ_k for elastic, composite excitation and ionisation collisions between electrons and neutral argon atoms as a function of the electron energy.

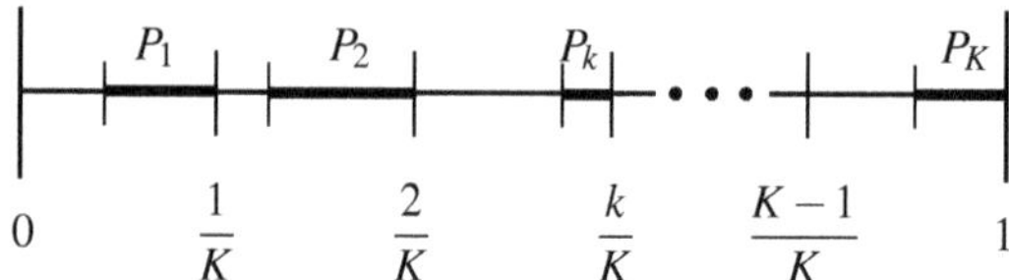

Figure 5.13: Collisional event sampling approach for electron-neutral interactions (cp. [Nan00]).

type is based on the method described in [Nan94] and [Nan00]. According to the NTC approach a collision takes place if,

$$U < P_{collision} = \frac{\sigma_T c_r}{(\sigma_T c_r)_{max}} \tag{5.71}$$

where $0 < U < 1$ is a random number. For the case of the collision between an electron and a neutral atom with three possible outcomes, the total collision probability $P_{collision}$ can be defined as,

$$P_{collision} = P_1 + P_2 + P_3 = \frac{\sigma_1 c_r + \sigma_2 c_r + \sigma_3 c_r}{(\sigma_T c_r)_{max}} = \frac{\sigma_T c_r}{(\sigma_T c_r)_{max}} \tag{5.72}$$

where σ_1, σ_2 and σ_3 correspond to the collision cross sections for the events 1, 2 and 3 respectively. In general, $P_k = \sigma_k c_r / (\sigma_T c_r)_{max}$ represents the probability of the kth collision event taking place. Note that the probabilities in Eq. 5.72 are normalized with

$(\sigma_T c_r)_{max}$ and the total collision probability is therefore $P_{collision} \leq 1$. In the method described in [Nan94], the distinction between individual collision types is achieved by dividing the range 0 to 1 into K equally spaced intervals, as shown in Fig. 5.13, with each of the intervals corresponding to one possible collisional event. Furthermore, each of the intervals is divided in two subintervals based on the value of the corresponding probability P_k. Hence, the sum of the individual subintervals on the right hand side equals the total collision probability $P_{collision}$. On the other hand, the sum of the subintervals on the left hand sides of each interval corresponds to the probability of the particles not colliding $1 - P_{collision}$. With this approach, if a random number $0 < U < 1$ falls into one of the subintervals on the right hand side, a collision will take place and the collision type is given by the label k of the subinterval. Similarly, if U falls into any of the subintervals on the left hand side, a collision event does not occur. Clearly, this approach is equivalent to randomly selecting one of the possible collision events for each of the considered particle pairs and subsequently evaluating whether the corresponding collision takes place. Although simple, the method has the disadvantage that all events are sampled with the same probability, an assumption which is implied in the equal spacing of the intervals representing each collisional event. This weakness can be illustrated by examining the case of an electron with an energy of 10 eV interacting with a neutral atom. As evident from Fig. 5.12, an elastic collision or no collision at all are the only possible outcomes of this interaction. Furthermore, a collision takes place if $U < \sigma_1 c_r / (\sigma_T c_r)_{max}$ where σ_1 represents the cross section for an elastic collisional event. If, for instance, $\sigma_1 c_r = (\sigma_T c_r)_{max}$, the collision probability is $P_{collision} = 1$ and an elastic collision should take place. However, if elastic collisions correspond to the first interval in Fig. 5.13 and $U > 1/3$, either excitation or ionisation events are considered and checked by the algorithm. Since the corresponding excitation and ionisation cross sections for an electron energy of 10 eV equal zero, no collision takes place. For this reason, a modified version of the method in [Nan94] is implemented in *dsmcPlasmaFoam*. Instead of dividing the range 0 to 1 in Fig. 5.13 into three equally large intervals, the intervals defined in this work are of variable size. The first interval $k = 1$ starts at the zero position and ends at the corresponding probability value P_1. Each of the following intervals starts at the end of the prior one as shown in Fig. 5.14. This way, the no-collision zone is located at the right hand side of the range. If the random number $0 < U < 1$ falls into this region, no collision takes place. On the other hand, if U falls into one of the intervals on the left side, the event corresponding to the interval containing U takes place. As with the NTC method, the collision determination step in *dsmcPlasmaFoam* does not require the evaluation of all possible particles pairs in the control volume. Instead, only a few collision candidates are preselected and assessed. Furthermore, the algorithm requires only one random value U in order to determine the actual collision type as well as whether a collision takes place. At the same time, the implemented approach avoids a potentially unrealistic sampling balancing between the possible collisional events.

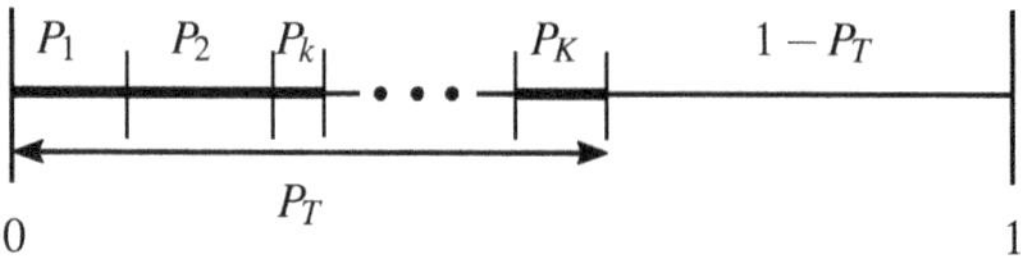

Figure 5.14: Collisional event sampling for electron-neutral interactions implemented in *dsmcPlasmaFoam*.

After the collision type has been determined, the next step in the developed plasma model involves the calculation of the particles post-collisions velocities. In the special case of an ionisation event, an additional step involving the creation of an electron-ion pair as well as the *deletion* of the neutral atom is required. The procedures for the three electron-neutral collisional events implemented in *dsmcPlasmaFoam* are:

- **Elastic collision:** From a numerical point of view, the case of an elastic collision between an electron and a neutral atom is relatively simple and very similar to a classic elastic sphere collision in typical DSMC applications (S. Section 3.2.2). However, compared with a classic elastic collision implementation in DSMC, the mass of the involved particles in an electron-neutral interaction differ by several orders of magnitude. If the electrons move slowly, the mass difference results in the scattering angles of electrons being much larger than those of their collision partners. In fact, for very low values of the electron energy ε, the electrons scattering angle becomes isotropic, which agrees with a classic DSMC collisional approach. However, if the electrons move with a very high energy ε, the probability of a head-on impact with a neutral atom drops, the majority of the interactions are grazing collisions with small deflection angles and forward electron scattering dominates. In order to determine the deflection angle of an electron after a collision with a neutral atom, a random sample of the scattering angle χ is obtained based on the electron energy and the following expression (cp. [Nan00]):

$$\cos\chi = \frac{1}{\varepsilon}[\varepsilon + 2 - 2(1+\varepsilon)^U] \tag{5.73}$$

where $0 < U < 1$ represents a random number. Note that the scattering angle χ

obtained from Eq. 5.73 obeys the probability distribution function,

$$g(\chi) = \frac{\varepsilon}{4\pi[1+\varepsilon\sin^2(\chi/2)]\ln(1+\varepsilon)} \tag{5.74}$$

which is based on the so-called screened Coulomb scattering. The reader is referred to [SGJ90] for additional information on the probability distribution function. The azimuthal scattering angle ψ is given by,

$$\psi = 2\pi U \tag{5.75}$$

Furthermore, the relative pre-collision velocity is defined as $g = v - V$ where v stands for the pre-collision electron velocity and V, for the pre-collision neutral velocity. The post-collision velocities v' and V' are obtained from the following expressions,

$$v' = v - \frac{M}{M+m}\left[g(1-\cos\chi) + h\sin\chi\right] \tag{5.76}$$

$$V' = V + \frac{m}{M+m}\left[g(1-\cos\chi) + h\sin\chi\right] \tag{5.77}$$

where the components of h are given by

$$\begin{aligned} h_x &= g_\perp \cos\psi \\ h_y &= -(g_x g_y \cos\psi + g g_z \sin\psi)/g_\perp \\ h_z &= -(g_x g_z \cos\psi + g g_y \sin\psi)/g_\perp \end{aligned} \tag{5.78}$$

In Eqs. 5.78, g stands for the magnitude of the pre-collision relative velocity and,

$$g_\perp = (g_y^2 + g_z^2)^{1/2} \tag{5.79}$$

- **Electronic Excitation:** In the case of electronic excitation, the collision between the electron and the neutral atom leads to the transfer of one (or more) electrons in the atom from their original electronic orbits to higher energy levels. Hence, the neutral atom gains an amount of energy equal to the threshold energy of excitation E for the specific excitation state. At the same time, the electron involved in the collision loses the same amount of energy E. As already mentioned, up to 25 different forms of electronic excitation states can be considered for argon. However, only one composite excitation collision type is implemented in the solver *dsmcPlasmaFoam*. The energy gained by the neutral atom through the excitation process is released when the electron returns to its original electronic state. This can happen either through excess energy transfer to another particle or via photon emission. Since these phenomena are not expected to have any macroscopic effect on the behaviour of a typical electric propulsion system, no new particle type representing the excited neutral atom is implemented in the present work. Therefore, the mathematical treatment of electronic excitation in *dsmcPlasmaFoam* involves the reduction of the pre-collision electron energy by the amount E, followed by the determination of the post-collision velocities using the same approach as for elastic electron-neutral collisions. The reduced electron pre-collision velocity is given by,

$$\widetilde{v} = v\left(1 - \frac{E}{\varepsilon}\right)^{1/2} \tag{5.80}$$

where $\varepsilon = mv^2/2$ stands for the electron pre-collision energy and E, for the threshold energy of excitation. The pre-collision velocity v is replaced by the reduced electron velocity $\widetilde{v}$ in Eq. 5.76 and the determination of the post-collision velocities follows the same approach as for the elastic-neutral collision case. In the present work, the threshold energy of excitation for argon is defined as $E = 11.55$ eV.

- **Ionisation:** During an ionisation collision, the interaction between the electron and the neutral atom leads to the creation of an electron-ion pair and the disappearance of the neutral. The ionisation process can be represented as follows:

$$e_1(v) + A((V)) \rightarrow e_1(v') + e_2(v'') + A^+(V') \tag{5.81}$$

where e_1 is the primary or parent electron, e_2 the secondary or progeny electron, A represents the neutral atom and the terms in parenthesis correspond to the particles velocities. Assuming that v' and v'' are much larger than the pre-collision centre of mass velocity, as well as $v \gg V$ and $M \gg m$ where m stands for the electron mass and M for the neutral mass, the following expression for the energy before and after the ionisation collision can be formulated:

$$\frac{1}{2}mv^2 - \frac{1}{2}mv'^2 = E + \varepsilon_p = \Delta\varepsilon \tag{5.82}$$

Eq. 5.82 implies that the energy of the parent electron decreases by an amount equal to the sum of the threshold energy of ionisation E and the post-collision energy of the progeny electron ε_p. The ionisation approach described in [Nan00] and implemented in *dsmcPlasmaFoam* is, in part, similar to the case of electronic excitation. First, the pre-collision energy of the parent electron is reduced by E

and ε_p. The corresponding reduced velocity is assumed to be the electron's pre-collision velocity and this value is used for the computation of the post-collision properties. However, this approach requires the energy of the progeny electron ε_p to be determined beforehand. The probability density function for ε_p is given in [Nan00] as,

$$g(\varepsilon_p) = \frac{a}{(\varepsilon_p - \varepsilon_0)^2 + a^2}\left(\tan^{-1}\frac{\varepsilon_1}{a} + \tan^{-1}\frac{\varepsilon_0}{a}\right)^{-1} \tag{5.83}$$

with

$$\varepsilon_0 = 2 - \frac{100}{\varepsilon + 10} \tag{5.84}$$

$$\varepsilon_1 = \frac{\varepsilon - E}{2} - \varepsilon_0 \tag{5.85}$$

$$a = 10.3 \tag{5.86}$$

where $\varepsilon = mv^2/2$ corresponds to the energy of the parent electron and the units of $\varepsilon, \varepsilon_0, \varepsilon_1$ and ε_p are eV. A random sample of ε_p obeying the probability distribution function in Eq. 5.83 can be calculated with the following expression,

$$\varepsilon_p = \varepsilon_0 + a\tan\left[U\left(\tan^{-1}\frac{\varepsilon_1}{a} + \tan^{-1}\frac{\varepsilon_0}{a}\right) - \tan^{-1}\frac{\varepsilon_0}{a}\right] \tag{5.87}$$

where $0 < U < 1$ represents a random number. Once the energy of the progeny electron has been sampled, the pre-collision velocity of the parent electron is updated as follows:

$$\widetilde{v} = v\left(1 - \frac{E + \varepsilon_p}{\varepsilon}\right)^{1/2} \tag{5.88}$$

The pre-collision velocity of the parent electron v is replaced by the reduced velocity $\widetilde{v}$ in Eq. 5.76 and the determination of the post-collision velocities v' and V' is performed in the same way as for the elastic electron-neutral collision case. The post-collision velocity of the progeny electron v'' is calculated by assuming a virtual collision between the progeny electron and the heavy neutral. Since the progeny electron has no actual pre-collision velocity $\widetilde{v}_p$, the velocity vector is assumed to have the same direction as that of the parent electron and the magnitude corresponding to the kinetic energy of the progeny electron. Hence,

$$\widetilde{v}_p = \frac{v}{v}\left(\sqrt{\frac{2\varepsilon_p}{m}}\right) \tag{5.89}$$

Finally, v is replaced by $\widetilde{v}_p$ in Eq. 5.76 and the post-collision velocity of the progeny electron v'_p is computed. In order for energy and momentum to be conserved with this approach, the neutral's post-collision velocity must be updated twice. The first velocity update corresponds to the collision of the neutral with the parent electron. The resulting post-collision velocity of the heavy atom is used as pre-collision value for the second update performed for the virtual collision with the progeny electron. If the parent electron and the neutral atom have the same numerical weight W, the neutral is deleted and an ion with the post-collision velocity of the heavy particle is created after the velocity updates are performed. The case of different particle weights is described below. In the present work, the threshold energy of ionisation for argon is defined as $E = 15.76$ eV.

In the developed solver, different weight values W are used for charge carriers (ions and electrons) and neutral particles. The reason behind this is the associated capability to use different numerical resolutions for the considered species, which is advantageous if large differences in the number densities of the species are expected. In *dsmcplasmaFoam*, when a neutral numerical particle of weight W' interacts with an electron of weight W and $W' \gg W$, the neutral particle is divided into two sub-particles, one with the same weight W as the electron and one of weight $(1-W)$. The neutral sub-particle of weight W collides with the electron and after the post-collision velocities are determined, both neutral sub-particles are merged at the end of the collision step. If the collision leads to ionisation, the sub-particle of weight W is deleted, the corresponding ion is created and the neutral sub-particle of weight $(1-W)$, not involved in the collision, remains unchanged throughout the interaction.

5.3.5.2 Implementation in OpenFOAM

The electron-neutral collisions model is implemented in the solver *dsmcPlasmaFoam* in the *collisions* function of the *evolve* method. The considered particles are ions, primary (parent) electrons and secondary (progeny) electrons. The particles properties are defined in the *dsmcProperties* dictionary. This dictionary also contains the threshold energy of ionisation and threshold energy of excitation for the neutral species. The collision cross sections for elastic collisions, composite excitation and ionisation, depicted in Fig. 5.12 as a function of the electron energy, are included in the dictionary *plasmaProperties* as look-up tables. The electron-neutral collision procedure is performed in a cell-wise manner after the Coulomb collisions step.

In order to calculate the number of potential collision candidates with the NTC method, the value $(\sigma_T c_r)_{max}$ must be estimated at the start of the simulation. In the developed model, this is done before the simulation as part of the *dsmcPlasmaInitialise* function. This function, which is executed before the actual solver, uses the initial macroscopic particles temperature defined in the *dsmcInitialiseDict* dictionary as well as the species mass to estimate the most probable particle velocity c from the corresponding Maxwellian velocity distribution function. The most probable electron energy is subsequently computed from this velocity as $mc^2/2$. The collision cross section is initialised using the most probable electron energy and the tabulated data for elastic electron-neutral collisions. The value $(\sigma_T c_r)_{max}$ is subsequently estimated as the product of the initialised cross section and the Maxwellian most probable speed. $(\sigma_T c_r)_{max}$ is stored for each cell and updated over the course of the simulation if a higher value is encountered.

After the number of potential collision candidates has been estimated with Eq. 5.67, particles pairs are randomly selected. As in the Coulomb collisions step, the particle selection is performed prioritizing interactions between electrons and neutrals contained in the same subcell. After two particles potentially involved in a collision are chosen, the following procedure is applied:

- Using one random number and the sampling approach depicted in Fig. 5.14, the collision type, if any, is obtained.
- If the collision is determined to be elastic, the post-collision velocities are calculated using Eqs. 5.76 and 5.77. These equations may be used regardless of possible differences in the particles weights. However, it should be noted that if the numerical weight of the neutral atom W' is higher than the weight of the electron W, the post-collision velocity of the neutral atom obtained from Eq. 5.77 applies only to the fraction W/W' of the test particle. This implies that only a fraction of the neutral DSMC test particle interacts with the electron. Since a DSMC particle represents a much larger number of real particles, the assumption that only a fraction of the neutral particle interacts with the electron does not introduce any physical inconsistencies. Let V'' represent the post-collision velocity obtained from Eq. 5.77 and valid for the fraction W/W' of the neutral particle. The post-collision velocity of the complete neutral test particle of weight W' can be defined as the combination of V'' and the pre-collision velocity V of the neutral's fraction which does not interact with the electron:

$$V' = \left(1 - \frac{W}{W'}\right)V + \frac{W}{W'}V'' \tag{5.90}$$

- If an excitation event takes place, the threshold energy of excitation is subtracted from the pre-collision electron energy and the same procedure as for the elastic collision case is followed.

- If the collision leads to an ionisation event, the energy of the progeny electron is first estimated with Eq. 5.87. This value and the threshold energy of ionisation are subtracted from the pre-collision energy of the parent electron and the post-collision velocities of the parent electron and heavy particles are obtained from Eqs. 5.76 and 5.77. Next, the energy of the progeny electron is used to estimate an imaginary pre-collision velocity and the second collision with the heavy particle is performed using again Eqs. 5.76 and 5.77. After the post-collision velocities are computed, an ion with the post-collision velocity of the heavy neutral and the numerical weight of the electron is created in the cell. If the colliding electron and neutral particles have the same numerical weight, the original neutral is deleted from the numerical grid. If the weight of the neutral atom is higher than that of the electron, the fraction $1-W/W'$ of the original neutral particle survives the ionisation process. This implies that this particle fraction does not interact with the electron during the ionisation process. Therefore, the post-collision velocity of the surviving neutral fraction corresponds to its pre-collision value and the neutral's numerical weight is reduced to compensate for the fraction of the test particle becoming a new ion.

The validation of the implemented model for the interactions between electrons and neutral atoms is presented in Section 6.5.

5.3.6 Recombination

5.3.6.1 Model description

In a plasma, recombination may occur through a high number of simultaneous reaction mechanisms. The dominating recombination mechanism is the result of several factors like the species present in the plasma, their number densities and the electron temperature. A comprehensive summary of different recombination mechanisms and their study is provided by Bates in [Bat75]. As described in [Bio17], recombination mechanisms may be categorized into two-body and three-body processes. One example of a two-body process is the *dissociative recombination* (DR) represented by the following general reaction:

$$XY^{+}+e^{-} \xrightarrow{\alpha} X^{*}+Y \tag{5.91}$$

In Eq. 5.91, X^{*} represents an electronic excited state and α, the recombination rate coefficient. During the dissociative recombination process, one free electron is captured by a diatomic ion leading to the creation of an excited diatomic molecule. This molecule may dissociate producing two neutral atoms, one of them being in an electronic excitation state. An example of a three-body process is the *collisional-radiative recombination* mechanism (CRR) schematically represented as follows:

$$X^+ + e^- + e^- \rightleftharpoons X^* + e^- \tag{5.92}$$

In this case, an additional electron facilitates the formation of the stable neutral reaction products by absorbing excess energy from the colliding electron. Hence, this mechanism involves the interaction between an ion and two electrons. The recombining electron enters the electronic cloud of the ion in a radiation-less manner. The energy released through this process, which equals the threshold energy of ionisation, is absorbed by the second electron leading to an increase of its kinetic energy. A comparable mechanism may employ a neutral atom as stabilizing third species if the number density of the neutrals is sufficiently high. The neutral stabilized mechanism can be schematically represented as follows:

$$X^+ + e^- + N \rightleftharpoons X^* + N \tag{5.93}$$

In the processes shown in Eqs. 5.91, 5.92 and 5.93, the energy level of the excited neutral product must decrease if the neutral is to remain stable. Otherwise, the captured electron can easily escape and the original electron-ion pair would reappear. One typical mechanism for the removal of the excess energy is radiation.

For two-body recombination processes, the recombination coefficient is given by,

$$\alpha = \langle \sigma_r c_e \rangle \tag{5.94}$$

where σ_r stands for the recombination cross section, c_e for the electron velocity and the brackets represent the average over the electrons velocity distribution. For three-body mechanisms, the effective recombination coefficient is given by,

$$\alpha_{eff} = K n_s \tag{5.95}$$

where K represents the three-body recombination coefficient and n_s, the number density of the stabilizing particle. The three-body coefficient for single charged ions and electron stabilized collisional-radiative recombination is defined in [Bio17] as follows:

$$K_{cr}(\mathrm{cm}^6/\mathrm{s}) \sim 10^{-19}[T_e(\mathrm{K})/300]^{-9/2} \tag{5.96}$$

For neutral stabilized recombination, the corresponding three-body coefficient is given by,

$$K_{ns}(\mathrm{cm}^6/\mathrm{s}) \sim 10^{-26}[T_e(\mathrm{K})/300]^{-9/2} \tag{5.97}$$

Note that $K_{cr} = 1 \times 10^{-19} [T_e(\mathrm{K})/300]^{-9/2}$ is employed in the study described in [Bog09].

The recombination algorithm implemented in *dsmcPlasmaFoam* includes the consideration of the electron stabilized collisional-radiative mechanism. Two-body dissociative recombination is not considered because the number densities of the diatomic argon species Ar_2 and Ar_2^+ are expected to be negligible in the INGA III arcjet thruster. Based on the *no time counter method* (NTC) described in Section 3.2.2, the number of potential collision candidates for an **electron stabilized three-body recombination** interaction is determined in *dsmcPlasmaFoam* as follows:

$$Candidates = \frac{1}{2} N_e N_i W (K_{cr,max}) n_e \frac{\Delta t}{V} \tag{5.98}$$

where N_e and N_i stand for the number of electrons and ions in the considered control volume and $K_{cr,max}$ represents the maximum value of the three-body recombination coefficient. The value $K_{cr,max}$ is estimated for each cell at the beginning of the simulation and updated if the value is exceeded over the course of the numerical run. Note that the developed solver uses the same numerical weight W for all charge carriers. Hence, no special considerations concerning W are necessary for the estimation of the number of potential collision candidates with Eq. 5.98. After the number of collision candidates has been determined, electron-electron-ion triples are randomly selected prioritizing particles located in the same sub-cell. Since the recombination probability decreases with electron velocity, the electron with the lower kinetic energy is defined as the recombining electron. The corresponding K_{cr} is calculated based on Eq. 5.96 as $K_{cr} = 1 \times 10^{-19} [T_e(\mathrm{K})/300]^{-9/2}$ and the three-body recombination interaction takes place with the probability:

$$P_{collision} = \frac{K_{cr}}{K_{cr,max}} \tag{5.99}$$

If recombination takes place, the simplest approach for the treatment of the excess energy transfer between the electrons is to assume a Coulomb collision between them prior to the actual recombination (cp. [Bio17]). This step is performed in *dsmcPlasmaFoam* using mainly the same Coulomb collision procedure outlined in Section 5.3.4. The only difference is the increase of the pre-collision kinetic energy of the stabilizing electron by an amount equal to the threshold energy of ionisation (15.76 eV for argon). The modified pre-collision velocity is therefore given by,

$$\widetilde{v} = v \left(1 + \frac{E}{\varepsilon}\right)^{1/2} \tag{5.100}$$

where $\varepsilon = mv^2/2$ stands for the original electron energy and E, for the threshold energy of ionisation. Note that the threshold energy is not to be subtracted from the kinetic energy of the recombining electron because E corresponds to the difference in energy levels between the ionised state and the excited neutral state.

After the energy transfer and Coulomb collision steps are performed, the electron and ion recombine to create a neutral. Strictly speaking, the position of the created neutral equals the position of the centre the mass of the two particles. However, since the mass of the ion is much larger than the mass of the electron, the original ion's position is used as the neutral's location in *dsmcPlasmaFoam*. Let v_e represent the pre-recombination velocity of the electron, V_i the pre-recombination velocity of the ion, V_n the post-recombination velocity of the created neutral, m the electron mass and M the mass of the heavy particle. Applying conservation of momentum, the post-recombination velocity of the created neutral can be computed as follows:

$$m_e v_e + MV_i = MV_n \rightarrow V_n = V_i + \frac{m}{M} v_e \tag{5.101}$$

where the mass of the ion and neutral particles is assumed to be the same. After the position and velocity of the new neutral are determined, the corresponding particle can be created and the electron-ion pair is deleted from the numerical grid and computer memory.

Recall that different numerical weights for charge carriers and initialised neutral particles are used in *dsmcPlasmaFoam*. In order to avoid an exponential increase in the amount of neutral particles of very low numerical weight, the developed solver includes an additional step in the algorithm that verifies whether additional neutrals are present in the cell during the recombination step. If additional neutrals are found, instead of creating a new neutral test particle, the algorithm increases the numerical weight of one randomly selected neutral particle in the cell. Note that nearby neutral atoms are prioritized in the selection process through the use of sub-cells. The position of the selected existing neutral particle is corrected so that it agrees with the centre of mass of the pair consisting of the new and the existing neutrals. In a similar way, the velocity is updated so that conservation of momentum is guaranteed. With this approach the amount of neutral numerical test particles is kept at roughly the same level throughout the simulation regardless of the amount of recombination events.

5.3.6.2 Implementation in OpenFOAM

The electron stabilized collisional-radiative recombination model is implemented in the solver *dsmcPlasmaFoam* in the new *recombination* function of the *evolve* method. The considered particles are ions, electrons and neutrals. The particles properties are stored in the *dsmcProperties* dictionary. This dictionary also contains the threshold energy of ionisation necessary for the implementation of the collisional radiative model. The factor for the recombination rate coefficient is stored in the *plasmaProperties* dictionary with a default value of 1×10^{-19} as in [Bog09]. The parameters required for the computation of the stabilizing electron-electron collision with Nanbu's Coulomb collisional model (s. Section 5.3.4 for details) are also stored in the *plasmaProperties* dictionary.

The electron-neutral collision procedure is performed in a cell-wise manner in *dsmcPlasmaFoam*. In order to calculate the number of potential collision candidates with the NTC method and Eq. 5.98, the value $K_{cr,max}$ must be estimated at the start of the simulation. In the developed model, this is done prior to running the actual simulation using the *dsmcPlasmaInitialise* function.

By examining Eqs. 5.96 and 5.97 it is obvious that the recombination coefficients increase with lower values of the electron temperature. Therefore, a reasonably value for the *minimum* electron energy must be assumed at the start of the simulation. In *dsmcPlasmaFoam*, the initial minimum electron energy is defined as $0.2E$, where E stands for the threshold energy of ionisation. Hence, the initial value for $K_{cr,max}$ is estimated in the *dsmcPlasmaInitialise* function as follows:

$$K_{cr,max}(\mathrm{cm}^6/\mathrm{s}) = 1 \times 10^{-19} [T_{e,min}(\mathrm{K})/300]^{-9/2} \tag{5.102}$$

with,

$$T_{e,min} = \frac{0.2E(\mathrm{J})}{k_B} \tag{5.103}$$

The value $K_{cr,max}$ is stored for each cell and updated over the course of the simulation if a higher value is encountered.

For a given cell of volume V, the number of potential collision candidates is first estimated with Eq. 5.98 and the current cell value $K_{cr,max}$. In the next step, electron-electron-ion triples are randomly selected in the cell. The particle selection is performed prioritizing interactions between particles contained in the same subcell. After the particles potentially involved in a recombination collision are chosen, the following procedure is applied:

- From the selected electrons, the one with the lower energy is assumed to be the recombining particle. The electron with the higher energy and hence, lower recombination probability, is defined as the stabilizing particle.
- The three-body coefficient for electron stabilized collisional-radiative recombination is calculated using Eq. 5.96. Here, the cell-wise or *swarm* electron temperature is used. The recombination interaction takes place with the probability:

$$P_{recombination} = \frac{K_{cr}}{K_{cr,max}} \tag{5.104}$$

- If recombination occurs, a Coulomb collision between the recombining and the stabilizing electron is performed next. To this end, the threshold energy of ionisation E is added to the energy of the stabilizing electron. Using the modified pre-collision velocity, a Coulomb collision between the stabilizing and the recombining electrons is computed using Nanbu's approach as described in Section 5.3.4.
- The velocity of the recombining electron is updated after the Coulomb collision. If the cell does not contain neutral particles, a new neutral is created and the ionising electron and ion are erased from the numerical grid. The position of the new neutral is assumed to be the same as the ion's position. Furthermore, the velocity of the new neutral is obtained from momentum conservation considerations as follows:

$$V' = \left(V + \frac{m}{M} v \right) \tag{5.105}$$

  where V' stands for the velocity of the new neutral, V for the velocity of the recombining ion and v for the velocity of the recombining electron.
- As already mentioned, *dsmcPlasmaFoam* verifies the presence of additional neutrals in the considered cell during the recombination step. If any are found, instead of creating a new test particle, the algorithm increases the numerical weight of one randomly selected neutral particle in the cell. The position of the selected existing neutral particle is corrected so that it agrees with the centre of mass of the pair consisting of the new and the existing neutrals. In addition, the velocity is updated so that conservation of momentum is guaranteed.

5.3.7 Boundary conditions in *dsmcPlasmaFoam*

A standard simulation with the solver *dsmcPlasmaFoam* employs a kinetically implemented, neutral background gas. The number of DSMC neutral particles in the numerical grid depends on the macroscopic number density and the numerical weight factor W of the gas. Furthermore, the particles velocities depend on the macroscopic initial temperature. Both the neutral number density and macroscopic temperature at the start of the simulation are defined in the *dsmcPlasmaInitialise* directory. The neutral background gas may be the only species present in the numerical grid at the start of a simulation with *dsmcPlasmaFoam*.

The electric current flowing through the ionisation chamber of an arcjet thruster like the INGA III device is implemented in *dsmcPlasmaFoam* as a kinetic electron flow from cathode to anode. The collisions between electrons and the background neutral gas trigger ionisation events as described in Section 5.3.5. This way, additional electrons and new ions are generated.

The electron generation at the cathode boundary surface is defined in the *dsmcPlasmaproperties* dictionary as the number of electrons per unit area and unit time to be added to the numerical domain at the cathode's surface. The number of electrons initialised at the cathode's surface can be set to agree with any desired electric current value between cathode and anode. For instance, a current of 30 A over a cathode area of 1×10^{-7} m^2 (typical values for the INGA III thruster) results in the following boundary condition in *dsmcPlasmaFoam*:

$$N_{e,Cath} = 30\text{A}\left(\frac{\text{C}}{\text{As}}\right)\left(\frac{1\ \text{electron}}{1.6 \times 10^{-19}\text{C}}\right)\left(\frac{1}{1 \times 10^{-7}\text{m}^2}\right) = 1.88 \times 10^{27}\frac{\text{electrons}}{\text{sm}^2} \tag{5.106}$$

Note that if a three-dimensional domain is simplified by using, for instance, a wedge geometry, the value for the electron generation boundary condition must be reduced by the corresponding area factor. The velocity of the new electrons is computed based on the fixed cathode wall temperature. Apart from electrons on the cathode wall, no further particles are generated via boundary conditions in *dsmcPlasmaFoam*. However, it should be noted that collisions of ions with metallic walls might lead to the production of secondary electrons at the metallic surface. This process is described in more detail later in this section.

Let us examine the interactions between particles in the numerical domain and the boundary surfaces. During their motion step, particles in the numerical grid might encounter one of the following boundaries in a typical *dsmcPlasmaFoam* simulation:

- *Cathode*: Where electrons are introduced into the numerical grid.
- *CathodeWall*: Surface wall adjacent to Cathode without electron generation.
- *Anode*: Opposite surface to cathode. Allows electrons to exit the numerical grid.
- *inletNeutral*: Symmetry plane in *dsmcPlasmaFoam*. Reserved for inlet boundary surface with neutral generation in future solver developments.
- *outletNeutral*: Symmetry plane in *dsmcPlasmaFoam*. Reserved for outlet boundary surface with neutral deletion in future solver developments.

The boundary surfaces in a typical arcjet thruster simulation with *dsmcPlasmaFoam* are depicted in Fig. 5.15. Here, the left image shows the convergent and throat section of the INGA III arcjet thruster with the red rectangles illustrating the ionisation ring section handled by *dsmcPlasmaFoam*. In addition, the right image provides details on the ring section and the *Anode* (blue), *Cathode* (red), *inletNeutral* (yellow) and *outletNeutral* (gray) boundary surfaces.

In the current version of *dsmcPlasmaFoam*, the neutral flow between the *inletNeutral* and *outletNeutral* boundaries is not considered because of the numerical challenges resulting from the highly different motion velocities of neutrals and electrons. Hence, the neutral gas is confined inside the numerical grid between the *Cathode*, *Anode*, *inletNeutral* and *outletNeutral* boundaries. To achieve this, standard symmetry or periodic boundary conditions may be applied to the *inletNeutral* and *outletNeutral* surfaces. The implementation of a macroscopic neutral flow between the *inletNeutral* and *outletNeutral* boundary surfaces is part of future solver developments.

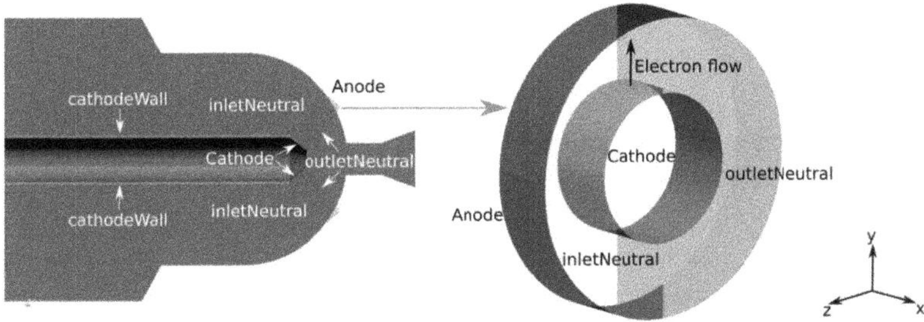

Figure 5.15: Boundary surfaces in a typical *dsmcPlasmaFoam* simulation. Left: Convergent and throat section of the INGA III arcjet thruster. Right: Detail on ring section and boundary surfaces.

The interaction types between the modelled particles and boundary surfaces are listed in Table 5.8 and can be summarized as follows:

- Electrons: The interaction between an electron and the *cathode* or *cathodeWall* boundaries is handled as a standard diffusive wall collision similar to DSMC applications as described in Section 3.2.3. With this approach, any electrons moving towards the cathode are reflected by the surface and sent back into the numerical domain. This way, the electric current boundary condition at the cathode's surface is fulfilled. On the *inletNeutral* and *outletNeutral* surfaces, symmetry boundary conditions are typically implemented leading to specular reflection of electrons at the surfaces. Considering this, electrons can exit the numerical domain only

	Cathode	CathodeWall	Anode	inletNeutral	outletNeutral
Electron	Wall	Wall	Anode BC	Symmetry	Symmetry
Ion	Neutralize $+e^{-}Yield$	Neutralize $+e^{-}Yield$	Neutralize $+e^{-}Yield$	Symmetry	Symmetry
Neutral	Wall	Wall	Wall	Symmetry	Symmetry

Table 5.8: Overview of interactions between particles and boundary surfaces in *dsmcPlasmaFoam*.

through the anode and a net electric current between cathode and anode arises. *dsmcPlasmaFoam* includes two different approaches for the treatment of electrons reaching the anode. With the simplest approach, the solver uses the value P_{ref} which describes the probability of electrons being reflected at the anode. P_{ref} is defined by the user in the *dsmcPlasmaProperties* dictionary and compared with a random number every time an electron reaches the anode in order to determine whether the electron is reflected back or deleted from the numerical domain. Note that $P_{ref} = 0$ effectively allows all electrons to exit the domain. The second approach implemented in *dsmcPlasmaFoam* employs a specific electric current value at the anode defined by the user in the *dsmcPlasmaProperties* dictionary. This current is used by the solver to dynamically determine the exact number of electrons that are allowed to exit the grid through the anode. The electrons leaving the domain are randomly selected and subsequently deleted. With this approach, it is possible to reach an agreement between the electric current into and out of the numerical domain, effectively replicating the work performed by the regulating power supply unit of the INGA III arcjet thruster.

- Ions: Similar to the electrons case, symmetry boundary conditions are implemented at the *inletNeutral* and *outletNeutral* boundaries leading to specular reflection of ions at these surfaces. Ions encountering the metallic surfaces *Cathode*, *CathodeWall* or *Anode* are neutralized and in the process, secondary electrons are emitted. Details on the neutralization and electron emission mechanisms can be found in [LL05]. Ion neutralization in *dsmcPlasmaFoam* is performed by a simple change of particle type. Furthermore, an ion's collision with a boundary surface is handled as a standard diffusive wall collision as described in Section 3.2.3. For secondary electron emission driven by ions interactions with metallic walls, electron yield factors describing the number of electrons created by one ion-metal collision are defined in the *dsmcPlasmaProperties* dictionary for the

Cathode, *CathodeWall* and for the *Anode* surfaces. In *dsmcPlasmaFoam* these values are effectively used as electron generation probabilities. If the ion-metal collision leads to the emission of one electron, the position of the new particle equals the location of the ion-wall collision. In addition, the electron's velocity is computed based on the wall's temperature. As described in [PP99], the electron yield per ion collision is a function of several factors like the energy of the incident ion, the specific metallic material and the surface's roughness and cleanness and its value lies between 0.01 and 0.8. For moderate energy values of the incident ions below 1,000 eV, the electron yield per ion collision is only weakly influenced by the ion's energy and can be assumed to be constant.

- Neutrals: The interaction between neutrals and the *cathode*, *cathodeWall* and *anode* boundaries corresponds to a standard diffusive wall collision as used in DSMC applications and described in Section 3.2.3. On the *inletNeutral* and *outletNeutral* surfaces, where symmetry boundary conditions are implemented, specular reflection takes place.

The interactions between particles and boundary surfaces described in this section are implemented in the file *particleTemplates.C* of *dsmcPlasmaFoam*.

5.3.8 Numerical aspects

5.3.8.1 Dynamic weighting

Because of the potential high differences in number densities, *dsmcPlasmaFoam* employs different weight factors W for neutral particles and charge carriers. Regarding charge carriers, only the primary electrons generated at the cathode are present at the start of a typical simulation with *dsmcPlasmaFoam*. At this point, a very low weight factor W is necessary in order for an appropriate number of electrons to be generated each time step. As ionisation collisions take place, secondary electrons and ions appear in the numerical domain. The new electrons are, in turn, accelerated towards the anode and collide with further neutrals. Further ionisation events lead to the appearance of more electrons and ions. The process concludes in an electron avalanche as expected in a typical electric discharge and the total number of charge carriers in the numerical domain increases exponentially.

The resulting exponential increase in computational requirements is addressed in *dsmcPlasmaFoam* through the implementation of a dynamic weighting approach for the charge carriers. Besides the initial weight factor W, the parameters *nElectronMax* and *reductionFactor* are defined by the user before the start of the simulation in the *dsmc-*

Plasmaproperties dictionary. The default value for the maximum number of electrons *nElectronMax* is 60,000 while the reduction factor is set to 0.8.

At the start of the *evolve* function, *dsmcPlasmaFoam* compares the total number of electrons in the numerical domain with the value *nElectronMax*. If the maximum number of electrons is exceeded, the function *reduceParticles* is called. The first step in this function involves the calculation of a new weight factor as follows:

$$W_{new} = W_{old} \frac{N_{elec}}{reductionFactor \cdot N_{elec,max}} \tag{5.107}$$

Next, the number of particles to be deleted is determined for both electrons and ions. Note that the number of electrons and ions to be deleted is calculated so that the weight factor W_{new} after the particles reduction is the same for both species:

$$N_{elec,del} = N_{elec} - reductionFactor \cdot N_{elec,max} \tag{5.108}$$

$$N_{ion,del} = N_{ion} - W_{old} \frac{N_{ion}}{W_{new}} \tag{5.109}$$

The next step involves the random selection of particles to be deleted from the numerical domain. Note that it is not possible to address individual particles in *dsmcPlasmaFoam* unless they are allocated to a specific cell. To perform the particles deletion step in a cell-wise environment is, however, extremely inefficient. For this reason, the *reduceParticles* function relies on the creation of a list of deletion candidates labels in a random manner. The entire particles cloud is then called by *reduceParticles* and the function counts through the particles cloud deleting a particle each time the internal particle counter reaches a value contained in the random deletion list. After the deletion process is finalized, the numerical weight of the remaining particles is updated and the simulation continues. An example of the dynamic weighting functionality is shown in Fig. 5.16.

Note that parallelization is supported by both *OpenFOAM®* and *dsmcPlasmaFoam*. For this reason, additional considerations are necessary for the correct implementation of the deletion candidates selection process. Specifically, the selection is performed independently by each core involved in the computation. Therefore, the *reduceParticles* function must take into account the number of particles in its own processor as well as in all other processors in order to guarantee a homogeneous selection over all computational cores. The reader is referred to the source code in the file *Cloud.C* of *dsmcPlasmaFoam* for details on the candidates selection process for parallel applications.

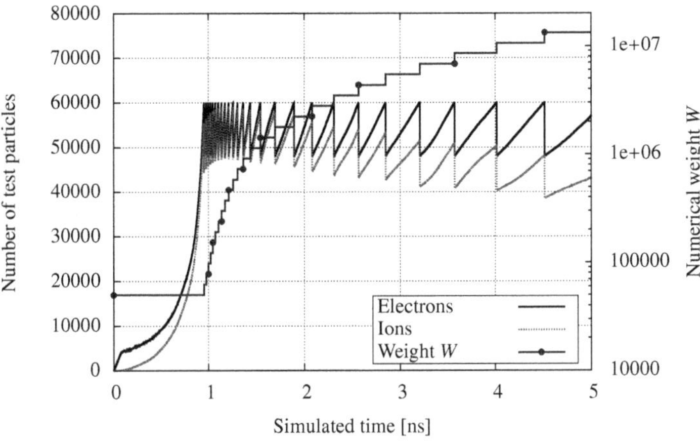

Figure 5.16: Dynamic weighting functionality on *dsmcPlasmaFoam*.

5.3.8.2 Neutral particles merging

During the recombination process described in Section 5.3.6, a new neutral particle with the same numerical weight W_{plasma} of the parents electron and ion is created. As already described in this chapter, the weights of charge carriers W_{plasma} and original neutrals not created through ionisation W_{neut} are usually different from each other in order to achieve an adequate numerical resolution with acceptable computational requirements. Hence, in a typical simulation, $W_{neut} \gg W_{plasma}$.

During the course of a simulation, a high number of ionisation events might take place. At the same time, the resulting higher number density of charge carriers will increase the probability of recombination events also occurring and therefore, a high number of low weight neutral particles might accumulate in the numerical domain. In order to avoid an exponential increase in the number of neutral test particles, a neutral merging function is implemented in *dsmcPlasmaFoam*. The function, called *mergeNeutrals*, is executed as part of the *evolve* method in *dsmcPlasmaFoam*.

In the *mergeNeutrals* procedure, the numerical cells are scanned for neutrals of standard (high) weight W_{neut} and of low numerical weight W_{plasma}. If a low weight neutral is found in the cell and the cell contains at least one additional neutral particle, both neutrals are merged together. If more than one additional neutral is available, the merging candidate selection prioritizes particles in close proximity by applying a sub-cell based procedure. Let P represent the low weight neutral particle and Q, its merging partner. The position of the merged neutral equals the position of the centre of mass of the particles P and Q. Furthermore, the velocity of the merged neutral V'_Q is obtained by applying conservation of momentum as follows:

$$W_P v_P + W_Q V_Q = (W_P + W_Q) V'_Q \rightarrow V'_Q = \frac{W_P v_P + W_Q V_Q}{(W_P + W_Q)} \tag{5.110}$$

where W_P and W_Q represent the weight of the particles P and Q and v_P and V_Q, their pre-merging velocities. After the position and velocity of the merged neutral are determined, the values are applied to the neutral particle Q, its weight is increased to $W_P + W_Q$ and the low weight particle Q is erased from computational memory.

5.3.9 Global model implementation in OpenFOAM

The complete algorithm implemented in *dsmcPlasmaFoam* is shown schematically in Fig. 5.17. A short description of the steps followed during the execution of *dsmcPlasmaFoam* as well as the corresponding main source files are provided in Table 5.9.

5.4 Summary

Electric propulsion systems for spacecraft applications are often designed based on empirical models and experimental data. The reasons behind this include the high complexity of plasma phenomena as well as the high computational requirements associated with numerical approaches for plasma modelling. With this in mind, a kinetic PIC-MCC solver, referred to as *dsmcPlasmaFoam*, has been developed in the frame of the present work. The described numerical model is a first step in an attempt to develop a hybrid PIC-MCC-FVM solver at ZARM aiming to tackle the numerous challenges concerning plasma modelling for electric spacecraft propulsion systems. Besides the obvious benefits in terms of thruster design and optimization, the development of an accurate hybrid numerical model has the potential to provide new insights into plasma phenomena, energy exchange and transport mechanisms not only in electric propulsion devices but also in further plasma-related applications like plasma arc welding and electron beam processing.

The described model is based on kinetic numerical methods and includes, among others, classic PIC elements, short-range Coulomb charge collisions as well as ionisation and recombination interactions. Additional features like dynamic weighting have been conceived and implemented in the solver over the course of its development which might prove beneficial in future numerical studies concerning not only plasma phenomena, but any research field heavily relying on kinetic particle models. The solver *dsmcPlasma-Foam* is expected to serve as a basis for the development of a hybrid kinetic-fluid model at ZARM able to accurately describe not only the INGA III thruster in hot-gas operation mode, but also the main electric propulsion concepts currently in service.

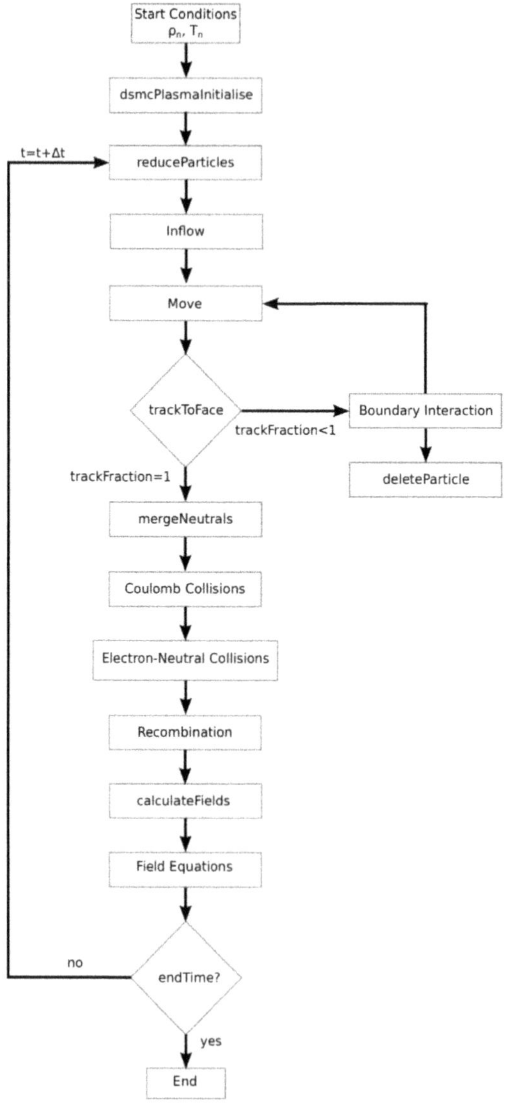

Figure 5.17: Schematic representation of the *dsmcPlasmaFoam* algorithm.

Step	Main files	Description
Initialisation	*dsmcInitialiseDict*, *DsmcCloud.C*	Generation of initial test particles in numerical domain using macroscopic values for T, n and the macroscopic fluid velocity u (s. Sections 5.3.2 and 5.3.5.2).
reduceParticles	*plasmaProperties*, *Cloud.C*	Random deletion of a predetermined number of electrons and ions in the numerical domain in order to avoid exponential particles increase linked to ionisation events. The procedure is associated with a change of the particles numerical weight (s. Section 5.3.8.1).
Inflow	*dsmcProperties*	Particle generation at numerical open boundaries. The *Inflow* function is also used for creation of electrons at the cathode surface in *dsmcPlasmaFoam* (s. Sections 5.3.2 and 5.3.7).
Move	*DsmcParcel.C*	Particles motion step. For charge carriers, the PIC force weighting step and the PIC Lorentz solver are implemented and executed at this point (s. Sections 3.2.1, 5.3.3.2 and 5.3.3.3).
Boundary Interactions	*particleTemplates.C*, *plasmaProperties*	During the motion step, particles interactions with boundary surfaces are evaluated. If a particle encounters a boundary, the particle's properties after the interaction as well as potential electron generation for ion-wall collisions are determined in this step. (s. Sections 3.2.3 and 5.3.7).
mergeNeutrals	*DsmcCloud.C*	After recombination, low weight neutrals appear in the numerical domain. In the *mergeNeutrals* step, the solver scans the numerical cells for such particles and if present, merges them with other neutrals in order to maintain a low number of test particles (s. Sections 5.3.6.2 and 5.3.8.2).
Coulomb Collisions	*DsmcCloud.C*, *plasmaProperties*	Coulomb interactions between charge carriers are computed in this step based on the theory of cumulative small-angle collisions developed by Kenichi Nanbu (s. Section 5.3.4).
Electron-Neutral collisions	*DsmcCloud.C*, *plasmaProperties*	Collisions between electrons and neutral background particles. The possible outcomes are elastic collision, electronic excitation and ionisation of the neutral collision partner (s. Section 5.3.5).
Recombination	*DsmcCloud.C*, *plasmaProperties*	In this step, electron stabilized collisional-radiative three-body recombination is evaluated. (s. Section 5.3.6).
calculateFields	*DsmcCloud.C*	The fields necessary for the determination of macroscopic properties for a numerical iteration are computed in this step. The function includes the PIC particle weighting step for determination of the number density from the charge carriers distribution (s. Sections 5.3.2 and 5.3.3.2).
Field Equations	*dsmcPlasmaFoam.C*	In this step, the PIC Maxwell solver, in which the electric field E is obtained from the total charge density, is executed (s. Section 5.3.3.1).

Table 5.9: Overview of main functions, source files and dictionaries in the *dsmcPlasmaFoam* solver.

Chapter 6

Validation of dsmcPlasmaFoam

In this chapter, the main components of the solver *dsmcPlasmaFoam* are analysed and validated. They include the Maxwell solver for the determination of the electric field from a particle distribution, the Lorentz solver describing the charge carrier's motion in an electric field as well as the weighting steps for the computation of the particle density from the particles distribution and the force acting on the charge carriers. Furthermore, the validation of the Coulomb interactions model as well as the electron-neutral collisional approach are presented.

6.1 Maxwell solver

The Maxwell solver constitutes one of the main elements in the classic PIC algorithm and is described in detail in Section 5.3.3.1. In *dsmcPlasmaFoam*, magnetic effects are neglected and only Gauss's law for electric fields requires consideration. Therefore, the equation solved by the Maxwell solver in *dsmcPlasmaFoam* corresponds to Poisson's equation:

$$\nabla^2 \phi = -\frac{\rho}{\varepsilon_0} \tag{6.1}$$

with $\rho = q_e n_e + q_i n_i$ being the total charge density. The solver validation is based on the approach presented in [Sti15]. To this end, numerical results for the electric potential ϕ and electric field E obtained from *dsmcPlasmaFoam* are compared with analytic values for selected test scenarios.

In the first considered case, the elementary charge $q = -1.6 \times 10^{-19}$ C is distributed across the central region of a cubic channel geometry of volume $V = L^3$, with $L = 4$ mm

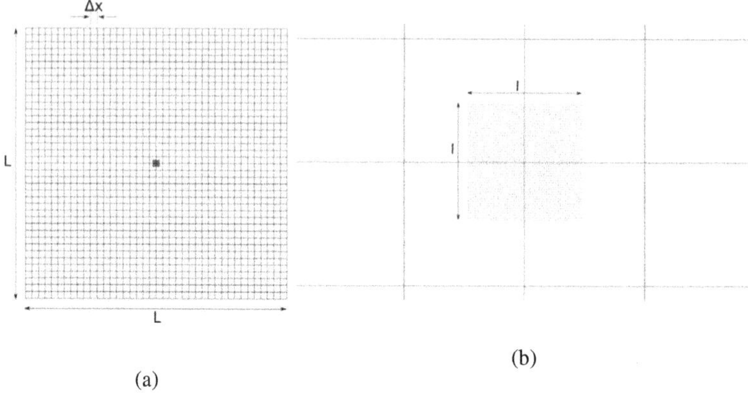

Figure 6.1: Numerical setup for Maxwell solver validation. (a) Channel geometry with a cell resolution of 40 cells per Cartesian coordinate, $\Delta x = 0.1$ mm; (b) Detail on charge distribution in the central mesh region with charges shown as blue dots.

as performed in [Sti15] and shown in Fig. 6.1. Note that the origin of the coordinate system $r_0 = \{0,0,0\}$ is located at the centre of the channel geometry. The elementary charge is distributed homogeneously in the centre of the geometry as an arrangement of $N = 50 \times 50 \times 50$ test particles inside the volume l^3, with $l = 0.1$ mm. Hence, the charge value q/N is assigned to each sub-particle. For the numerical tests, the particles placement is performed using a custom version of the *dsmcPlasmaInitialise* function in *dsmcPlasmaFoam*. In addition, the six boundary surfaces of the numerical channel are treated as standard open surfaces with a fixed value for the electric potential $\phi = 0$ V. The analytic results for ϕ are calculated as the sum of the individual particle contributions to the electric potential over the total number of particles N as follows:

$$\phi = \sum_{i=1}^{N} \frac{1}{4\pi\varepsilon_0} \frac{q_i}{|r - r_i|} \tag{6.2}$$

where ε_0 stands for the vacuum permittivity, r_i for the position of the particle i and r represents the location where the potential is calculated. In a similar way, the component E_x of the electric field E at the location $r = \{x, y, z\}$ is analytically obtained using the expression:

$$E_x = \sum_{i=1}^{N} = \frac{q_i}{4\pi\varepsilon_0} \frac{x - x_i}{|r - r_i|^3} \tag{6.3}$$

The numerical simulations on the channel geometry are performed with different mesh resolutions as listed in Table 6.1. Here, the test cases are named according to the number of cells employed in each Cartesian coordinate. For cases with mesh grading, the cell resolution increases gradually towards the centre of the geometry where the test particles are located. With this approach, an increase in the resolution in the critical regions is achieved without impacting the computational requirements of the computation. Specifically, the 400Grd1 test case employs a two-fold increase in the number of cells in the geometry's central region while the 400Grd2, a four-fold increase.

Test Case	Resolution	Total Cells	Grading
40	$40 \times 40 \times 40$	6.4×10^4	n/a
80	$80 \times 80 \times 80$	5.12×10^5	n/a
160	$160 \times 160 \times 160$	4.096×10^6	n/a
200	$200 \times 200 \times 200$	8×10^6	n/a
400	$400 \times 400 \times 400$	6.4×10^7	n/a
400Grd1	$400 \times 400 \times 400$	6.4×10^7	$2\times$
400Grd2	$400 \times 400 \times 400$	6.4×10^7	$4\times$

Table 6.1: Numerical test cases for Maxwell validation in channel geometry.

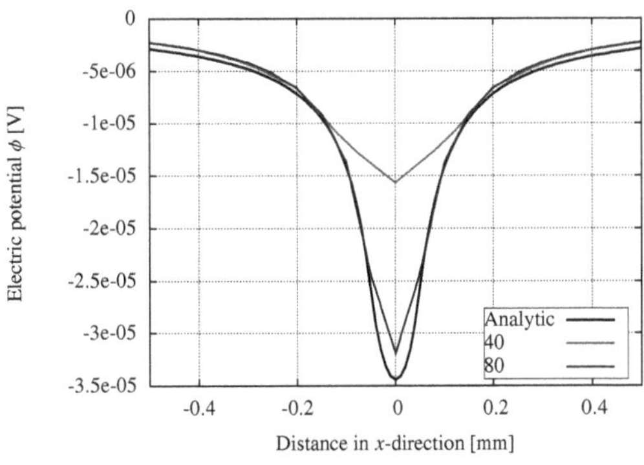

Figure 6.2: Analytic and numerical results for the electric potential ϕ along the x axis $r = \{x, 0, 0\}$. Numerical results are depicted for the test cases 40 and 80.

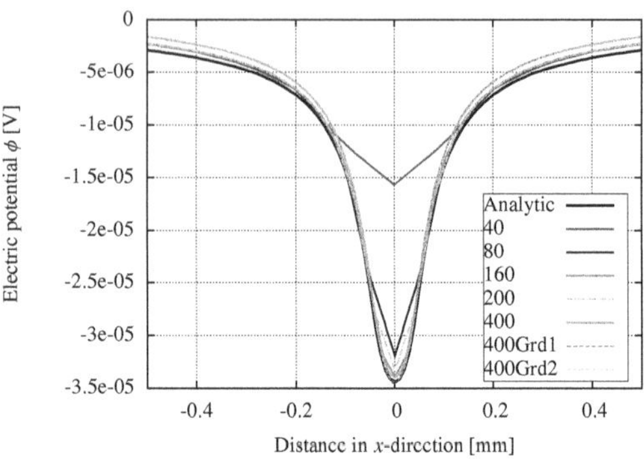

Figure 6.3: Analytic and numerical results for the electric potential ϕ along the x axis $r = \{x, 0, 0\}$. Numerical results for all considered test cases.

Figure 6.2 shows the analytic and numerical results for the electric potential ϕ along the x axis. Here, only the numerical results for the test cases with 40 and 80 cells per axis are depicted. As can be seen, the grid resolution plays a decisive role on the quality of the results. The low number of cells in the central region of the mesh makes it impossible for the solver to adequately resolve the high gradients resulting from the concentrated location of charged particles in an otherwise vacuum environment. The clear quality improvement with test case 80 highlights the importance of an adequate grid resolution. The results for all tests cases listed in Table 6.1 are shown in Fig. 6.3. As can be seen, a resolution of 400 cells per axis produces the best results if no grading is used. However, the results of the test cases 160 and 200 can be considered to be acceptable while having considerably lower computational requirements. The overall best results are obtained with the grading test cases which agree extremely well with the analytic profiles.

The analytic and numerical results for the components E_x and E_y of the electric field E are shown in Fig. 6.4. As for the electric potential, the results for E_x obtained with 400 cells per Cartesian coordinate and using mesh grading agree extremely well with the analytic results. Furthermore, the results with 400 cells per axis and no grading show a very good agreement with the analytic data while the values obtained with a resolution of 160 and 200 cells per coordinate can be considered to be acceptable. Because of the symmetry of the test particle distribution and the location of its centre of mass at the origin of the coordinate system, the perpendicular component of the electric field E_y along the x axis equals zero. Note that the same applies for the component E_z. In summary, the results produced by the Maxwell solver in *dsmcPlasmaFoam* show a very

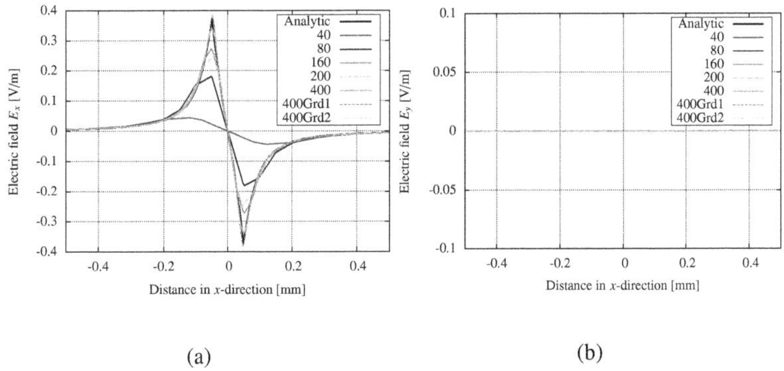

(a) (b)

Figure 6.4: Analytic and numerical results for the components E_x (a) and E_y (b) of the electric field E along the x axis $r = \{x, 0, 0\}$.

good agreement with the analytic data. This is particularly the case if a high mesh resolution is used, either through a high number of cells in the domain or by applying a local resolution increase via mesh grading. Note that for a future simulation campaign of the INGA III thruster, the identification of regions with high gradients of the electric potential inside the ionisation chamber is not a straightforward task because ionisation can, in theory, take place anywhere in the numerical domain. As a consequence, mesh regions with high gradients of charge number density are difficult to predict and the use of a high and homogeneous numerical resolution is expected to be necessary.

Next, the behaviour of the Maxwell solver is examined for a cylindrical mesh more closely related to the INGA III ionisation chamber geometry. To this end, the Maxwell solver of *dsmcPlasmaFoam* is used to compute the electric potential and electric field produced by a charge distribution placed inside a cylinder of 8 mm in length and 4 mm in radius, dimensions that are representative of the ionisation chamber in the INGA III thruster. For the validation in the cylindrical geometry, charge carriers are homogeneously distributed in a ring shape with 20 test particles in both the x and radial directions and a total of 1,440 particles in the angular coordinate. Hence, the elementary charge 1.6×10^{-19} C is distributed among a total of 576,000 particles. In the x coordinate, the particles are placed in the range $x = 0 \pm 0.05$ mm. In the radial direction, the range for the particles positions is $r = 3 \pm 0.05$ mm while in angular direction, the particles are distributed homogeneously between 0 and 2π. As in the channel geometry, the analytic results for ϕ and E are obtained using Eqs. 6.2 and 6.3. The employed cylindrical geometry is shown in Fig. 6.5.

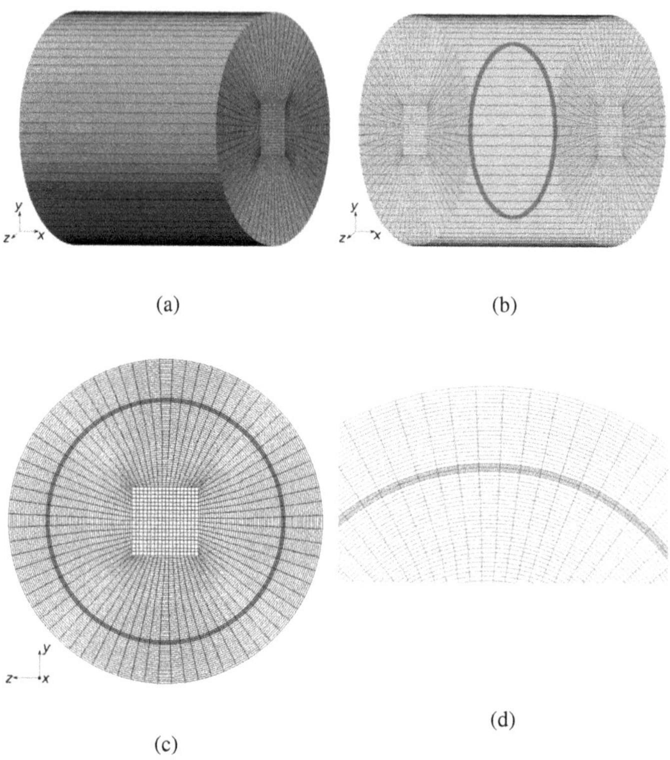

Figure 6.5: Cylindrical geometry for validation of Maxwell solver. (a) Surface view; (b) and (c) Internal view with test particles shown as red dots; (d) Detailed view of test particles.

Regarding the boundary conditions for the cylindrical tests, the flat surfaces at $x = -4$ mm and $x = 4$ mm are modelled using the fixed boundary value $\phi = 0$ V. Note that pre-tests showed the 4 mm distance to be large enough for the fixed value $\phi = 0$ V not to influence the results in the critical central regions of the cylinder. The curved boundary surface, which technically corresponds to the anode of the thruster, requires the definition of a fixed value for ϕ in order for the Maxwell solver to accurately solve Poisson's equation. To this end, the electric potential ϕ at the location $r = 4$ mm is calculated using Eq. 6.2 and the result $\phi_{Anode} = -4.3794 \times 10^{-7}$ V is used as fixed boundary condition. As for the channel geometry, simulations are performed using different mesh resolutions. Specifically, a total of 80 cells in x direction and 72 cells in angular direction are used for all test cases. Furthermore, the number of cells in radial direction varies according to the values given in Table 6.2. Here, the radial resolution corresponds to the number of cells in the radial direction implemented in the critical mesh region 2 mm$< r <$ 4 mm. Note that pre-tests showed the use of numerical grading factors to be unnecessary for the cylindrical geometry.

Test Case	Radial Resolution	Total Cells
40	40	7.43×10^5
80	80	1.43×10^6
160	160	2.816×10^6
200	200	3.508×10^6
400	400	6.964×10^6

Table 6.2: Numerical test cases for Maxwell validation in cylindrical geometry.

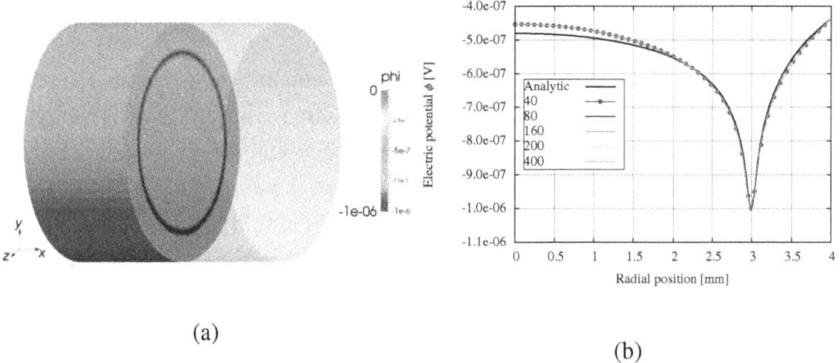

(a)

(b)

Figure 6.6: Results for the electric potential ϕ. (a) Potential field for test case 40; (b) Analytic and numerical results for ϕ as a function of radial position r with $x = y = 0$.

Figure 6.6 (a) shows a slice of the cylindrical geometry at the position $x = 0$ where the centre of the particles distribution lies. As can be seen, a ring-shaped area of homoge-

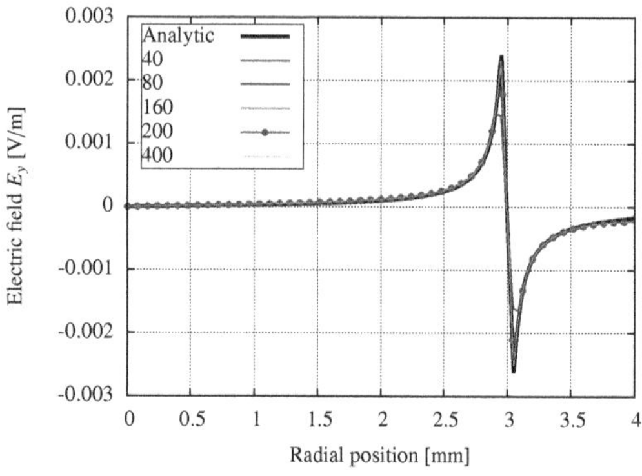

Figure 6.7: Results for the component E_y of the electric field E as a function of radial position r with $x = y = 0$.

neously low electric potential $\phi = -1 \times 10^{-6}$ V appears. The numerical and analytic results for the electric potential ϕ as a function of the radial position and with $x = y = 0$ are shown in Fig. 6.6 (b). Here, the numerical results show a very good agreement with the analytic values in the electron-rich region $r \approx 3$ mm, while the solver slightly overpredicts ϕ in the region $r \approx 0$ mm. The results for the electric field E as a function of the radial position are shown in Fig. 6.7. As can be seen, the simulations with a lower resolution (40 and 80 cells in the mesh region 2 mm$< r < 4$ mm) exhibit difficulties reproducing the electric field peaks near the electron-rich region. However, the simulations with higher resolution do not have the same issues as they agree extremely well with the analytic profiles. In summary, the Maxwell solver produces very good results for the studied cylindrical geometry. However, the quality of the results is highly influenced by the accuracy of the boundary conditions employed. The results presented in Figs. 6.6 and 6.7 are only achievable if the cylinder dimensions in the x coordinate are large enough for the boundary condition $\phi = 0$ V to be applicable at the flat boundary surfaces. At the same time, the determination of the anode boundary condition $\phi_{Anode} = -4.3794 \times 10^{-7}$ V and its implementation in the numerical simulations are essential for achieving accurate results.

In order to verify the solver behaviour in a simplified geometry with rotational symmetric boundary conditions, additional simulations with the same configuration and the same fixed electric potential values as in the cylindrical geometry are performed in a wedge mesh compressing an angular section of 5° of the original cylinder. For this particular analysis, two radial resolutions (40 and 400) are considered. The two new boundary surfaces of the wedge geometry are modelled with the *OpenFOAM*® boundary condition "*symmetryPlane*". The results are depicted in Fig. 6.8 for both ϕ and E. As can be seen, the wedge approach with *symmetryPlane* boundary conditions produces results of the same high quality as with a full three-dimensional cylindrical mesh, while at the same time, strongly reducing the computational requirements of the simulations.

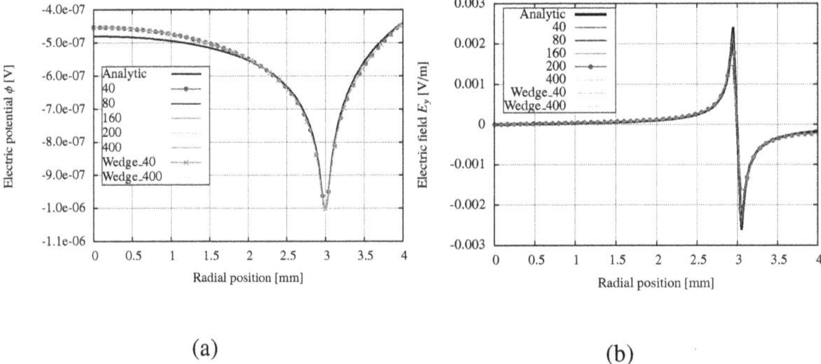

(a) (b)

Figure 6.8: Analytic and numerical results for the electric potential (a) and electric field component E_y (b) for the cylindrical and simplified wedge geometry as a function of radial position.

Finally, the behaviour of the Maxwell solver is verified in the ring-shaped geometry shown in Fig. 6.9 which includes cathode and anode boundary surfaces and strongly resembles the ring-shaped ionisation chamber of the INGA III thruster. As for the cylindrical geometry, the ring-shaped numerical mesh has a total length of 8 mm in the longitudinal x coordinate. The internal radius of the ring is defined as 2 mm and the external, as 4 mm. The test particles are distributed in the same manner as for the cylindrical geometry. Furthermore, the fixed value boundary condition for electric potential $\phi = 0$ V is implemented in the flat surfaces at $x = -4$ mm and $x = 4$ mm. The electric potential at the external curved boundary surface (anode) is defined as $\phi_{Anode} = -4.3794 \times 10^{-7}$ V while the potential at the internal curved surface (cathode) is set as $\phi_{Cathode} = -5.5302 \times 10^{-7}$ V. Both values are obtained analytically from Eq. 6.2. As for the cylindrical mesh, several test cases with different radial resolutions ranging from 40 to 400 cells are considered.

The analytic and numerical results for the electric potential ϕ and the electric field component E_y obtained with the ring-shaped geometry are shown in Fig. 6.10. As can be seen, the results for ϕ are not strongly influenced by the radial mesh resolution and all

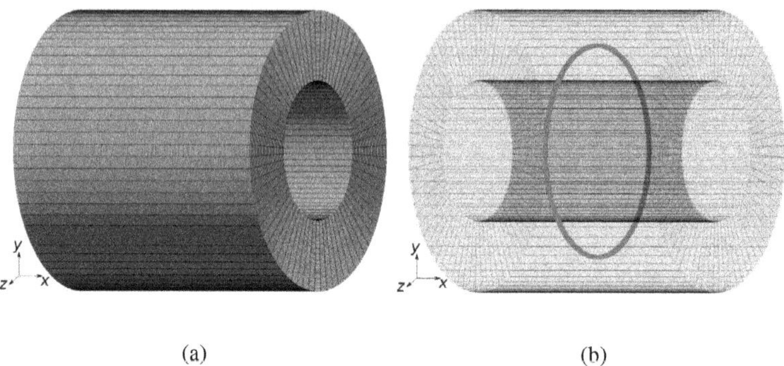

(a) (b)

Figure 6.9: Ring-shaped geometry for Maxwell validation. (a) Surface view; (b) Internal view with test particles shown in red.

considered tests produce nearly the same results. On the other hand, the electric field results with the coarsest mesh have difficulties reproducing the E_y peaks in the regions near the particles cloud. However, fine mesh configurations with at least 160 cells produce high quality results in this region. As evident from Fig. 6.10 (a), the numerical results for ϕ consistently lie under the analytically obtained values. This deviation can be explained by a fundamental difference between the analytic and numerical approach. In the numerical simulations, the solid region $r < 2$ mm inside the cathode is not modelled in any way. This limitation effectively transforms the solid cathode into a region free of any electric potential or field. Hence, the fields generated by the particles distribution cannot penetrate the cathode boundary surface and they are, as a consequence, confined to the discretized mesh region. On the other hand, the analytic approach does not include any particular treatment for the region $r < 2$ mm and hence, allows the ϕ produced by the test particles to be present in this zone and to be distributed in a larger volume. As a consequence, the numerical electric potential field is stronger than the analytically obtained one.

Note that in reality, the region inside the cathode is not isolated from the electric fields produced by charge carriers inside the ionisation chamber. Considering this, improved results are to be expected if the solid cathode region $r < 2$ mm is also discretized and modelled accordingly. However, because of the considerable differences in the phenomena taking place in the ionisation chamber and solid region and the resulting high differences of time scales in the fluid and solid phases, this approach is not straightforward and hence, not considered in the present work. Nevertheless, the results shown in Fig. 6.10 are in good agreement with the analytic values and the Maxwell solver is con-

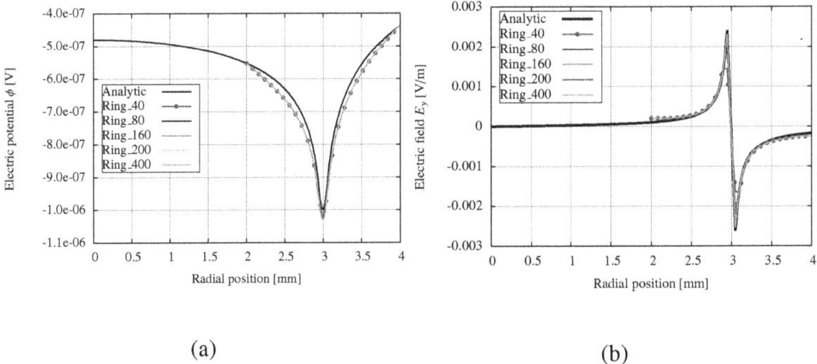

(a) (b)

Figure 6.10: Analytic and numerical results for the electric potential (a) and electric field component E_y (b) for the ring-shaped geometry as a function of radial position.

sidered to be validated for all geometries relevant for the present work. This is also the case if the three-dimensional ring geometry is replaced by a simplified wedge mesh as demonstrated by the high quality results with a wedge approach depicted in Fig. 6.11.

6.2 Lorentz solver

The Lorentz solver, described in Section 5.3.3.3, constitutes one of the main steps in the classic PIC algorithm and is responsible for the modelling of the charge carriers motion inside electric fields in *dsmcPlasmaFoam*. In order to examine its behaviour, the motion of one electron inside an electric field generated by a single ion is analysed in this section. The validation approach is based on the strategy described in [Sti15]. Let us imagine an ion of charge Q fixed at the position $R = \{0,0,0\}$ at the centre of a channel geometry of dimensions 4 mm $\times$4 mm $\times$0.07 mm, where the smallest dimension corresponds to the z coordinate. In this setup, an electron with initial velocity $v = \{6.2 \times 10^6, 0, 0\}$ m/s is placed 1 mm away from the ion at the position $r = \{0, 0.001, 0\}$. In order for the electron to move around the ion in a perfect circular orbit at a distance $r = 1$ mm, an specific centripetal force must be exerted by the ion and its electric field on the electron. Note that at the start of the electron's motion, the direction of the electric field is in the positive y direction, pointing from the ion's position at the origin of the coordinate system towards the electron. With this information, the ion charge Q, for which the electron will move in a perfect circular orbit, can be calculated based on the assumption that the magnitudes of the Coulomb force F and the centripetal

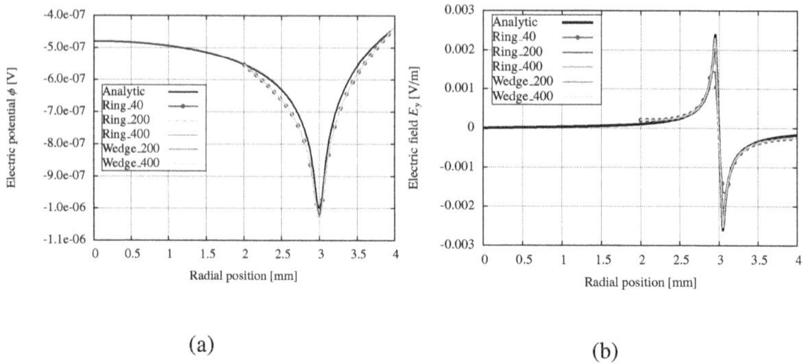

(a)

(b)

Figure 6.11: Analytic and numerical results for the electric potential (a) and electric field component E_y (b) for the three-dimensional ring-shaped and the simplified wedge geometry as a function of radial position.

force F_c are equal. Hence,

$$
\begin{aligned}
F = qE &= m\frac{v^2}{r} = F_c \\
&= q\frac{Q}{4\pi\varepsilon_0}\frac{r}{|r^3|} = m\frac{v^2}{r}
\end{aligned}
\tag{6.4}
$$

where q stands for the electron charge, m for its mass, v for its velocity and r, for the distance between the electron and ion at the start of the analysis. The ion charge for which the electron will move in a circular orbit at the constant linear velocity 6.2×10^6 m/s and at the constant distance $r = 1$ mm is therefore given by:

$$
Q = mv^2 r\frac{4\pi\varepsilon_0}{q} = 2.4352 \times 10^{-11}\text{C}
\tag{6.5}
$$

In order to evaluate the solver behaviour and its accuracy, several simulations for the case described above are performed using *dsmcPlasmaFoam* and the obtained electron orbits are compared with the expected perfect circular motion. The considered cases and numerical results are described in the following.

6.2.1 Solver behaviour without implementation of the Leapfrog algorithm

The first part of the numerical analysis is performed without implementation of the Leapfrog algorithm. As a consequence, the electron is advanced in time without any time-centred scheme and the *old* and *new* values of the considered variables are computed at equal time levels t_n and $t_n + \Delta t$ respectively. Hence, the *old* values at the start of the simulation t_0 are given by:

$$
\begin{aligned}
x_0 &= \{0; 0.001; 0\}[\text{m}] \\
v_0 &= \{6.2 \times 10^6; 0; 0\}[\text{m/s}] \\
E_0 &= E_{x=x_0} = \frac{Q}{4\pi\varepsilon_0} \frac{x_0}{|x_0|^3} = \{0; 218,865; 0\}[\text{N/C}] \\
F_0 &= qE_0
\end{aligned}
$$

Moreover, the updated positions and velocities for $t_1 = t_0 + \Delta t$ are obtained from Eqs. 5.12 and 5.13 as follows:

$$v_{t_1} = F_0 \frac{\Delta t}{m} + v_0 \tag{6.6}$$

$$x_{t_1} = v_{t_1} \Delta t + x_0 \tag{6.7}$$

Two different strategies regarding the electric field estimation are used in this analysis. The approaches as well as the results are presented in the following sections.

6.2.1.1 Electric field computed at the electron's location

For the first considered strategy, the electric field generated by the ion and acting on the electron is obtained analytically at each time level and at the exact electron's position using the expression in Eq. 6.3. This approach has the advantage that potential deviations caused by the Maxwell solver (s. Section 5.3.3.1) and the force weighting step (s. Section 5.3.3.2) are excluded from the analysis. Hence, the electron's motion and the quality of the numerical results are directly linked to the employed time step Δt. Note that the *trackToFace* method (s. Section 5.3.2) partially mitigates the negative impacts of very large time steps by automatically reducing Δt to a value for which the electron cannot cross more than one cell at a time. However, the electron's position before changing cells (i.e., at the cells interface) is estimated based on a predicted trajectory computed using the original, large Δt value. Therefore, part of the error introduced by very large time steps is not compensated by the *trackToFace* method. This aspect is described in more detail in Section 6.2.1.2. Note that the analytic calculation of the electric field at the exact electron's position is implemented in *dsmcPlasmaFoam* by deactivating the Maxwell solver and introducing Eq. 6.3 into the *move* function of *dsmcPlasmaFoam*.

In order to assess the impact of the time step on the quality of the results, simulations for the electron's motion are performed with Δt values ranging from 1×10^{-10} s to 1×10^{-13} s. The numerically obtained electron trajectories are shown in Fig. 6.12 along with the perfect circular orbit $r = 1$ mm. For the lowest considered temporal resolution, the high time step $\Delta t = 1 \times 10^{-10}$ s leads to the electron being pulled too strongly towards the centre of the numerical mesh at the start of the Lorentz algorithm. From a numerical point of view, the excessive pull towards the grid's centre is the result of the overly long exposure of the electron to the electric field at the starting position prior to the first position and velocity update. While the electron moves inwards and changes cells, the *trackToFace* method kicks in several times before the time step $\Delta t = 1 \times 10^{-10}$ s is finally covered. Since the electric field generated by the ion becomes stronger towards the geometry's centre, the Coulomb force acting on the electron increases as the electron moves inwards. Finally, very close to the centre of the geometry, the singularity of the electric field created by the point charge manifests itself as an extremely high electric force acting on the electron. This, combined with the high time step, results in the electron being strongly pulled past the ion and accelerated out of the numerical mesh. For this particular case, the motion depicted in Fig. 6.12 takes place over a time of 0.2 ns (two full Δt iterations).

A reduction of the time step by a factor of 10 down to $\Delta t = 1 \times 10^{-11}$ s leads to clear improvements of the numerical results. As evident from Fig. 6.12, the electron is now able to perform several revolutions around the ion. However, the distance between the particles drops constantly as a result of the still excessive exposure of the electron to old values of the electric field before position and velocity updates take place. As for $\Delta t = 1 \times 10^{-10}$ s, the electron is eventually accelerated out of the numerical mesh by the electric field singularity at the geometry's centre. In this particular case, the electron motion shown in Fig. 6.12 takes place over a time interval of approximately $t = 1.6$ ns.

For an increased time resolution with $\Delta t = 1 \times 10^{-12}$ s, the electron is now able to orbit the ion in a trajectory which clearly resembles the expected circular motion. Slight deviations from the analytic orbit are, however, still evident as the separation between the electron and ion oscillates around the theoretical distance $r = 1$ mm. With the smallest considered time step $\Delta t = 1 \times 10^{-13}$ s, the results show very good agreement with the theoretical circular trajectory. Based on this analysis, the Lorentz algorithm of *dsmcPlasmaFoam* is considered to be able to produce accurate results for the charge carriers motion inside electric fields even without implementation of the Leapfrog algorithm. However, it should be noted that the high temporal resolution associated with the time step $\Delta t = 1 \times 10^{-13}$ s also results in high computational requirements. This aspect is discussed in more detail in Section 6.2.2.

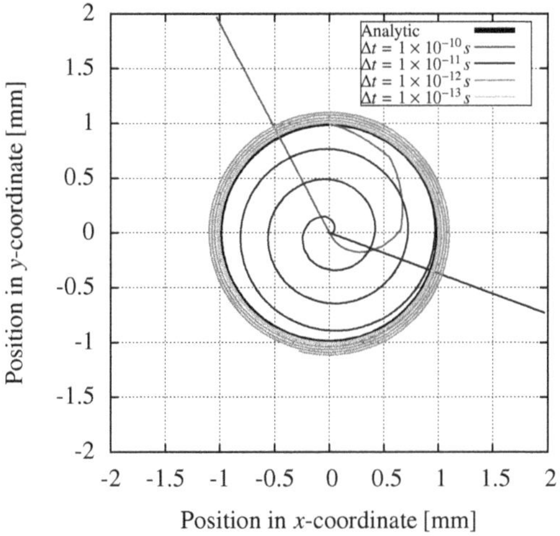

Figure 6.12: Electron trajectories for different values of Δt without Leapfrog algorithm and with E obtained at the electron's location.

6.2.1.2 Electric field computed at the cell centres

In Section 6.2.1.1, the electric fields are calculated at the exact electron's location. As a consequence, an infinite number of E values are available during the simulation and the electric field resolution is, effectively, much higher than the actual grid resolution. However, in a typical numerical simulation with the Finite Volume Method, only one field value per cell is actually available. In order to examine the Lorentz solver's behaviour in a more realistic numerical setup, the tests in Section 6.2.1.1 are re-run using a single E value per cell calculated with Eq. 6.3 at the coordinates of the cells centres. The corresponding numerical simulations are performed with a grid resolution of $161 \times 161 \times 1$ cells. The obtained electron trajectories with this numerical setup as well as the relative errors compared to the analytic orbit $r = 1$ mm are shown in Fig. 6.13.

As mentioned in Section 6.2.1.1, errors introduced by the use of very large time steps

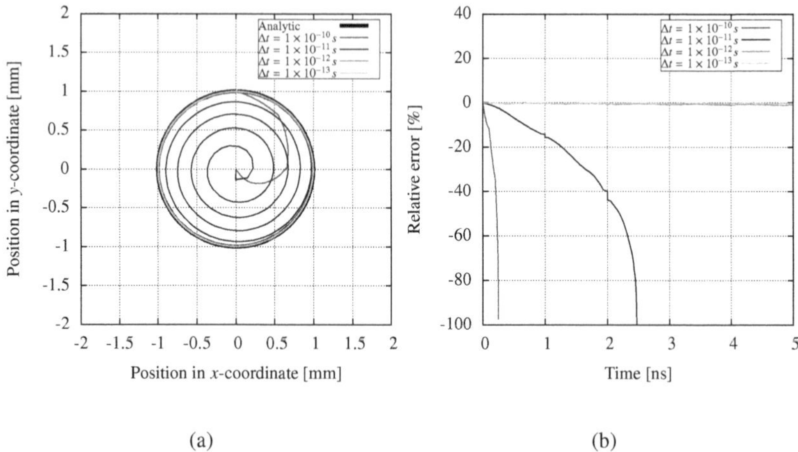

Figure 6.13: Electron trajectories for different values of Δt without Leapfrog algorithm and with E obtained at the cells centres (a) and relative deviations to analytic orbit (b).

are partially mitigated in *dsmcPlasmaFoam* by the *trackToFace* method. With this function, a very large time step is automatically reduced to the value for which the tracked particle is estimated to encounter a cell face. The motion step is then performed until this point and the algorithm is repeated until Δt is completely covered. As a consequence, the *trackToFace* method effectively limits the time step of the Lorentz solver to a value closely linked to the spatial grid resolution. Regarding the results in Fig. 6.13 for the two largest considered time steps $\Delta t = 1 \times 10^{-10}$ s and $\Delta t = 1 \times 10^{-11}$ s, it should be noted that both values are considerably higher than the average time needed for the electron to cross a cell. Therefore, the effective time step values for both cases are mostly equal as they are, in fact, dictated by the *trackToFace* method. However, as evident from Fig. 6.13, the results for the trajectories are extremely different. This deviation is explained by the procedure used for the estimation of the location where the particle encounters the cell's boundary in the *trackToFace* method. As shown schematically in Fig. 6.14, the method uses the original large Δt to estimate the final particle's position before checking for cell faces located between the final and the initial position. If a cell interface is found, the vector connecting the initial and estimated final position is scaled down so that it ends at the identified cells interface. The corresponding scaling factor is subsequently used to reduce Δt, as the particle is advanced in time and placed at the intersection between the estimated particle's path and the cell's face. Although simple, this approach is only accurate if the particle's velocity is constant during the motion step. This assumption is usually correct for typical DSMC applications, as the particles are assumed not to collide during the motion step and no external accelerating force acts on them. However, as can be seen in Fig. 6.14 representing the initial conditions of the simulations in this section, the acceleration term associated with the electric field will introduce deviations of the estimated intersection coordinates if different values of Δt are initialised. In the depicted case, the estimated particle's position X_1 at the cell's interface using Δt as initial time step is clearly closer to the geometry's centre than its counterpart X_1' obtained with the smaller time step $\Delta t/2$. Although the particles are advanced in time by almost the same time step, the errors introduced by the estimation of the particles position at the cell's interface using the original Δt accumulate over the course of the simulation, inducing the much stronger drift towards the grid's centre visible in Fig. 6.13 for $\Delta t = 1 \times 10^{-10}$ s compared to $\Delta t = 1 \times 10^{-11}$ s.

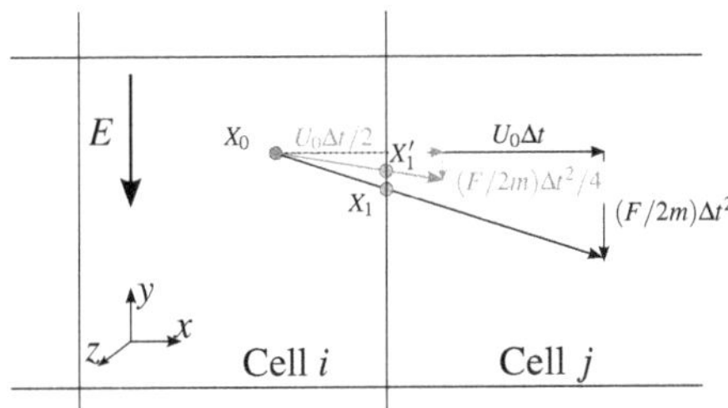

Figure 6.14: Schematic representation of the estimated particle's position X_1 at the cell's interface using the *trackToFace* method with black vectors representing the estimation using Δt as initial time step and red vectors, the result for $\Delta t/2$.

As can be seen in Fig. 6.13, the electron's trajectory for the simulation with the lowest temporal resolution and highest time step $\Delta t = 1 \times 10^{-10}$ s is very similar to the one obtained in Section 6.2.1.1 and the electron reaches the geometry's centre after approximately 0.2 ns. However, instead of the electron exiting the numerical grid, the simulation ends with a numerical crash once the distance between the ion and the electron (i.e., the cell's centre) effectively becomes zero and the electric force, infinite. Although the behaviour for the simulation with $\Delta t = 1 \times 10^{-11}$ s and E obtained at the cell's centre is very similar to the case with E obtained at the exact electron's location, a closer look to the results reveals that the electron performs a higher number of revolutions around the ion for the former case (4) compared to the latter configuration (3). This behaviour is explained by the fact that the calculation of the electric field at the electron's location leads to a constant increase in the field and force magnitude as the electron gradually moves towards the grid's centre. In contrast, for E computed at the cell's centre, the electric field remains constant and hence, weaker during the electron's inwards motion as long as a cell change does not take place. As a consequence, since the location and associated electric field errors introduced by large time steps are partially mitigated by the constant nature of the electric field inside each cell, the results with the electric field obtained at the cells centres show better agreement with the analytic case for high values of Δt compared to the electric field estimation at the electron's location. This smoothing effect is also evident from the improvement of the results for $\Delta t = 1 \times 10^{-12}$ s in Fig. 6.13 compared to Fig. 6.12. However, it should be noted that this conclusion is valid for the numerical setup used for the Lorentz validation only because the electron is initialised at a location, with a velocity and under an electric field which already match the problem's solution. Therefore, instead of advancing towards a result, the performed simulations attempt to conserve the initial equilibrium. In a different numerical setup, the smoothing effect might have the opposite effect, effectively slowing down the computation and in the worst case, compromising the quality of the simulation results.

6.2.2 Solver behaviour with implemented Leapfrog algorithm

In Section 6.2.1, a Lorentz solver approach without use of time-centred schemes is discussed. This procedure is compared in this section with a Leapfrog based algorithm. Based on the general Leapfrog procedure described in Section 3.3.1 and its implementation in *dsmcPlasmaFoam* as discussed in Section 5.3.3.3, the particle's position as well as the electric and force fields are computed at the time levels t_n and $t_n + \Delta t$, while the velocities are evaluated at the intermediate time levels $t_n - \Delta t/2$ and $t_n + \Delta t/2$. Hence, the *old* values at the start of the simulation t_0 for the electron's position x, the electric field E and the force acting on the electron F are given by:

$$x_{old} = x_0 = \{0; 0.001; 0\}[\mathrm{m}]$$

$$E_{old} = E_0 = E_{x=x_0} = \frac{Q}{4\pi\varepsilon_0} \frac{x_0}{|x_0|^3} = \{0; 218,865; 0\}[\mathrm{N/C}]$$

$$F_{old} = F_0 = qE_{old}$$

Furthermore, in order to estimate v_{old}, the initial velocity $v_0 = \{6.2 \times 10^6, 0, 0\}$ valid for t_0 is pushed back to the time level $t_0 - \Delta t/2$ using the force F_{old} as follows:

$$v_{old} = v_0 - F_{old}\frac{\Delta t}{2m} \tag{6.8}$$

In the next step, the leapfrog velocity v_{new}, valid for the time level $t + \Delta t/2$, is calculated using Eq. 5.12:

$$v_{new} = v_{old} + F_{old}\frac{\Delta t}{m} \tag{6.9}$$

Finally, the updated positions for the next time level $t_1 = t_0 + \Delta t$ are obtained using the updated velocity v_{new} and Eq. 5.13:

$$x_{new} = x_{t_1} = v_{new}\Delta t + x_{old} \tag{6.10}$$

The fields E and F are subsequently updated at the electron's new position and the results are used to advance the velocity to the next intermediate time level $t_1 + \Delta t/2$. The steps are repeated until the end of the simulation.

As mentioned in Section 5.3.3.3, the approach described above effectively assumes the velocity to be constant during the particle's motion step, i.e., between t_n and $t_n + \Delta t$. Hence, if a particle crosses a cell boundary during the motion step, it will retain the trajectory and velocity initially estimated based on the electric field of its original cell. Therefore, the particular electric field E in the new cell has, at first, no impact on the particle's motion and the new value of E will play a role only during the next velocity update (next iteration). As a consequence, this procedure does not include force or velocity updates after cell changes. In this chapter, this approach is referred to as *standard Leapfrog* ($v = const.$).

In order to assess the implemented standard Leapfrog procedure, the numerical tests described in Section 6.2.1.2 are repeated using the approach described above. Here, one single value per cell of the electric field E, calculated with Eq. 6.3 at the coordinates of the cells centres, is considered. The corresponding numerical simulations are performed with a grid resolution of $161\times161\times1$ cells and with time step values in the range $1\times10^{-13} < \Delta t < 1\times10^{-10}$ s. The obtained electron trajectories with the standard Leapfrog algorithm are depicted in Fig. 6.15 (a). In addition, the relative deviations from the theoretical orbit for the Lorentz solver with standard Leapfrog as well as without Leapfrog (from Section 6.2.1.2) are shown in Fig. 6.15 (b).

The results depicted in Fig. 6.15 (a) highlight the advantages of a time-centred scheme. Specifically, the orbits obtained using low temporal resolutions show substantial improvements compared to the results from the simulations with the same numerical set-up and without Leapfrog implementation (s. Fig. 6.13). Instead of the electron being

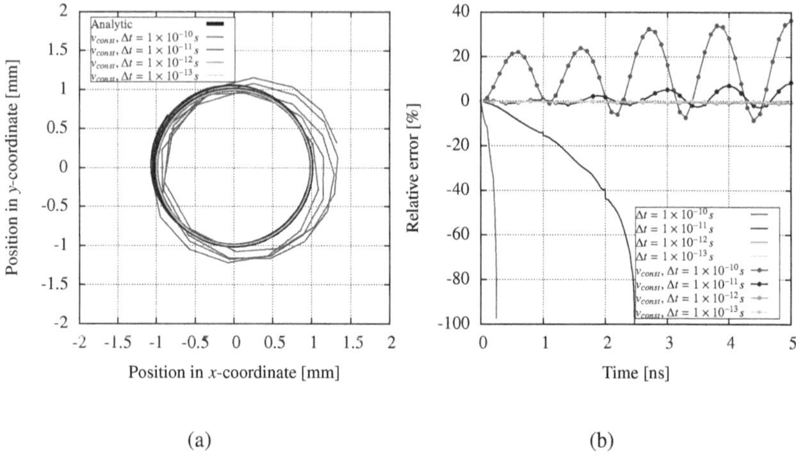

Figure 6.15: (a) Electron trajectories for different values of Δt with the Leapfrog algorithm and with E obtained at the cells centres; (b) Relative deviations from analytic orbit for the Lorentz solver with standard Leapfrog (v_{const}) and without Leapfrog algorithm.

pulled strongly towards the geometry's centre, the time-centred computation with the highest time step $\Delta t = 1 \times 10^{-10}$ s shows the electron still orbiting around the geometry's centre at the end of the simulation, $t = 5$ ns. Although the accuracy of the orbit is clearly low and the large Δt is visible as a succession of straight lines, the resulting trajectory partially resembles the expected circular orbit. However, as seen in Fig. 6.15 (b), the relative deviation from the theoretical orbit gradually increases and the electron is expected to eventually exit the numerical domain. A reduction of the time step to $\Delta t = 1 \times 10^{-11}$ s greatly improves the quality of the results and the obtained orbit is in good agreement with the theoretical trajectory. In comparison, the non-time-centred scheme, used to produce the results for the same time step and shown in Fig. 6.13, ends with the electron reaching the geometry's centre and the simulation crashing. For numerical tests with sufficiently low values of Δt, the results with and without Leapfrog implementation are in very good agreement with the theoretical orbit and the relative deviations are negligible. Based on these results and the highly improved accuracy for low temporal resolutions, the Leapfrog algorithm is shown to be clearly superior to the simplified, non-time-centred strategy presented in Section 6.2.1.

The *trackToFace* functionality included in *dsmcPlasmaFoam* makes it possible to consider a further adaptation of the Leapfrog method in which force and velocity updates are performed each time a particle changes cells. As described in Section 5.3.3.3, this approach involves partially advancing the particle in time using the fraction of the original time at which the particle encounters a cell face. Therefore, if a cell face is located between the particle's old position x_{old} and its estimated final position x_{new}, the particle is first moved forward to the interface location with coordinates x'_{old} and the corresponding time step fraction is then given by,

$$trackFraction = \frac{|x'_{old} - x_{old}|}{|x_{new} - x_{old}|} \tag{6.11}$$

Let F^i_{old} and F^j_{old} represent the force acting on the particle as a result of the electric field at the cells i and j respectively. In order to consider the effects of the electric field in the new cell, the original estimation of the leapfrog velocity v_{new} is corrected using $trackFraction$ so that it equals the linear combination of the contributions of the electric fields in the old and new cells. Hence,

$$v_{new} = v_{old} + (trackFraction)F^i_{old}\frac{\Delta t}{m} + (1 - trackFraction)F^j_{old}\frac{\Delta t}{m} \tag{6.12}$$

The particle's motion step through the new cell is subsequently performed starting at the interface with coordinates x'_{old} and using the new estimation of the leapfrog velocity v_{new}. In order to assess the implemented Leapfrog procedure with velocity updates at cell interfaces, the numerical tests performed for the standard Leapfrog approach are repeated using the same numerical set-up. The obtained electron trajectories are depicted in Fig. 6.16 (a). In addition, the relative deviations from the theoretical orbit for the Lorentz solver with modified Leapfrog method (velocity updates at cell interfaces) as well as with the standard Leapfrog algorithm ($v = const$) are shown in Fig. 6.16 (b).

As can be seen in Fig. 6.16, the results from the Leapfrog approach with velocity updates are comparable with the results with the constant velocity strategy for small values of Δt. This is to be expected because for small time steps, the electron's trajectory through the numerical grid is dominated by a very large number of small position and velocity updates inside the cells and in comparison, the number of velocity updates triggered by cell changes is negligible. For very large values of Δt, the opposite is the case and the number of velocity updates triggered by cell changes can become higher than the number of updates performed as part of a standard Leapfrog iteration. For instance, with the highest considered time step $\Delta t = 1 \times 10^{-10}$ s, the electron crosses several cell interfaces during the very first iteration and does so with an inwards-biased trajectory. Note that this initial inwards trajectory is the result of the large Δt as explained in Section 6.2.1.2 and Fig. 6.14. For the Leapfrog approach with velocity updates, the velocity re-estimation at the cells interfaces during this first iteration employs each time a different and, because of the inwards-biased initial trajectory, stronger

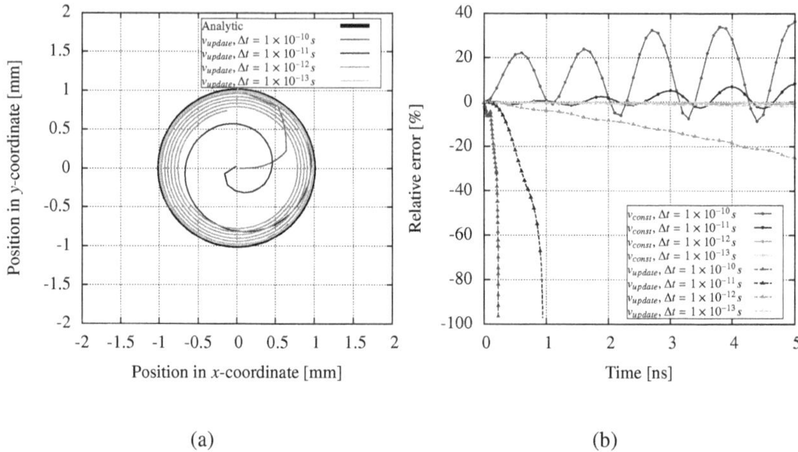

Figure 6.16: (a) Electron trajectories for different values of Δt with the Leapfrog algorithm, E obtained at the cells centres and F updated after cell changes; (b) Relative deviations from analytic orbit for approaches with Leapfrog algorithm, with and without v update.

E field. As a consequence, the electron is strongly pulled towards the domain's centre reaching it after less than three full time steps. Based on this behaviour, it can be concluded that for large values of Δt, the error introduced during the initial estimation of the particle's final position is amplified by the updates of E, F and v_{new} performed each time the electron changes cells. Despite this, it should be noted that the Leapfrog procedure with velocity updates allows for the particles to follow the electric field lines during iterations involving cell changes more closely, while at the same time conserving the time-centred nature of the Leapfrog algorithm. However, for large values of Δt, the algorithm has limitations similar to the ones observed in Section 6.2.1.2 for the Lorentz validation approach with E calculated at the exact particle's position.

The maximum relative deviations from the theoretical orbit using the approaches described in this section are summarized in Table 6.3 as a function of the time step Δt. Here, an error of 100% represents a computation in which the electron either exits the numerical domain or is captured in its centre before the simulation ends. For comparison and based on the initial velocity of the electron $|v| = 6.2 \times 10^6$ m/s, as well as on the cells size $\Delta x \approx 0.025$ mm, the average transit time of the electron inside a cell in the performed numerical tests is estimated as 4×10^{-12} s. As can be seen in Table 6.3, all considered approaches produce accurate results with the very small time

Time step Δt	No Leapfrog, E at exact pos.	No Leapfrog, E at cell centres	Leapfrog, E at cell centres v_{const}	Leapfrog, E at cell centres v_{update}
1×10^{-10} s	100%	100%	36.16%	100%
1×10^{-11} s	100%	100%	8.39%	100%
1×10^{-12} s	11.41%	1.12%	1.03%	25.16%
1×10^{-13} s	1.86%	0.42%	0.44%	2.05%

Table 6.3: Maximum relative deviation from theoretical orbit for considered Lorentz solver approaches and different values of Δt.

step $\Delta t = 1 \times 10^{-13}$ s. Furthermore, for $\Delta t = 1 \times 10^{-12}$ s, the Lorentz solver approach with E obtained at the cell centres and without Leapfrog implementation, as well as the approach with leapfrog implementation and constant velocity both produce accurate results. For $\Delta t = 1 \times 10^{-11}$ s, the time step becomes larger than the estimated average transit time of the electron and only the results with implemented Leapfrog and constant velocity can be considered to be acceptable. Based on the overall good accuracy of the numerical results with appropriate small values of Δt, the Lorentz solver in *dsmcPlasmaFoam* is considered to be validated. Furthermore, because of its higher accuracy and robustness for large values of Δt, the Leapfrog approach with constant velocity during cell changes is implemented as the standard numerical scheme in *dsmcPlasmaFoam*.

6.3 Particle and force weighting

The particle and force weighting procedures described in Section 3.3.3 are responsible for the assignment of particle quantities to grid points as well as for the application of the Coulomb force to charge carriers inside a plasma. As described in Section 5.3.3.2, a zero-order weighting approach, also known as nearest-grid-point NGP, is implemented in *dsmcPlasmaFoam*. Hence, the cell's number density n of a particular species is given by:

$$n = \frac{W}{V} \sum_i N_i \tag{6.13}$$

where W stands for the species numerical weight, V for the cell's volume and the sum corresponds to the amount of particles located inside the cell. As already discussed, the electric field E is estimated by the Maxwell solver based on the number densities of the charge carriers and the result is valid for the cells centres. In *dsmcPlasmaFoam*, the force acting on a charge carrier located in a given cell is computed using the electric field at the corresponding cell's centre. Therefore, the particle's position relative to the cell's centre and the electric field values in the neighbour cells are not considered. The approach has low computational requirements because the handling of field values on additional cells is avoided.

Considering the successful validation of the Maxwell and Lorentz solvers described in Sections 6.1 and 6.2 and in order to examine the quality of the selected zero-order weighting approach, the complete PIC algorithm of *dsmcPlasmaFoam* is tested for the well-documented case of a plasma wave. This approach allows the validation of the zero-order weighting approach as well as of the interactions between the individual components of the PIC solver. Note that the plasma wave validation strategy used in this work is based on the approaches employed in [Sti15] and [Sto13].

In order to validate the weighting steps and to analyse the overall behaviour of the PIC algorithm, a channel geometry with dimensions x =50 mm, y =0.5 mm and z =0.05 mm is discretized employing a resolution of 1,000 cells in the x direction, 10 cells in the y direction and 1 cell in the z coordinate. Hence, the plasma wave tests are performed in a two-dimensional numerical setup. Periodic boundary conditions are applied to each of the six boundary faces of the channel geometry allowing the particles in the numerical domain to be representative of an infinite-sized plasma. The numerical grid is populated with the same number of both motion-less ions and electrons. Since the numerical weight of both species is equal, the plasma can be considered to be macroscopic neutral. The ions are distributed uniformly in the numerical grid while the electrons are initialised with slight deviations from the uniform distribution in the x coordinate defined as follows (cp. [Sti15], [Sto13]):

$$\Delta x_e = A \sin(k x_i) \tag{6.14}$$

Here, x_i represents the position of one ion (uniformly distributed in the grid), $A = 0.1$ mm the amplitude or maximum possible deviation from the uniform distribution for the electrons and $k = 4\pi/50$ mm, the phase for which the numerical mesh contains two full cycles of the electrons distribution function. As an example, a section of the resulting particles distribution for a numerical setup containing 10,000 ions and 10,000 electrons is shown in Fig. 6.17.

The deviation in the electrons position in the x coordinate creates an unbalance in the system which manifests itself as electric potential and therefore, as electric force. The

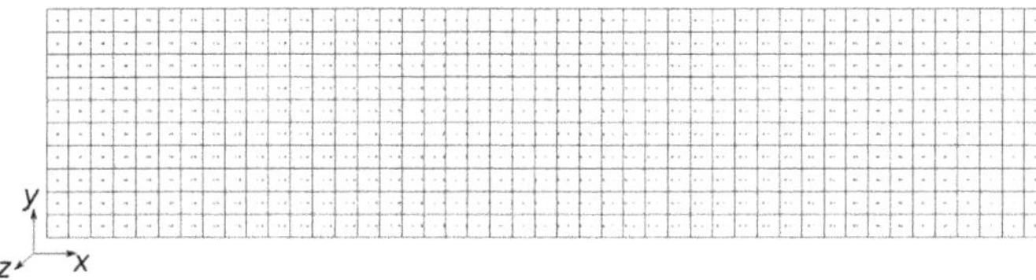

Figure 6.17: Particles distribution for a plasma wave numerical setup containing 10,000 ions and 10,000 electrons. The ions (red dots) are located in the cells centres while the electrons positions (blue dots) relative to the ions are a function of x.

total potential energy E_{pot} stored as electric force is given by,

$$E_{Pot} = \int\limits_V \frac{1}{2}\varepsilon_0 |E|^2 \mathrm{d}V = \sum_n \frac{1}{2}\varepsilon_0 |E_n|^2 V_n \tag{6.15}$$

where E_n represents the electric field in the cell n, V_n, the cell's volume and the summation is performed over all cells in the numerical grid. The potential energy E_{Pot} is transformed into kinetic energy at the start of the simulation as the electrons, initialised with no velocity, start to move as a result of the electric force. The total relativistic kinetic energy of the electrons in the system corresponds to the sum of the individual energy contributions and can be calculated as follows:

$$E_{kin} = \sum_{i=1}^{N} \left(\frac{mc^2}{\sqrt{1 - v_i^2/c^2}} - mc^2 \right) \tag{6.16}$$

where c stands for the speed of light in vacuum, m is the electron's mass, v the magnitude of the electron's velocity and the summation is performed over all electrons in the numerical grid. In their attempt to restore neutrality, the electrons overshoot the equilibrium point and electron oscillations, also known as a plasma wave, appear. For a system containing 10,000 numerical electrons, each representing 1,250 real particles and with the channel dimensions defined above, the number density of the electrons becomes $n_e = 1 \times 10^{16}\ \mathrm{m}^{-3}$. The analytic plasma frequency for the simulated system is a function of the electron number density and is given by,

$$\omega_{pe} = \sqrt{\frac{n_e e^2}{m\varepsilon_0}} = 5,633.61\,[\mathrm{rad/s}] \tag{6.17}$$

where m stands for the electron mass and e, for the elementary charge. Therefore, the theoretical angular plasma frequency f_{pe} and plasma period T_{pe} for the simulated system are estimated as follows:

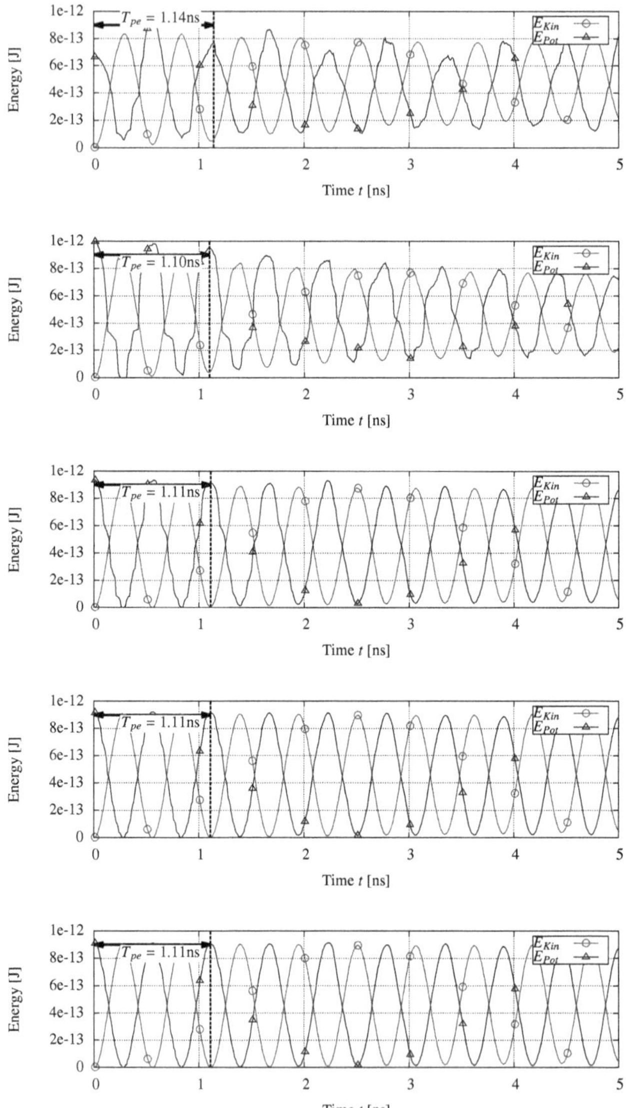

Figure 6.18: Potential and kinetic energy in the simulated plasma wave as a function of time. From top to bottom with 5,000, 10,000, 20,000, 30,000 and 40,000 numerical electrons.

$$f_{pe} = \frac{\omega_{pe}}{2\pi} = 896.62\,\text{MHz}$$
$$T_{pe} = \frac{1}{f_{pe}} = 1.115\,\text{ns} \tag{6.18}$$

The analytic values for f_{pe} and T_{pe} can be used to evaluate the quality of the results obtained with the PIC algorithm implemented in *dsmcPlasmaFoam*. To this end, numerical simulations are performed with 5,000, 10,000, 20,000, 30,000 and 40,000 electrons. Note that the total number of ions is always the same as that of the electrons. Furthermore, the numerical weight W of the ions and electrons is also adapted so that the number density $n_e = 1 \times 10^{16}$ m^{-3} remains constant across all tests. Simulations are run for 5 ns using the time step $\Delta t = 1 \times 10^{-11}$ s and the results for the potential and kinetic energy as a function of time are shown in Fig. 6.18.

As can be seen in Fig. 6.18, the results obtained with 5,000 electrons, the lowest amount of particles considered, exhibit highly irregular energy profiles specially for the potential energy. Specifically, the potential energy shows strong variations of amplitude with the maximum value $E_{pot} = 9.48 \times 10^{-13}$ J at $t = 0.57$ ns being over 40% higher than the value obtained after the first numerical iteration. Furthermore, a considerable level of numerical noise can be observed in the profiles. The irregularities in the energy profiles can be explained by the low amount of particles in the system combined with the zero-order weighting approach implemented in *dsmcPlasmaFoam*. Specifically and as described in Section 3.3.3, a particle changing cells with the zero-order weighting approach will lead to a sudden jump of the number density and hence, to strong oscillations of the electric field. The problem becomes evident if one considers that this particular numerical setup contains, on average, only one numerical electron per two cells. Despite these issues, the obtained plasma period $T_{pe} = 1.14$ ns, estimated as the time between the start of the simulation (first peak on potential energy) and the time the electrons complete a full oscillation (third peak on potential energy as the electrons return to their initial position) is a good approximation of the analytic value as it deviates only by 2.24 %.

By doubling the number of electrons to 10,000 the irregularities in the energy profiles become much less pronounced while the plasma period estimation $T_{pe} = 1.14$ ns gains considerably in accuracy. Further increases of the number of electrons lead to additional improvements in the quality and regularity of the temporal energy profiles. With 40,000 numerical electrons in the numerical grid and an average of four electrons per cell, the jump effects caused by electrons changing cells are almost completely smoothed out. The stabilising effect associated with a higher number of particles is also evident from the comparison of the total energy in the system shown in Fig. 6.19 for the simulations with 10,000 and 40,000 electrons. The plasma period values obtained for the simulations with 20,000, 30,000 and 40,000 electrons $T_{pe} = 1.11$ ns are very close to the analytic result and exhibit a relative deviation of only 0.45 %. However, because of the employed time step $\Delta t = 1 \times 10^{-11}$ s, the numerical setup does not allow for the simulation to capture the analytic result $T_{pe} = 1.115$ ns exactly and therefore, an additional accuracy improvement can be expected if the temporal resolution is increased. The results of the performed simulations are summarized in Table 6.4. As can be seen, the computational requirements are moderate and *dsmcPlasmaFoam* is considered to be able to model basic plasma phenomena in an accurate way using a standard workstation and some level of parallelization for the simulations.

Considering that individual charges are assigned directly to a given cell in *dsmcPlasmaFoam* and no form function for particles and force weighting is employed, conservation of charge is given and a detailed analysis on this topic is not necessary. Based on the results described in this section, the zero-order approach implemented in *dsmcPlasmaFoam* is considered to be able to deliver accurate results for basic plasma phenomena, provided that an adequate number of numerical particles is employed. Furthermore, the quality of the results for the plasma wave modelling combined with the positive validation results for the Maxwell and Lorentz solvers in Sections 6.1 and 6.2 show that, if the spatial and temporal domains are discretized in an appropriate way, the implemented PIC algorithm can deliver accurate results for plasma-related phenomena. Hence, this component of the solver *dsmcPlasmaFoam* is considered to be validated.

Number of test electrons	Frequency f_{pe} [MHz]	Period T_{pe} [ns]	Error [%]	Computational time [s]
5,000	877.2	1.14	2.24%	68.5
10,000	909.1	1.10	-1.35%	166
20,000	900.9	1.11	-0.45%	173
30,000	900.9	1.11	-0.45%	239.5
40,000	900.9	1.11	-0.45%	476

Table 6.4: Summary of results for numerical simulation of a plasma wave. Each of the computations is performed in a parallel manner using 4 cores.

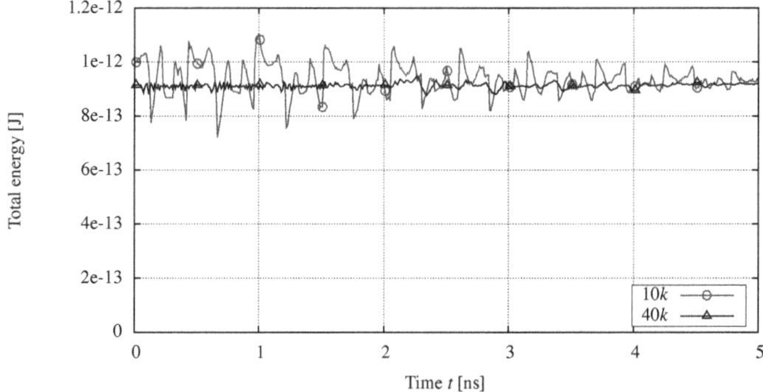

Figure 6.19: Comparison of total energy in simulated plasma system as a function of time using 10,000 and 40,000 test electrons.

6.4 Coulomb collisions

The PIC component, discussed in the previous sections, is mainly responsible for the handling of long-range interactions between charge carriers. In order for *dsmcPlasmaFoam* to take into account short-range Coulomb collisions inside a plasma, the algorithm is combined with a Monte-Carlo-Collision approach (MCC). As described in Section 5.3.4, the implemented MCC algorithm is based on the theory of cumulative small-angle collisions developed by Kenichi Nanbu (s. [Nan97] and [Nan00]). The phenomena modelled by this component of *dsmcPlasmaFoam* is the result of forces acting over distances of the order of the Debye length λ_D, which are, in turn, associated with small-angle changes of the particles velocity vectors. The validation of the implemented MCC algorithm presented in this section is based on the study cases examined in [Nan97].

In the first numerical test, the thermalization of an electron beam inside a field of fixed ions is modelled and the obtained results are compared with theoretical data. Let the subscript α represent the electrons and β, the ions. For this numerical test, a field of ions with number density $n_\beta = 1 \times 10^{16}$ m^{-3} and homogeneous velocities $v_{\beta 0} = \{1, 1, 1\}$ m/s is initialised in a numerical domain of dimensions 5 mm $\times$5 mm$\times$0.05 mm. With the numerical weight defined as $W = 125$, the initialised ions field is comprised of 100,000 numerical particles. The ions interact with an electron beam with the following initial velocity in the x coordinate,

$$v_{\alpha 0,x} = \sqrt{\frac{3k_B T_\alpha}{m_\alpha}} = 6.74 \times 10^5 \text{m/s} \tag{6.19}$$

which corresponds to $T_\alpha = 10,000$ K. Note that at the start of the thermalization process, the electrons velocities have only the component in the x coordinate and $v_{\alpha0,y} = v_{\alpha0,z} = 0$ m/s. Furthermore, $v_{\beta0} = \{1,1,1\}$ m/s is used as fixed ion velocity instead of $v_{\beta0} = \{0,0,0\}$ m/s in order to avoid numerical singularities at the start of the simulations when the y and z components of the relative pre-collision velocities $g = v_{\alpha0} - v_{\beta0}$ equal zero. As for the ions field, the beam electrons are initialised with the number density $n_\alpha = 1 \times 10^{16}$ m^{-3} and the numerical weight $W = 125$. Hence, the simulation comprises 100,000 test electrons. Furthermore, the numerical domain is discretized using 100 cells in the x, 100 cells in the y and 1 cell in the z coordinates. As a result, an average of 10 electrons and 10 ions can be found in each of the numerical cells. Periodic boundary conditions arc applicd to all boundary surfaces.

In the thermalization test, the initialised electrons interact with the field ions through Coulomb collisions only and the temporal development of the electrons velocities is examined. Since the initial velocities of the particles making up each of the species ensembles are equal and considering the homogeneous distribution in the numerical domain as well as the periodic character of the applied boundary conditions, the computation of the particles motion through the numerical mesh is not required. Hence, the Lorentz solver is deactivated. Moreover, as the numerical test focuses on short-range interactions between charge carriers, the Maxwell solver is also deactivated. Note that only collisions between electrons α and ions β are considered, i.e., electron-electron and ion-ion interactions are not modelled. For the described numerical setup, the Debye length can be estimated as follows:

$$\lambda_D = \sqrt{\frac{\varepsilon_0 k_B T_\alpha}{n_\alpha q_e^2}} \tag{6.20}$$

where q_e stands for the electron charge. The result $\lambda_D = 6.9 \times 10^{-5}$ m is of the same order of magnitude as the spatial mesh resolution 5×10^{-5} m. In addition, the magnitude of the pre-collision relative velocity g can be assumed to agree with the magnitude of the initial velocity of the electrons. Hence, the parameter b_0 can be estimated using Eq. 5.25 as follows:

$$b_0 = \frac{|q_\alpha q_\beta|}{2\pi\varepsilon_0 m_\alpha g^2} = 1.114 \times 10^{-9}\text{m} \tag{6.21}$$

Furthermore, the reference time τ_0 for the described numerical setup can be calculated using the following expression (cp. [Nan97]):

$$\frac{1}{\tau_0} = \frac{n_\beta q_\alpha^2 q_\beta^2 \ln\Lambda}{8\pi\sqrt{2}\varepsilon_0^2 m_\alpha^{1/2} \varepsilon_{\alpha0}^{3/2}} \tag{6.22}$$

where,

$$\ln \Lambda = \ln \frac{\lambda_D}{b_o} \tag{6.23}$$

corresponds to the Coulomb logarithm and $\varepsilon_{\alpha 0} = m_\alpha v_{\alpha 0}^2/2$. Based on Eqs. 6.20 through 6.23, the reference time for the numerical test is estimated as $\tau_0 = 3.447 \times 10^{-6}$ s.

The electron beam, comprised of particles with initial velocity only in the x coordinate, interacts with the fixed ions in the numerical domain leading to a change in the electrons velocity distribution over the course of the simulation. Since the ions velocities are held constant, the electrons scatter over time, eventually reaching an isotropic state. The expected values of the nondimensional electrons velocity in the x coordinate $\langle \hat{v}_x \rangle$ as well as the expectation of the sum of the nondimensional squared perpendicular components (i.e., in the y and z coordinates) at the start of the thermalization process $\langle \hat{v}_\perp^2 \rangle$ are given by the following analytic expressions (cp. [Nan97]):

$$\langle \hat{v}_x \rangle = 1 - \hat{t} \tag{6.24}$$

$$\langle \hat{v}_\perp^2 \rangle = 2\hat{t} \tag{6.25}$$

where $\hat{t}$ stands for the nondimensional time t/τ_0, the expected values, denoted by the angle brackets, correspond to the average over the electrons ensemble and the velocities are normalized using the initial electrons velocity $v_{\alpha 0}$. The results for $\langle \hat{v}_x \rangle$ and $\langle \hat{v}_\perp^2 \rangle$ as well as the kinetic energy Cartesian distribution obtained with the time step $\Delta\hat{t} = 0.003$ are depicted in Fig. 6.20.

As evident from Fig. 6.20 (a), the numerical results for the early stages of the thermalization process show excellent agreement with the analytic expressions in Eqs. 6.24 and 6.25. At the start of the simulation, the electrons velocity vectors have only a positive x component and their average velocity in x equals $v_{\alpha 0,x}$. During the course of the simulation, the electrons interact with the fixed ions field. As they do so, the positive x component of the velocity distribution becomes less dominant because of the cumulative electron-ion collisions and the associated electron scattering. At the end of the simulation, $\langle \hat{v}_x \rangle$ approaches zero as the electrons in the ensemble have the same probability of moving in the positive as in the negative x direction. The ensemble average of the sum of the squared perpendicular components $\langle \hat{v}_\perp^2 \rangle = \langle \hat{v}_y^2 + \hat{v}_z^2 \rangle$ gradually increases during the thermalization process, as the electrons kinetic energy, initially present only in the x dimension, is distributed among all spatial coordinates. At the end of the simulation, $\langle \hat{v}_\perp^2 \rangle$ reaches approximately 2/3, which is in agreement with the electrons kinetic energy being distributed equally in all coordinates. At this point, the electron beam has become fully thermalized and an isotropic velocity distribution is reached. The energy equalization process depicted in Fig. 6.20 (b) also shows the decrease in kinetic energy in the x direction from 1 to 1/3. Because of the nature of Nanbu's cumulative collisions model, the results for the velocity distributions are expected to be mainly independent of the employed time step. This can be confirmed based on the results for $\langle \hat{v}_{\perp}^2 \rangle$ obtained with different values of $\Delta\hat{t}$ and depicted in Fig. 6.21. Here, only the relatively large time step $\Delta\hat{t} = 0.15$ results in some level of deviation compared to other time steps. Nevertheless, the thermalization process also ends with an isotropic velocity distribution.

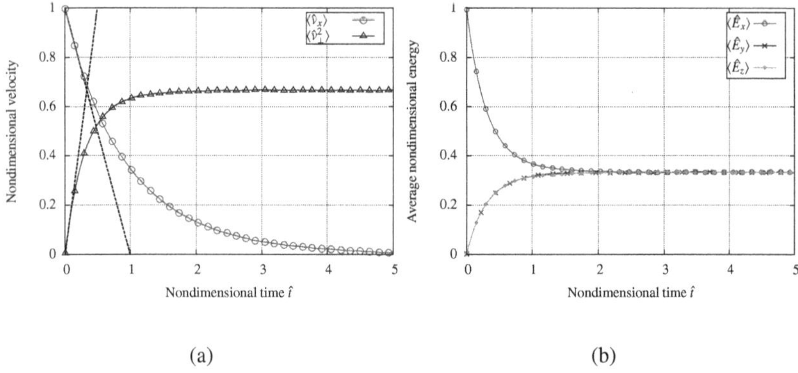

Figure 6.20: Numerical results for electron beam thermalization in a field of fixed ions. (a) $\langle \hat{v}_x \rangle$ and $\langle \hat{v}_{\perp}^2 \rangle$ as a function of time with Eqs. 6.24 and 6.25 shown as dashed lines; (b) Normalized kinetic energy components.

In the next study case, an electron beam is thermalized as it interacts with a field of electrons in equilibrium (cp. [Nan97]). The numerical setup is similar to the one used in the previous example. Beam electrons (species α) and field electrons (species β) are initialised in the same numerical domain as before, using the same number of cells and subjected to the same boundary conditions. Furthermore, $n_\alpha = n_\beta = 1 \times 10^{16}$ m^{-3} and the numerical weights are defined as $W_\alpha = W_\beta = 125$, resulting in 100,000 test particles for each species. The initial temperature of the beam electrons, set to 100 eV, is implemented as kinetic energy in the positive x coordinate only. In addition, the field electrons are initialized with a temperature of 2 eV and the initial velocities obey the Maxwell velocity distribution with no macroscopic net velocity.

As before, particle trajectories and long-range interactions play no role and the Lorentz

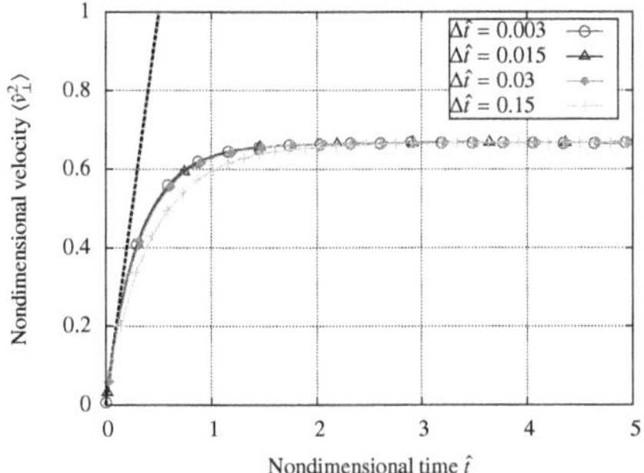

Figure 6.21: Numerical results for $\langle \hat{v}_\perp^2 \rangle$ during electron beam thermalization in a field of fixed ions using different time steps $\Delta \hat{t}$.

and Maxwell components of *dsmcPlasmaFoam* are deactivated. Using Eqs. 6.20 through 6.23 with $g \approx v_{\alpha 0}$, the reference time of the numerical test is estimated as $\tau_0 = 2.62 \times 10^{-3}$ s. During the simulations, the field electrons velocities are fixed. This corresponds to the assumption of a much higher number density of field electrons compared to beam electrons, which effectively pushes the beam electrons equilibrium conditions towards those of the initial field electrons. In the performed tests, particles are allowed to collide with any of the particles located in the same cell regardless of their species.

Let $\eta_{\alpha\beta} = \varepsilon_{\alpha 0}/k_B T_\beta = 50$ represent the initial energy ratio between the electron beam and the electron field. The analytic solutions for the early thermalization stages are (cp. [Nan97]):

$$\langle \hat{v}_x \rangle = 1 - 2\hat{t} \tag{6.26}$$

$$\langle \hat{v}_\perp^2 \rangle = (2 - \eta_{\alpha\beta}^{-1})\hat{t} \tag{6.27}$$

Furthermore, $\langle \hat{v}_\perp^2 \rangle$ in equilibrium is defined as follows (cp. [Nan97]):

$$\langle \hat{v}_\perp^2 \rangle = \frac{1}{\eta_{\alpha\beta}} = 0.02 \tag{6.28}$$

Comparison between $\langle \hat{v}_\perp^2 \rangle$ for the early stages in Eq. 6.27 and the equilibrium result in Eq. 6.28 reveals that the beam electrons are initially scattered in the perpendicular direction. However, the corresponding energy is not conserved in the species. Through further interactions with the field particles of fixed velocity, the beam electrons lose

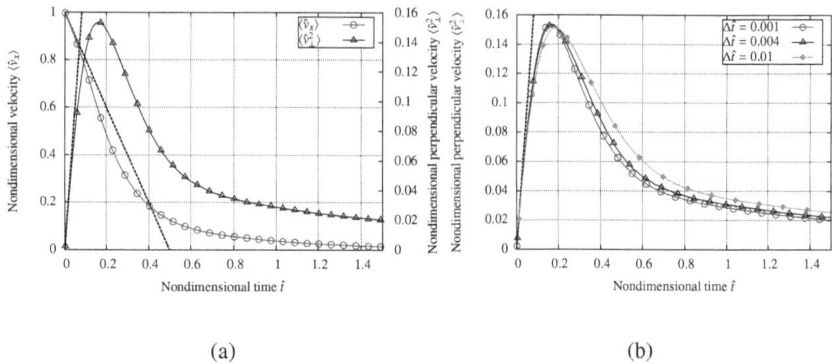

Figure 6.22: Numerical results for electron beam thermalization in a field of fixed velocity electrons. (a) $\langle \hat{v}_x \rangle$ and $\langle \hat{v}_\perp^2 \rangle$ as a function of time with Eqs. 6.26 and 6.27 shown as dashed lines; (b) Results for $\langle \hat{v}_\perp^2 \rangle$ as a function of the dimensionless time step $\Delta\hat{t}$.

energy and cool down. As a consequence, $\langle \hat{v}_\perp^2 \rangle$ reaches a maximum before its value starts to decay towards the equilibrium result $\langle \hat{v}_\perp^2 \rangle = 0.02$.

The results for $\langle \hat{v}_x \rangle$ and $\langle \hat{v}_\perp^2 \rangle$ as a function of time are shown in Fig. 6.22 (a). As can be seen, the expected maximum for $\langle \hat{v}_\perp^2 \rangle$ is reproduced by the model implemented in *dsmcPlasmaFoam* and its late stage value agrees with the analytically expected result $\eta_{\alpha\beta}^{-1} = 0.02$. In addition, the results for both $\langle \hat{v}_x \rangle$ and $\langle \hat{v}_\perp^2 \rangle$ at the start of the beam thermalization process are in very good agreement with Eqs. 6.26 and 6.27. Furthermore, the average beam electrons velocity in the x direction falls over time and approaches zero as the electrons are scattered in all directions. As evident from Fig. 6.22 (b), the results using different time steps are comparable, although lower $\Delta\hat{t}$ values lead to the equilibrium solution being reached slightly faster.

In the final beam thermalization case, the beam electrons α are initialised with the exact same conditions as in the previous example. However, the field medium consists now of an argon plasma in equilibrium consisting of both electrons β and argon ions β'. The field electrons β are initialised with the same parameters as in the previous test. On the other hand, the field ions initial temperature is defined as $T_{\beta'} = 0.02$ eV, their number density as $n_{\beta'} = 1 \times 10^{16}$ m^{-3} and the numerical weight is $W = 125$. The particles velocities of both field species are held constant over the course of the simulation. All other numericals settings as well as the reference time $\tau_0 = 2.62 \times 10^{-3}$ s remain unchanged compared to the previous example.

For the case of a plasma field, the solutions for the early thermalization stages are given by the following expressions (cp. [Nan97]):

$$\langle \hat{v}_x \rangle = 1 - 3\hat{t} \tag{6.29}$$

$$\langle \hat{v}_\perp^2 \rangle = (4 - \eta_{\alpha\beta}^{-1})\hat{t} \tag{6.30}$$

As evident from the numerical results depicted in Fig. 6.23, the implemented Nanbu cumulative collision model delivers good quality results for the electron beam thermalization in an argon plasma. The profiles for both $\langle \hat{v}_x \rangle$ and $\langle \hat{v}_\perp^2 \rangle$ agree with the analytic solutions obtained with Eqs. 6.29 and 6.30 in the initial phase of the thermalization process. Moreover, the numerical results approach the late stage equilibrium value $\eta_{\alpha\beta}^{-1} = 0.02$ over the course of the simulation and previous observations regarding the results accuracy as a function of the employed time step are confirmed.

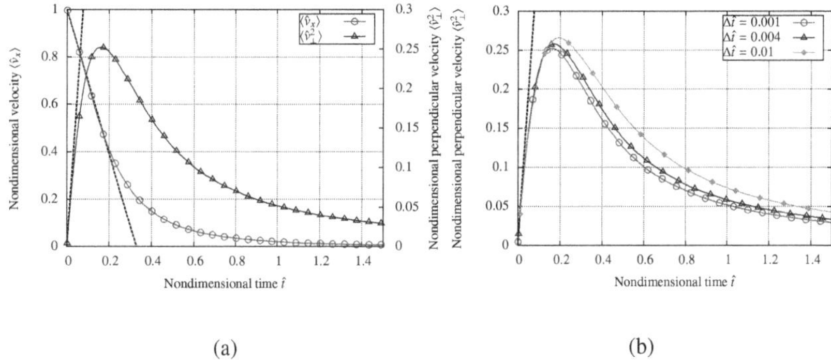

(a) (b)

Figure 6.23: Numerical results for electron beam thermalization in argon plasma. (a) $\langle \hat{v}_x \rangle$ and $\langle \hat{v}_\perp^2 \rangle$ as a function of time with Eqs. 6.29 and 6.30 shown as dashed lines; (b) Results for $\langle \hat{v}_\perp^2 \rangle$ as a function of the dimensionless time step $\Delta\hat{t}$.

The last considered study case for the Nanbu collisional model corresponds to the relaxation of the velocity distribution function in an electron particles ensemble. Here, a field of electrons with number density $n_\alpha = 1 \times 10^{16}$ m^{-3} and numerical weight $W = 12.5$ is initialised in the same numerical grid employed for the previous examples. Hence, the numerical test comprises a total of 100,000 test electrons. Periodic boundary conditions are applied to all boundary surfaces and the Lorentz and Maxwell solvers are deactivated. In order to analyse the relaxation process, the electrons are initialised with different energy values in the spatial coordinates. Specifically, $T_x = 1.3T_y = 1.3T_z$. The different temperature values imply an ellipsoidal form of the Maxwell velocity distribution as initial condition. This is implemented by defining the initial velocity components as follows:

$$\begin{aligned} v_{\alpha 0,x} &= \sqrt{\frac{1.3k_B T_0}{m_\alpha}} Z_1 \\ v_{\alpha 0,y} &= \sqrt{\frac{k_B T_0}{m_\alpha}} Z_2 \\ v_{\alpha 0,z} &= \sqrt{\frac{k_B T_0}{m_\alpha}} Z_3 \end{aligned} \tag{6.31}$$

where Z_1,Z_2 and Z_3 are random numbers following a standard normal distribution and $T_0 = 2$ eV. Hence, the overall electron temperature T_e is given by,

$$T_e = \frac{1}{3}T_x + \frac{2}{3}T_y = 2.2\,\text{eV} \tag{6.32}$$

The reference time for the relaxation of the velocity function is obtained from the following expression (cp. [Nan97]):

$$\frac{1}{\tau_0} = \frac{n_\alpha q_\alpha^4 \ln \Lambda}{8\pi\sqrt{2}\varepsilon_0^2 m_\alpha^{1/2} (k_B T_e)^{3/2}} \tag{6.33}$$

Furthermore, with $\Delta T = T_x - T_e$, $(\Delta T)_0 = \Delta T$ for $t = 0$ and $\hat{t} = t/\tau_0$, the relaxation process is analytically defined by the following expression (cp. [Nan97]):

$$\Delta T = (\Delta T)_0 \exp\left(-\frac{8}{5\sqrt{2\pi}}\hat{t}\right) \tag{6.34}$$

Fig. 6.24 shows the numerical and analytic results for the relaxation process. As can be seen, the normalized energy in the x component drops over the course of the simulation, as the excess kinetic energy in x is redistributed via Coulomb interactions to the y and z spatial components. As a consequence, the depicted energy in the y component increases with $\hat{t}$. Although the numerical relaxation process advances slower than the analytic prediction using Eq. 6.34, the energy components clearly approach the equilibrium state towards the end of the simulation. Considering this, the results

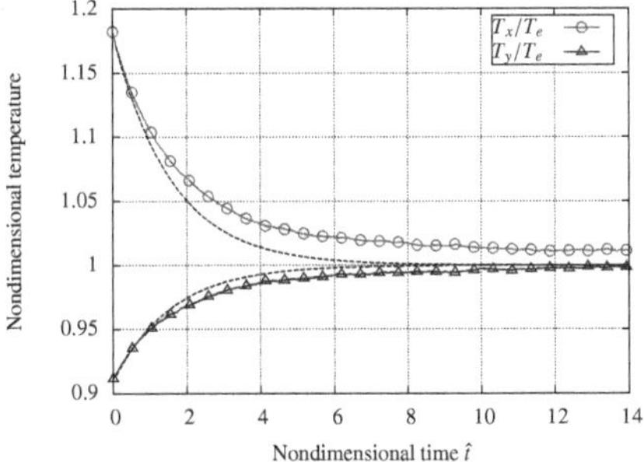

Figure 6.24: Results for T_x/T_e and T_y/T_e for the relaxation of the velocity distribution function in an electron field as a function of time with Eq. 6.34 shown as dashed lines.

for the relaxation of the velocity distribution function can be considered to be in good agreement with the theoretically obtained profiles.

The case studies for the implemented cumulative collisional model in *dsmcPlasmaFoam* presented in this section deliver highly satisfactory results as the solver is able to accurately reproduce basic short-range Coulomb collisional phenomena in a plasma. Based on these results, this component of *dsmcPlasmaFoam* is considered to be validated.

6.5 Electron-neutral collisions

The interactions between electrons and neutrals in *dsmcPlasmaFoam* are modelled using the collisional approach described in Section 5.3.5.1. As already mentioned, the three possible outcomes of a collision between an electron and a neutral particle are elastic collision, electronic excitation and ionisation. In order to assess the implemented model, a parametric study focused on the macroscopic or drift velocity v_d of an electron ensemble through a neutral argon particles field as well as on the electron impact ionisation coefficient α is presented in this section. The validation approach is based on the study described in [NK97].

In the parametric study, a field of neutral argon atoms is initialised between a cathode and an anode in a numerical domain of dimensions $d = 50$ mm, $L_1 = x$ mm and $L_2 = 0.2$ mm. Here, d corresponds to the separation between cathode and anode and two different values for the channel height $L_1 = 50$ mm (for simulations with low computational requirements) and $L_1' = 10$ mm (for cases with high computational requirements) are employed. The domain is discretized using 250 cells in the d dimension (x coordinate), 250 cells for $L_1 = 50$ mm, 50 cells for $L_1' = 10$ mm (y coordinate) and 1 cell for L_2 (z coordinate). The argon neutral gas is initialised with a temperature of 323 K and following a Maxwell distribution without macroscopic net velocity. Furthermore, an electric potential of 500 V is applied between cathode and anode of the geometry. Hence, the electric field magnitude between cathode and anode equals 10,000 V/m. In the numerical simulations, electrons are dynamically initialised at the cathode in a way that approximately 100 test electrons enter the domain each iteration for $L_1 = 50$ mm. Furthermore, for the cases with reduced channel height $L_1' = 10$ mm, the amount of initialised electrons per iteration equals 20. With the time step $\Delta t = 1 \times 10^{-10}$ s, the cathode area $L_1 \times L_2 = 1 \times 10^{-5}$ m^2 and $W_e = 2.05 \times 10^5$ as numerical weight for the electrons, the boundary condition for the real particles flux through the cathode becomes 2.05×10^{22} particles /sm^2 and the value is independent of the used channel height.

E/N_A [Td]	N_A [1/m^3]
10,000	1×10^{21}
5,000	2×10^{21}
1,000	1×10^{22}
500	2×10^{22}
100	1×10^{23}
50	2×10^{23}
10	1×10^{24}
5	2×10^{24}
1	1×10^{25}

Table 6.5: Parametric study for validation of electron-neutral collisional model.

In the numerical setup, the electrons initialised each iteration at the cathode's surface are accelerated by the external, constant electric field towards the anode, where they exit the domain. During their motion through the channel geometry, the electrons collide with the neutral test particles leading to changes in the electrons velocities as well as to electronic excitation and ionisation events. Since all potential collisional outcomes result in scattering of the velocity vectors, the numerically obtained average electron drift velocity from cathode to anode can be compared with literature data and used to assess the quality of the approach implemented in *dsmcPlasmaFoam*. In a similar way, the ionisation coefficient, estimated as the ensemble average of the number of ion-electron pairs created by each moving electron per unit length due to collisions with neutrals, is compared to literature values. Note that internal electric forces generated by the electronic distribution in the domain are not considered and the Maxwell solver is deactivated. Except for the cathode and anode surfaces, all domain boundaries are defined using periodic boundary conditions. Since the amount of neutrals in the domain and the associated collision frequency are key parameters for both the electrons drift velocity and the ionisation coefficient, the simulations are performed for a wide range of argon number densities as listed in Table 6.5. Note that the different number densities for the neutral gas are implemented together with a change of the numerical weight so that approximately 8 test particles per cell are present in the domain for all considered cases. This equals a total of roughly 500,000 for $L_1 = 50$ mm and 100,000 for $L_1' = 10$ mm. In order to keep the computational requirements low and considering the low velocity and high inertia of the heavy particles, the Lorentz solver is activated only for the electrons motion. Furthermore, the tests are run for simulation times ranging from 1×10^{-8} s (10 ns) up to 3×10^{-7} s (300 ns) according to the convergence speed of the particular test. The obtained results for the electrons drift velocity in the x direction v_d and the ionisation coefficient α are averaged over the last 50 time steps of the numerical run and the results obtained from the parametric study are shown in Figs. 6.25 and 6.26.

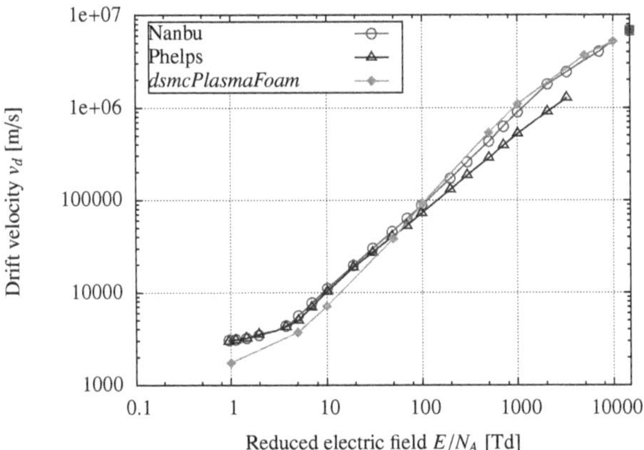

Figure 6.25: Electron drift velocity between cathode and anode v_d obtained with *dsmcPlasmaFoam* as a function of the reduced electric field and comparison with numerical results from Nanbu [NK97] and Phelps [Phe].

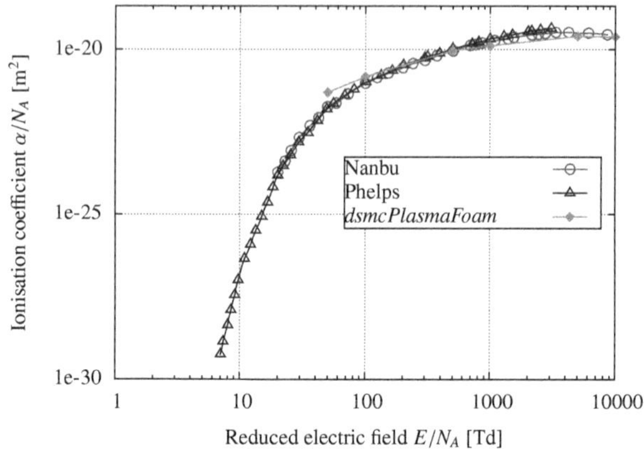

Figure 6.26: Ionisation coefficient results obtained with *dsmcPlasmaFoam* as a function of the reduced electric field and comparison with numerical results from Nanbu [NK97] and Phelps [Phe].

As evident from Fig. 6.25, the obtained results for the drift velocity of the electrons v_d are in very good agreement with the reference numerical data. For the lowest considered argon number density $N_A = 1 \times 10^{21}$ m^{-3}, which corresponds to a reduced electric field force $E/N_A = 10,000$ Td, the electrons cannot collide very often with the neutrals and the macroscopic drift velocity from cathode to anode is dominated by the acceleration induced by the external electric force. This point becomes clear from the direct comparison of the corresponding average drift velocity $v_d \approx 5.19 \times 10^6$ m/s with the average electron velocity without neutral background gas $v \approx 6.76 \times 10^6$ m/s valid for $E/N_a = \infty$ and shown as a lone red square on the RHS edge of Fig. 6.25.

The gradual increase in the number density of the neutrals results in higher collision frequencies between electrons and argon atoms. As a result, the velocity scattering process becomes more prominent and the net macroscopic velocity in the x direction decreases with lower values of E/N_A. However, it should be noted that the drift velocity is not only a function of scattering angles and overall collision frequency. Indeed, the proportion in which each of the three collisional outcomes takes place and hence, the sampling scheme depicted in Fig 5.14, plays a decisive role on the quality of the results. This becomes evident by recalling the fact that for excitation and ionisation events, the threshold energies of excitation and ionisation are removed from the kinetic energy of the colliding electrons and hence, from the system. This has a direct effect on the overall kinetic energy of the electrons ensemble and is therefore closely related to the electrons net velocities. Considering this and based on the very high quality of the electron drift velocity results shown in Fig. 6.25, both the scattering process due to electron-neutral interactions as well as the implemented collisional sampling approach can be considered to be validated.

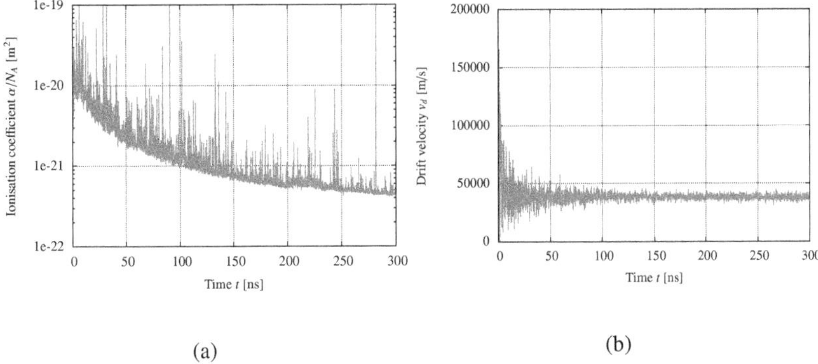

(a) (b)

Figure 6.27: Numerical results for the test case $E/N_A = 50$ Td as a function of time. (a) Ionisation coefficient α/N_A; (b) Electron drift velocity v_d.

The results for the ionisation coefficient α/N_a as a function of the reduced electric force are depicted in Fig. 6.26. As can be seen, *dsmcPlasmaFoam* produces high quality results in the range $E/N_A \geq 100$ Td or $N_A \leq 1 \times 10^{23}$ m^{-3}. On the other hand, the obtained results for high values of number density do not agree as well with the reference literature data and deviations for $E/N_A = 50$ Td can be observed. For this case, the established disagreement can be explained by an observed exponential increase in computational time needed for the ionisation coefficient results to reach stationary conditions with high values of N_A. As an example, for $E/N_A = 1{,}000$, overall stationary conditions are easily reached for a simulation time of $t = 1 \times 10^{-8}$ s or 10 ns. However, for $E/N_A = 50$ Td and as shown in Fig. 6.27 (a), the ionisation coefficient is still not completely stationary after 300 ns as the results still exhibit a slight decreasing tendency in the final stages of the computation. In comparison and as shown in Fig. 6.27 (b), the electrons drift velocity reach stationary conditions much faster. Note that since stationary results are far from being reached for $E/N_A < 50$ Td, the results are not included in Fig. 6.26. The low convergence time is not a particular surprise considering the typical limitations of a kinetic modelling approach for high number densities and conditions approaching the continuum regime. Based on these observations, additional improvements on the quality of the obtained results for α and high values of N_A can be expected if an increase in the amount of numerical test particles for both electrons and neutrals as well as an extension of the total simulation time are implemented. Nevertheless, the ionisation coefficients obtained for the range $E/N_A \geq 100$ Td are in very good agreement with literature data and the simulations exhibit relatively low computational requirements. Based on this fact and the overall good quality for the estimated drift velocity, the approach implemented in *dsmcPlasmaFoam* for the handling of interactions between electrons and neutral atoms can be considered to be validated and accurate, providing a powerful tool for the kinetic modelling of the important ionisation process in the plasma state.

6.6 Summary

In this chapter, the main components of the developed solver *dsmcPlasmaFoam* have been tested using a wide range of numerical study cases. The Maxwell solver, one of the standard elements of a classic PIC algorithm, has been shown to deliver accurate results for different geometrical setups and the importance of the used boundary conditions for the electric potential was highlighted in the process. Moreover, different approaches for the modelling of the charge carrier motion through an electric field using the Lorentz solver have been evaluated. The results clearly show the overall benefits of the time-centred Leapfrog algorithm as well as its accurate implementation in *dsmcPlasmaFoam*. The positive results for a typical plasma wave test case confirm not only the suitability of the zero-order or NGP approach but also, the proper interaction of all of the above key elements of the solver for the numerical modelling of long-range plasma phenomena.

The implemented cumulative collisional approach for Coulomb interactions has been shown to deliver extremely accurate results compared to analytic data for study cases like electron beam relaxation and velocity distribution thermalization. The associated MCC approach allows the solver to reproduce short-range plasma phenomena taking place over a distance of the order of the Debye length. For cases in which an established plasma is not the initial state of the simulation, the implemented electron-neutral collisional schemes and the ionisation procedure constitute a key tool for the consideration of plasma generation processes as expected in, among others, typical electric propulsion systems for spacecraft like the INGA III arcjet thruster. The results presented in this chapter confirm the solver *dsmcPlasmaFoam* as a powerful, flexible and accurate modelling tool covering a wide range of typical plasma phenomena. Furthermore, the solver is expected to serve as one of the two main submodels in the envisioned hybrid

PIC-MCC-FVM plasma model currently in development at the Center of Applied Space Technology and Microgravity ZARM.

Chapter 7

Conclusion

7.1 Summary

New trends regarding fundamental design approaches of orbital spacecraft have been developing in the space industry in recent years. They include an increased interest in miniaturized satellites with the ability to perform spacecraft formation flying as well as the partially linked general rise in the use of electric propulsion systems for orbit and attitude control. A number of challenges, including the appearance of rarefied conditions inside the nozzles of miniaturised propulsion devices as well as an increased need for new approaches regarding the numerical modelling of plasma phenomena inside electric propulsion systems arise from these trends. The present thesis has focused on both these topics.

Chapters 2 and 3 of this work present the theoretical and numerical concepts needed for the study of both rarefied gas flows as well as plasma phenomena. After introducing the Knudsen number and the different flow regimes, classic transport equations, applicable to flows in the continuum regime, have been developed and discussed. Furthermore, fundamental concepts of plasma physic, gas kinetics and the Laval nozzle have been described. With this background, numerical approaches for modelling flows in the continuum regime, under rarefied conditions as well as in the plasma state have been introduced and analysed in Chapter 3.

Based on the theoretical and numerical concepts presented in the early chapters of this work, rarefied conditions in transonic micronozzle flows have been examined in Chapter 4 both numerically and experimentally. The study focuses on the cold-gas expansion of several noble gases through a millimetre-scale Laval nozzle, with the experimental series taking advantage of the INGA III setup at ZARM. The Knudsen numbers of the considered cases lie in the range $0.02 < Kn < 0.33$. Hence, the flow conditions correspond to the slip-flow and transition regimes. Comparisons between the experimentally obtained pressure in the ionisation chamber, numerical Navier-Stokes results with no-slip boundary conditions and the DSMC method were performed in order to assess the applicability of the numerical approaches as well as the slip-behaviour of the studied flows.

The performed analysis has shown the DSMC and experimental results to be in good agreement in the range $Kn > 0.1$. Furthermore, the Navier-Stokes simulations produce better results for lower Knudsen numbers in the range $Kn < 0.05$ but still outside the continuum regime. For the higher values of Kn considered in the studies, a strong disagreement between the experimental data and the pressure values in the ionisation chamber estimated with continuum-based simulations is established as the relative deviations exceed 50%. In these cases, the fact that flow conditions are well inside the slip-flow and transition regimes combined with the use of a no-slip boundary condition for the Navier-Stokes simulations leads to a considerable overestimation of the pressure drop through the Laval nozzle. Based on this information, a relation describing the deviation of the pressure drop through the nozzle as a function of the Knudsen number has been developed. Analysis of the obtained Knudsen-function, which can be formulated with either gas-dependent or gas-independent accommodation coefficients, has shown that small collision cross sections are linked to strong deviations from the no-slip assumption under rarefied gas conditions, with the obtained relation being described by a second degree polynomial function. Furthermore, it has been shown that deviation from the no-slip behaviour becomes less pronounced with larger collisional cross sections, as the obtained Knudsen relation approaches a linear function.

Provided that a rigorous handling of the flow characteristics is not required, the developed Knudsen functions can be employed as a correcting tool for numerically obtained pressure results using continuum-based models, avoiding this way a numerical case distinction between continuum, slip-flow and transition regimes. This can be particularly helpful for large parametric studies with different propellant gases and across multiple mass flow rates and flow regimes. Considering that the accommodation coefficients are developed from a dimensionless analysis, the applicability of the Knudsen function to comparable Laval nozzle geometries is expected. As part of the analysis, thrust and specific impulse for the cold-gas thruster have been determined based on numerical results. The maximum thrust value reached in the study is 640 μN and the maximum specific impulse is 18 s, both achieved with argon at a mass flow rate of 3.568 mg/s. However, an increase of the thickness of the viscous boundary layer through the diffuser of the micronozzle has been observed, leading to a shock-less decrease of Mach number, flow velocity and hence, of thrust efficiency. The study has shown that the negative effects of the viscous boundary layer can be reduced by using higher mass flow rates (higher *Re* numbers) and through a reduction of the diffuser length. In particular, an 85% reduction of the diffuser length for xenon with $\dot{m} = 3.568$ mg/s leads to an improvement in thrust performance of roughly 55%. This demonstrates the importance of the characterization of the viscous losses for thrust optimization in propulsion systems operating with micronozzles.

Chapters 5 and 6 of this thesis are dedicated to the conception of a global modelling approach applicable to electric propulsion systems as well as to the development of a kinetic PIC-MCC solver for the description of the dynamic plasma processes taking place inside them. The global hybrid modelling concept is based on the coupling of a kinetic submodel for plasma phenomena with a fluid submodel for ohmic gas heating and neutral gas handling. The fully developed kinetic submodel referred to as *dsmcPlasmaFoam* includes, among others, classic PIC elements, short-range Coulomb collisions as well as ionisation and recombination interactions. Additional features like dynamic numerical weighting have been conceived and implemented in the solver over the course of its development which might prove beneficial in future numerical studies concerning not only plasma phenomena, but any research field heavily relying on kinetic particle models.

The validation results for the Maxwell solver, one of the standard elements of a classic PIC algorithm, have confirmed both its ability to deliver accurate results for different geometrical setups as well as the importance of the employed boundary conditions for the electric potential. Moreover, different approaches for the modelling of the charge carrier motion through an electric field using the Lorentz solver have been evaluated in this work. The performed tests have demonstrated the overall benefits of the time-centred Leapfrog algorithm as well as confirmed its accurate implementation in *dsmcPlasmaFoam*. The results obtained for a typical plasma wave problem have confirmed both the suitability of the zero-order or NGP weighting approach as well as the proper interaction of all of the typical PIC elements of the solver. Based on these results, the PIC components of *dsmcPlasmaFoam* are considered to be validated as the developed solver has been shown to accurately model typical long-range plasma phenomena.

The implemented cumulative collisional approach for Coulomb interactions has been shown to deliver extremely accurate results compared to analytic data for study cases like electron beam relaxation and velocity distribution thermalization. The associated MCC approach allows the solver to reproduce short-range plasma phenomena taking place over a distance of the order of the Debye length. For cases in which an established plasma is not the initial state of the simulation, the implemented electron-neutral collisional schemes and the ionisation procedure constitute a key tool for the consideration of plasma generation processes as expected in, among others, typical electric propulsion systems for spacecraft like the INGA III arcjet thruster. The results presented in the final chapter of this thesis confirm the solver *dsmcPlasmaFoam* as a powerful, flexible and accurate modelling tool covering a wide range of typical plasma phenomena. The developed solver is expected to serve as one of the two main submodels in the envisioned hybrid PIC-MCC-FVM plasma modelling approach currently in development at the Center of Applied Space Technology and Microgravity ZARM.

7.2 Conclusion and outlook

The study of gas rarefaction effects on transonic micronozzle flows has provided extremely interesting insights into the effects of high Knudsen numbers on pressure drop and general flow characteristics, as well as into the way flows under rarefied conditions deviate from the no-slip boundary condition. Furthermore, the effects of viscous losses caused by the Knudsen boundary layer constitute one key aspect to consider during the development of miniaturized propulsion systems in which the ratio of wall surface area to nozzle volume becomes very high. In this regard, the effects of the diffuser's opening angle and the nozzle throat diameter on the pressure drop deviations are expected to be the subject of future complementary studies. The transferability of the obtained coefficients to other propellants such as diatomic gases or the widely used hydrazine or ammonia gases in arcjet thrusters might also be explored in the future.

Regarding the numerical modelling of plasma phenomena inside electric propulsion systems, *dsmcPlasmaFoam* is considered to be a major first step in an attempt to develop a global hybrid PIC-MCC-FVM solver at ZARM, aiming to tackle the numerous challenges concerning plasma modelling for electric spacecraft propulsion systems. Besides the obvious benefits in terms of thruster design and optimization, the development of an accurate hybrid numerical model has the potential to provide new insights into plasma phenomena, energy exchange and transport mechanisms not only in electric propulsion systems but also in other plasma-related applications like plasma arc welding and electron beam processing. The next steps in this long-term project include the development and implementation of a mapping functionality for the one-way coupling between a fluid submodel and *dsmcPlasmaFoam*. Furthermore, the development of appropriate numerical techniques for the handling of the time scale differences expected between the phenomena modelled by *dsmcPlasmaFoam* and a fluid submodel is key for the success of the envisioned modelling concept. At the time of writing, work on these topics is already in progress at ZARM with support from the German Research Foundation (*Deutsche Forschungsgemeinschaft* DFG).

Appendix A

Mathematical Theorems

A.1 Divergence theorem

The volume integral of the divergence of a vector field inside a volume V equals the total of sources and sinks within V. At the same time, the result of this operation equals the flux of the vector field into or out of the volume through its boundaries. This relation is mathematically expressed by the divergence theorem:

$$\int_V \nabla \cdot F \,\mathrm{d}V = \int_{\delta V} (F \cdot n)\,\mathrm{d}A \tag{A.1}$$

where the LHS represents the volume integral of $\nabla \cdot F$ over the volume V, the RHS the surface integral over the surface δV with normal surface vector n and F is a continuously differentiable arbitrary vector field.

A.2 Reynolds transport theorem

Let ψ represent an arbitrary property. The rate of change of the volume integral of ψ is given by:

$$\frac{\mathrm{d}}{\mathrm{d}t}\int_V \psi \,\mathrm{d}V = \int_V \frac{\partial \psi}{\partial t}\,\mathrm{d}V + \int_{\delta V} \psi (u_b \cdot n)\,\mathrm{d}A \tag{A.2}$$

where V stands for the considered volume, u_b for the velocity of the differential area element $\mathrm{d}A$, n for the normal surface vector of the area element and the last term on the RHS constitutes the surface integral over the area δV.

Therefore, the rate of change of the property ψ in the control volume V equals the sum of the volume integral of its partial derivative with respect to time and an additional component which takes into account the temporal change caused by the deformation of the dynamic boundaries represented by $u_b(x,t) \cdot n$. For the case of continuum mechanics, the theorem is often applied to fluid or solid parcels assuming that no material enters or leaves the considered parcel. In this case the particles in the control volume move with the boundaries and,

$$u_b \cdot n = u \cdot n \qquad \text{(A.3)}$$

where u stands for the element's velocity. Note that ψ can be an scalar, vector or tensor value.

Appendix B

Source Code dsmcPlasmaFoam

For completeness, key elements of the source code in *dsmcPlasmaFoam* are presented in this section. They include the Maxwell and Lorentz solvers, the collisional approaches for Coulomb and electron-neutral interactions as well as the dynamic weighting scheme.

B.1 Maxwell solver

The Maxwell solver for the determination of the electric field based on the solution of Poisson's equation is implemented in the master solver file dsmcPlasmaFoam.C. Note that for the first iteration, the Maxwell solver component (lines 58:80) is executed prior to all other steps in the algorithm in order to obtain E_{old} and F_{old} for $t = 0$ and push the initial velocities back in time during the first iteration of the Lorentz solver. In all other time steps, the Maxwell solver is run after the *evolve* function (including Lorentz and collisional schemes) is executed (lines 88:101).

Listing B.1: File: dsmcPlasmaFoam.C - Maxwell solver.

```
24  Application
25      dsmcPlasmaFoam
26
27  Description
28      PIC - MCC Solver based on Direct simulation Monte Carlo (DSMC)
29
30  \*---------------------------------------------------------------------------*/
31
32  #include "fvCFD.H"
33  #include "dsmcCloud.H"
34  #include "simpleControl.H" //for laplacian equation
35
```

```
36 // * * * * * * * * * * * * * * * * * * * * * * * * * * * * * * * * * * * * * //
37
38 int main(int argc, char *argv[])
39 {
40     #include "setRootCase.H"
41     #include "createTime.H"
42     #include "createMesh.H"
43     simpleControl simple(mesh);
44     // * * * * * * * * * * * * * * * * * * * * * * * * * * * * * * * * * * //
45
46     Info<< nl << "Constructing dsmcCloud with dsmcPlasmaFoam" << endl;
47
48     dsmcCloud dsmc("dsmc", mesh);
49
50     Info<< "\nStarting time loop\n" << endl;
51     dimensionedScalar AsUnit ("zero",  dimensionSet(0, 0, 1, 0, 0, 1, 0), 1);
                    // Unit for charge
52
53     bool firstIteration = true;
54
55     //Start of solver
56     while (runTime.loop())
57     {
58         if (firstIteration == true)
59         {
60             //calculate density fields for initial particle distribution
61             dsmc.initialEvolve();
62
63             //calculate E Field for initial particle distribution
64             if (dsmc.MaxwellSolver == true)
65             {
66                 while (simple.correctNonOrthogonal())
67                 {
68                     solve
69                     (
70                             fvm::laplacian(dsmc.phi()) + dsmc.rhoNElektron()*
                                dsmc.constProps(0).charge()*AsUnit/
                                electromagnetic::epsilon0
71                             + dsmc.rhoNIon()*dsmc.constProps(2).charge()*AsUnit
                                /electromagnetic::epsilon0
72                     );
73                 }
74                 dsmc.phi().correctBoundaryConditions();
75                 dsmc.EInt()=-1.0*fvc::grad(dsmc.phi());
76                 dsmc.laplacetest()=-fvc::laplacian(dsmc.phi())*electromagnetic
                    ::epsilon0/(dsmc.constProps(0).charge()*AsUnit);
77             }
78
79             firstIteration = false;
80         }
81
82         //Evolve function including Lorentz and collisional solvers
83         dsmc.evolve();
84
85         Info<< "Time = " << runTime.timeName() << nl << endl;
86
87         // ********* Maxwell Solver *********  //
```

```
88          if (dsmc.MaxwellSolver == true)
89          {
90              while (simple.correctNonOrthogonal())
91              {
92                  solve
93                  (
94                          fvm::laplacian(dsmc.phi()) + dsmc.rhoNElektron()*dsmc.
                                constProps(0).charge()*AsUnit/electromagnetic::
                                epsilon0
95                          + dsmc.rhoNIon()*dsmc.constProps(2).charge()*AsUnit/
                                electromagnetic::epsilon0
96                  );
97              }
98              dsmc.phi().correctBoundaryConditions();
99              dsmc.EInt()=-1.0*fvc::grad(dsmc.phi());
100             dsmc.laplacetest()=-fvc::laplacian(dsmc.phi())*electromagnetic::
                    epsilon0/(dsmc.constProps(0).charge()*AsUnit);
101         }
102         // ********* End Maxwell Solver *********   //
103         dsmc.info();
104
105         runTime.write();
106
107         if (runTime.outputTime())
108                 {
109             dsmc.weightProperties.set("plasmaWt", dsmc.plasmaWt);
110             dsmc.weightProperties.Foam::regIOobject::write();
111
112                 }
113
114         Info<< nl << "ExecutionTime = " << runTime.elapsedCpuTime() << " s"
115             << "  ClockTime = " << runTime.elapsedClockTime() << " s"
116             << nl << endl;
117     }
118
119     Info<< "End\n" << endl;
120
121     return(0);
122 }
123 // ************************************************************************* //
```

B.2 Lorentz solver

The Lorentz solver is called by the *evolve* function of *dsmcPlasmaFoam* with the main steps included in the file DsmcParcel.C. Note that each particle carries two different velocities in *dsmcPlasmaFoam*: The Leapfrog velocity $p.ULeapfrog()$ valid for intermediate time steps and the velocity $U\_$ valid at normal, non-shifted time steps. If the particle enters the Lorentz solver for the very first time, its Leapfrog velocity is pushed back in time to the time level $t - \Delta t/2$ in lines 107:113 as required for the time-centred Leapfrog scheme. Leapfrog velocities are advanced to the intermediate time step $t+\Delta t/2$ in lines 116:122 and the *trackToFace* method is subsequently executed based on the particle's estimated final position. Finally, the velocity $U\_$ at the time level $t+\Delta t$ is computed in lines 137:148. Note that $U\_$ is used in collisional schemes and for writing of simulation results.

Listing B.2: File: DsmcParcel.C - Lorentz solver.

```
24 \*---------------------------------------------------------------------------*/
25
26 #include "DsmcParcel.H"
27 #include "meshTools.H"
28
29 #include <iostream>
30 #include <fstream>
31 #include <string>
32 #include <stdio.h>
33 #include <stdlib.h>
34
35 // * * * * * * * * * * * * * * * * Member Functions  * * * * * * * * * * * * * * //
36
37 template<class ParcelType>
38 template<class TrackData>
39 bool Foam::DsmcParcel<ParcelType>::move(TrackData& td, const scalar trackTime)
40 {
41     typename TrackData::cloudType::parcelType& p =
42         static_cast<typename TrackData::cloudType::parcelType&>(*this);
43
44     td.switchProcessor = false;
45     td.keepParticle = true;
46
47     const polyMesh& mesh = td.cloud().pMesh();
48     const polyBoundaryMesh& pbMesh = mesh.boundaryMesh();
49
50     /////////~~~~~~~~~~~~~~~~For new external Force
        -------------------~~~~~~~~~~~~~~~~~~~~~~~~~~~~///////////////
51
52     const constantProperties& constProps = td.cloud().constProps(typeId_);
53     const volVectorField EExt_ = td.cloud().EExt();
54     const volVectorField EInt_ = td.cloud().EInt();
55     /////////~~~~~~~~~~~~~~~~For new external Force
        -------------------~~~~~~~~~~~~~~~~~~~~~~~~~~~~///////////////
56
57     scalar tEnd = (1.0 - p.stepFraction())*trackTime;
58
59     if (typeId_ == 2 && td.cloud().IonMotion == false) //Ion skipped
60     {
61         tEnd = 0;
62     }
63
64     if (typeId_ == 1 &&  td.cloud().NeutralMotion == false) //Neutral skipped
65     {
66         tEnd = 0;
67     }
68
69     const scalar dtMax = tEnd;
```

```
70      scalar tracktoFaceResult;
71
72      // For reduced-D cases, the velocity used to track needs to be
73      // constrained, but the actual U_ of the parcel must not be
74      // altered or used, as it is altered by patch interactions an
75      // needs to retain its 3D value for collision purposes.
76      vector Utracking = U_;
77
78      scalar deltaTforLog = dtMax;
79      bool particleLeapfrogFirsIt = true;
80
81      while (td.keepParticle && !td.switchProcessor && tEnd > ROOTVSMALL)
82      {
83
84          // Apply correction to position for reduced-D cases
85          meshTools::constrainToMeshCentre(mesh, p.position());
86
87          // Set the Lagrangian time-step
88          scalar dt = min(dtMax, tEnd);
89
90
91          //////////////////   Determination of F_old at initial position of
                particle x_old     ////////////////////////////////////
92          vector ETot;
93          vector EInt;
94          vector EExt;
95          vector F;
96
97          if (particleLeapfrogFirsIt == true)
98
99          {
100             EExt = EExt_[p.cell()];
101             EInt = EInt_[p.cell()];
102             ETot = EExt+EInt;
103             F = constProps.charge()*ETot;
104         }
105
106         //Leapfrog velocity initialisation for very first iteration (velocity
                pushed back)
107         if (p.parFirstLeapFrog() == 1)
108         {
109             p.ULeapfrog() = U_ -0.5*F*dt / constProps.mass();
110             p.parFirstLeapFrog() = 0;
111             Utracking = p.ULeapfrog();
112             meshTools::constrainDirection(mesh, mesh.solutionD(), Utracking);
113         }
114
115         //Normal velocity push-forward to Delta + t/2
116         if (particleLeapfrogFirsIt == true)
117         {
118             p.ULeapfrog() += F*dt/constProps.mass();
119             Utracking = p.ULeapfrog();
120             meshTools::constrainDirection(mesh, mesh.solutionD(), Utracking);
121             particleLeapfrogFirsIt = false;
122         }
123
124         //Tracktoface returns fraction of deltaT until face/patch etc found
```

```
125            td.cloud().correctedTime = dt;  //For manual velocity update prior to
                   patch interactions inside trackToFace method
126            dt *= p.trackToFace(p.position() + dt*Utracking , td);
127            deltaTforLog = dt;
128            scalar leaptrackFraction = dt/td.cloud().correctedTime;
129
130            if (leaptrackFraction == 0 && typeId_ == 0) //Special Case at start of
                   sim in which particle changes cell without any motion. Next loop to
                   restart everything
131            {
132                p.parFirstLeapFrog() = 1;
133                particleLeapfrogFirsIt = true;
134            }
135
136            ///////////~~~~~~~~~~~~~~~~~~~~~~~~~~~~ Velocity update
                   ~ -----~~~~~~~~~~~~~~~~~~~~~~~~~~~~///////////////
137            if (td.cloud().correctVelocity == true ) //Update Vel if no patch has
                   been found --> Only at final position
138            {
139                ////// Force update at new position
140                EExt = EExt_[p.cell()];
141                EInt = EInt_[p.cell()];
142                ETot = EExt+EInt;
143                F = constProps.charge()*ETot;         //F_new for x_New
144
145                //Velocity advanced 0.5 dt --> same time level as Xnew and Fnew
                       ---> For writing purposes
146                U_ = p.ULeapfrog() + 0.5 * trackTime*F/constProps.mass(); // 1/2
                       deltaT forward to agree with time level of x and F.
147
148            }
149
150            tEnd -= dt;
151
152            p.stepFraction() = 1.0 - tEnd/trackTime;
153
154            if (p.onBoundary() && td.keepParticle)
155            {
156                if (isA<processorPolyPatch>(pbMesh[p.patch(p.face())]))
157                {
158                    td.switchProcessor = true;
159                }
160            }
161        }
162
163        return td.keepParticle;
164
165
166    }
```

B.3 Cumulative Coulomb collisions

The cumulative Coulomb collisional approach is implemented in *dsmcPlasmaFoam* in the *collisions* function of the *evolve* method, both located in the file DsmcCloud.C. As described in Section 5.3.4.1, all charge carriers interact with other carriers at least once and the concept of collision probability is not needed. With electrons and ions as the two considered species, the initial part of the algorithm (lines 803:970) determines the amount of collisions taking place between the charge carriers present in the cell. Subsequently, collision pairs are randomly selected in lines 988:1180 prioritizing particles in the same subcell and considering the fact that particles are to collide only once during the time step. In lines 1185:1238, several lists indicating the particles availability for additional collisions are updated. Nanbu's collisional parameter τ (s in Section 5.3.4.1) is estimated in the next steps and the value is used as input for the sub-function *binaryCollision().collide*, where the post-collision velocities are computed. Note that an update of the Leapfrog velocities of the charge carriers is performed based on the obtained post-collision velocities (lines: 1304:1310).

Listing B.3: File: DsmcCloud.C - Coulomb collision candidates selection.

```
799  /////////*********** Collision Candidates Selection,
         ********////////////////
800
801  //Coulomb Collisions only if more than one charge carrier in cell
802  // Coulomb Collisions only if more than one e- in cell
803   if (NElec >= 2 && NanbuCollisions == true)
804   {
805           // available Candidates for collision; available =1; already Selected = 0
806           List<label> availableCand(cellParcels.size(),1);
807
808           scalar CollElecIonB = NElec*rhoNIonColl/rhoNTotalPlasma;
809           //Number of Electron - Ion Collisions
810           label CollElecIon = NElec*rhoNIonColl/rhoNTotalPlasma;
811           scalar Rsd = CollElecIonB - CollElecIon;
812           if (Rsd >= 0.5)
813           {
814                   //CollElecIon rounded up
815                   CollElecIon += 1;
816           }
817                           //Number of Electron - Electron Collisions
818                           label CollElecElec;
819                           //Number of Ion - Ion Collisions
820                           label CollIonIon;
821
822                           //Number of Partners for Electron - Electron Collisions
823                           label NElecElec = NElec - CollElecIon;
824                           //Number of Partners for Ion - Ion Collisions
825                           label NIonIon = NIon - CollElecIon;
826
```

```
827                 if (NanbuElecElec == false)
828                 {
829                     NElecElec = 0;
830                 }
831 
832                 if (NanbuIonIon == false)
833                 {
834                     NIonIon = 0;
835                 }
836                 // True if only one electron available for
                        ElecElecCollisions
837                 bool OneElectron = (NElecElec == 1);
838                 // True if only one Ion available for ElecElecCollisions
839                 bool OneIon = (NIonIon == 1);
840                 label Eliminate=-1;
841                 scalar chargeEliminate;
842                 //Number of electrons in each subcell
843                 List<label> varsubCellsizeElec(8,0);
844                 //Number of ions in each subcell
845                 List<label> varsubCellsizeIon(8,0);
846 
847                 forAll(cellParcels, i)
848                 {
849                         const ParcelType& p = *cellParcels[i];
850 
851                         if (p.typeId()==0 || p.typeId()==3)
852                         {
853                                 //Number of electrons in each subcell
854                                 varsubCellsizeElec[whichSubCell[i]]+=1;
855                         }
856                         else if (p.typeId()==2)
857                         {
858                                 //Number of ions in each subcell
859                                 varsubCellsizeIon[whichSubCell[i]]+=1;
860                         }
861                 }
862 
863                 // If only one electron and one ion available --> Collide
864                 if (OneElectron && OneIon)
865                 {
866                         CollElecIon+=1;
867                         NElecElec = 0;
868                         NIonIon = 0;
869                 }
870                 // If only one electron available --> One electron
                        ignored
871                 else if (OneElectron)
872                 {
873                         NElecElec = 0;
874                         do
875                         {
876                                 Eliminate = rndGen_.integer(0, nC - 1);
877                                 ParcelType& EliminateP = *cellParcels[
                                        Eliminate];
878                                 chargeEliminate = constProps(EliminateP.
                                        typeId()).charge();
879                                 if (chargeEliminate >= 0)
```

```
880                         {
881                                 Eliminate = -1;
882                         }
883 
884                 }       while (Eliminate == -1);
885                 //One Electron "marked" as used
886                 availableCand[Eliminate]=0;
887                 //One Electron "marked" as used
888                 varsubCellsizeElec[whichSubCell[Eliminate]] -= 1;
889         }
890         // If only one ion available --> One ion ignored
891         else if (OneIon)
892         {
893                 NIonIon = 0;
894                 do
895                 {
896                         Eliminate = rndGen_.integer(0, nC - 1);
897                         ParcelType& EliminateP = *cellParcels[
                                Eliminate];
898                         chargeEliminate = constProps(EliminateP.
                                typeId()).charge();
899 
900                         if (chargeEliminate <= 0)
901                         {
902                                 Eliminate = -1;
903                         }
904 
905                 }       while (Eliminate == -1);
906                 //One Ion "marked" as used
907                 availableCand[Eliminate]=0;
908                 //One Electron "marked" as used
909                 varsubCellsizeIon[whichSubCell[Eliminate]] -= 1;
910         }
911         //Remaining Electrons (not yet involved in collisions)
912         label RemElecTotal = NElec;
913         //Remaining Ions (not yet involved in collisions)
914         label RemIonTotal = NIon;
915 
916         scalar totalCollElectrons = double(NElecElec)/2;
917         label   rCollElectrons(totalCollElectrons);
918         //switch for odd number of electrons in cell (if > 0)
919         scalar diffCollElec = totalCollElectrons - rCollElectrons
            ;
920 
921         scalar totalCollIons = double(NIonIon)/2;
922         label   rCollIons(totalCollIons);
923         //switch for odd number of electrons in cell (if > 0)
924         scalar diffCollIons = totalCollIons - rCollIons;
925 
926         //Total number of particles available for Collisions
927         scalar nC2 = NElec+NIon;
928 
929         //Corrected deltaT for odd number of electrons
930         scalar deltaTElecElec;
931         //Corrected deltaT for odd number of Ions
932         scalar deltaTIonIon;
933 
```

```
934                         if (diffCollElec > 0 && diffCollIons > 0)
935                         {
936                                 //1 additional collision with one particle
                                        colliding twice
937                                 CollElecElec = rCollElectrons+1;
938                                 //1 additional collision with one particle
                                        colliding twice
939                                 CollIonIon = rCollIons+1;
940                                 //deltaT correction for odd number of electrons
941                                 deltaTElecElec = NElecElec*deltaT/(NElecElec+1);
942                                 //deltaT correction for odd number of ions
943                                 deltaTIonIon = NIonIon*deltaT/(NIonIon+1);
944 
945                         }
946                         else if (diffCollElec > 0 && !diffCollIons > 0)
947                         {
948                                 //1 additional collision with one particle
                                        colliding twice
949                                 CollElecElec = rCollElectrons+1;
950                                 CollIonIon = rCollIons;
951                                 //deltaT correction for odd number of electrons
952                                 deltaTElecElec = NElecElec*deltaT/(NElecElec+1);
953                         }
954                         else if (!diffCollElec > 0 && diffCollIons > 0)
955                         {
956                                 CollElecElec = rCollElectrons;
957                                 //1 additional collision with one particle
                                        colliding twice
958                                 CollIonIon = rCollIons+1;
959                                 //deltaT correction for odd number of ions
960                                 deltaTIonIon = NIonIon*deltaT/(NIonIon+1);
961                         }
962 
963                         else
964                         {
965                                 // All Particles collide only once (even numbers)
966                                 CollElecElec = rCollElectrons;
967                                 CollIonIon = rCollIons;
968                         }
969                         //Total Collisions
970                         label nCandidates = CollElecIon+CollElecElec+CollIonIon;
971                         //list of already selected particles
972                         List<label> alreadySelected(nC2,-1);
973                         //Counter for list alreadySelected
974                         label selCounter=0;
975                         //List of electrons involved in ElecElec collisions
976                         List<label> alreadyElecElec(NElecElec,-1);
977                         //List of Ions involved in IonIon collisions
978                         List<label> alreadyIonIon(NIonIon,-1);
979                         label ElecElecCounter = 0;
980                         label IonIonCounter = 0;
981 
982 
983                         //////////////////////////////////////////////////////////
984                         ////////     Selection of Candidates       //////////////
985                         //////////////////////////////////////////////////////////
986 
```

```
987         for (label c = 0; c < nCandidates; c++)
988         {
989             // ------------------------------------
990             // subCell candidate selection procedure
991 
992             // Select the first collision candidate
993             label candidateP = -1;
994             scalar chargeP;
995 
996             do
997             {
998                 candidateP = rndGen_.integer(0, nC - 1);
999                 ParcelType& ProvparcelP = *cellParcels[
                        candidateP];
1000                chargeP = constProps(ProvparcelP.typeId()
                        ).charge();
1001
1002                if (chargeP == 0 || availableCand[
                        candidateP]==0)
1003                {
1004                    candidateP=-1;
1005                }
1006            }       while (candidateP == -1);
1007
1008            //Marked as already selected
1009            availableCand[candidateP]=0;
1010            //parcelP added to list as already used
1011            alreadySelected[selCounter] = candidateP;
1012            //counter for list alreadySelected
1013            selCounter+= 1;
1014
1015            ParcelType& parcelP = *cellParcels[candidateP];
1016            chargeP = constProps(parcelP.typeId()).charge();
1017
1018            if (chargeP > 0)
1019            {
1020                //Remaining Ions in Subcell
1021                varsubCellsizeIon[whichSubCell[candidateP
                        ]]-=1;
1022                //Remaining Ions in whole Cell
1023                RemIonTotal -= 1;
1024            }
1025            else if (chargeP < 0)
1026            {
1027                //Remaining Electrons in Subcell
1028                varsubCellsizeElec[whichSubCell[
                        candidateP]]-=1;
1029                //Remaining Electrons in whole Cell
1030                RemElecTotal -= 1;
1031            }
1032
1033            // Declare the second collision candidate
1034
1035            label candidateQ = -1;
1036            scalar chargeQ;
1037            List<label> subCellPs = subCells[whichSubCell[
                    candidateP]];
```

```
1038
1039                    //1 Electron collides a 2nd Time
1040                    if (chargeP < 0 && RemElecTotal == 0 &&
                            diffCollElec >0)
1041                    {
1042                            List<label> subCellPEE = subCellsElecElec
                                    [whichSubCell[candidateP]];
1043                            label nEESC = subCellPEE.size();
1044                            if (nEESC >= 1)
1045                            {
1046                                    do
1047                                    {
1048                                            candidateQ = subCellPEE[
                                                    rndGen_.integer(0,
                                                    nEESC - 1)];
1049                                    } while (candidateP == candidateQ
                                            );
1050                            }
1051                            else
1052                            {
1053                                    do
1054                                    {
1055                                            candidateQ =
                                                    alreadyElecElec[
                                                    rndGen_.integer(0,
                                                    NElecElec - 2)];
1056                                    } while (candidateP == candidateQ
                                            );
1057                            }
1058
1059
1060                    }
1061                    //1 Ion collides a 2nd Time
1062                    else if (chargeP> 0 && RemIonTotal == 0 &&
                            diffCollIons >0)
1063                    {
1064                            List<label> subCellPII = subCellsIonIon[
                                    whichSubCell[candidateP]];
1065                            label nIISC = subCellPII.size();
1066                            if (nIISC >= 1)
1067                            {
1068                                    do
1069                                    {
1070                                            candidateQ = subCellPII[
                                                    rndGen_.integer(0,
                                                    nIISC - 1)];
1071                                    } while (candidateP == candidateQ
                                            );
1072                            }
1073                            else
1074                            {
1075                                    do
1076                                    {
1077                                            candidateQ =
                                                    alreadyIonIon[rndGen_.
                                                    integer(0, NIonIon -
                                                    2)];
```

```
1078                    } while (candidateP == candidateQ
                          );
1079                }
1080 
1081            }
1082 
1083            //Normal declaration of Q without repetition of
                    partners
1084            else
1085            {
1086 
1087 
1088                bool enterSubCell = false;
1089                //For P = Ion -> If Ion-Ion Coll possible
                        and Ion in subcell
1090                bool C1 = (chargeP > 0 && CollIonIon > 0
                    && varsubCellsizeIon[whichSubCell[
                    candidateP]] > 0 && RemIonTotal-1 >=
                    CollElecIon);
1091                //For P = Ion -> If Ion-Elec Coll
                    possible and Elec in subcell
1092                bool C2 = (chargeP > 0 && CollElecIon > 0
                     && varsubCellsizeElec[whichSubCell[
                    candidateP]] > 0);
1093                //For P = Elc -> If Ion-Elec Coll
                    possible and Ion in subcell
1094                bool C3 = (chargeP < 0 && CollElecIon > 0
                     && varsubCellsizeIon[whichSubCell[
                    candidateP]] > 0);
1095                //For P = Elec -> If Elec-Elec Coll
                    possible and Elec in subcell
1096                bool C4 = (chargeP < 0 && CollElecElec >
                    0 && varsubCellsizeElec[whichSubCell[
                    candidateP]] > 0 && RemElecTotal-1 >=
                    CollElecIon);
1097 
1098                if (C1 || C2 || C3 || C4)
1099                {
1100                    enterSubCell = true;
1101                }
1102 
1103                /////////////////////////////////////
1104                if (enterSubCell)
1105                {
1106                    // If there are two or more
                        particle in a subCell, choose
1107                    // another from the same cell.
                        If the same candidate is
1108                    // chosen, choose again.
1109                    label nSC = subCellPs.size();
1110                    do
1111                    {
1112                        candidateQ = subCellPs[
                            rndGen_.integer(0, nSC
                            - 1)];
1113                        ParcelType& ProvparcelQ =
                                *cellParcels[
```

```
                                                        candidateQ];
1114                                                chargeQ = constProps(
                                                        ProvparcelQ.typeId()).
                                                        charge();
1115
1116                                                //Neutrals or already
                                                        selected
1117                                                if (chargeQ == 0 ||
                                                        availableCand[
                                                        candidateQ]==0)
1118                                                {
1119                                                        candidateQ=-1;
1120                                                }
1121                                                //Ion-Electron Collisions
1122                                                else if (chargeP*chargeQ
                                                        < 0 && CollElecIon
                                                        ==0)
1123                                                {
1124                                                        candidateQ=-1;
1125                                                }
1126                                                //Electron-Electron
                                                        Collision
1127                                                else if (chargeP < 0 &&
                                                        chargeQ < 0 && (
                                                        CollElecElec == 0 ||
                                                        RemElecTotal-1 <
                                                        CollElecIon))
1128                                                {
1129                                                        candidateQ=-1;
1130                                                }
1131                                                //Ion-Ion Collision
1132                                                else if (chargeP > 0 &&
                                                        chargeQ > 0 && (
                                                        CollIonIon == 0 ||
                                                        RemIonTotal-1 <
                                                        CollElecIon))
1133                                                {
1134                                                        candidateQ=-1;
1135                                                }
1136                                        }       while (candidateQ == -1);
1137                                }
1138                                //
                                        ////////////////////////////////////////

1139                                else
1140                                {
1141                                        // Select a possible second
                                                collision candidate from the
1142                                        // whole cell.  If the same
                                                candidate is chosen, choose
1143                                        // again.
1144
1145                                        do
1146                                        {
1147                                                candidateQ = rndGen_.
                                                        integer(0, nC - 1);
```

```
1148                                            ParcelType& ProvparcelQ =
                                                     *cellParcels[
                                                    candidateQ];
1149                                            chargeQ = constProps(
                                                    ProvparcelQ.typeId()).
                                                    charge();
1150
1151                                            //Neutrals or already
                                                    selected
1152                                            if (chargeQ == 0 ||
                                                    availableCand[
                                                    candidateQ]==0)
1153                                            {
1154                                                    candidateQ=-1;
1155                                            }
1156
1157                                            //Ion-Electron Collision
1158                                            else if (chargeP*chargeQ
                                                    < 0 && CollElecIon
                                                    ==0)
1159                                            {
1160                                                    candidateQ=-1;
1161                                            }
1162                                            //Electron-Electron
                                                    Collision
1163                                            else if (chargeP < 0 &&
                                                    chargeQ < 0 && (
                                                    CollElecElec == 0 ||
                                                    RemElecTotal-1 <
                                                    CollElecIon))
1164                                            {
1165                                                    candidateQ=-1;
1166                                            }
1167                                            //Ion-Ion Collision
1168                                            else if (chargeP > 0 &&
                                                    chargeQ > 0 && (
                                                    CollIonIon == 0 ||
                                                    RemIonTotal-1 <
                                                    CollElecIon))
1169                                            {
1170                                                    candidateQ=-1;
1171                                            }
1172                                    } while (candidateQ == -1);
1173                            }
1174
1175                            availableCand[candidateQ]=0;
1176                            //parcelQ added to list as already used
1177                            alreadySelected[selCounter] = candidateQ;
1178                            //counter for list alreadySelected
1179                            selCounter+= 1;
1180                    }
1181
1182                    ParcelType& parcelQ = *cellParcels[candidateQ];
1183                    chargeQ = constProps(parcelQ.typeId()).charge();
1184
1185                    bool ElecIonColl = false;
1186                    bool ElecElecColl = false;
```

```
1187                bool IonIonColl = false;
1188
1189                if (chargeQ > 0)
1190                {
1191                        //becomes negative for odd number of
                            particles
1192                        varsubCellsizeIon[whichSubCell[candidateQ
                            ]]-=1;
1193                        //becomes negative for odd number of
                            particles
1194                        RemIonTotal -=1;
1195                }
1196                else if (chargeQ < 0)
1197                {
1198                        //becomes negative for odd number of
                            particles
1199                        varsubCellsizeElec[whichSubCell[
                            candidateQ]]-=1;
1200                        //becomes negative for odd number of
                            particles
1201                        RemElecTotal -=1;
1202                }
1203
1204                if (chargeP*chargeQ < 0)
1205                {
1206                        CollElecIon -=1;
1207                        ElecIonColl = true;
1208                }
1209                else if (chargeP < 0 && chargeQ < 0)
1210                {
1211                        CollElecElec -=1;
1212                        ElecElecColl = true;
1213                        alreadyElecElec[ElecElecCounter]=
                            candidateP;
1214                        subCellsElecElec[whichSubCell[candidateP
                            ]].append(candidateP);
1215                        if (!(CollElecElec == 0 && diffCollElec
                            >0))
1216                        {
1217                                alreadyElecElec[ElecElecCounter
                                    +1]=candidateQ;
1218                                subCellsElecElec[whichSubCell[
                                    candidateQ]].append(candidateQ
                                    );
1219                        }
1220                        ElecElecCounter +=2;
1221
1222                }
1223                else if ((chargeP > 0 && chargeQ > 0))
1224                {
1225                        CollIonIon -=1;
1226                        IonIonColl = true;
1227                        alreadyIonIon[IonIonCounter]=candidateP;
1228                        subCellsIonIon[whichSubCell[candidateP]].
                            append(candidateP);
1229                        if (!(CollIonIon == 0 && diffCollIons >
                            0))
```

```
1230                {
1231                        alreadyIonIon[IonIonCounter+1]=
                            candidateQ;
1232                        subCellsIonIon[whichSubCell[
                            candidateQ]].append(candidateQ
                            );
1233                }
1234                IonIonCounter +=2;
1235        }
1236
1237
1238        scalar  expRelVel;
1239        deltaT = mesh().time().deltaTValue();
1240
1241        if (ElecElecColl)
1242        {
1243                if (diffCollElec > 0)
1244                {
1245                        deltaT = deltaTElecElec;
1246                }
1247                //Expected relative Velocity squared
1248                expRelVel = 3*physicoChemical::k.value()
                    *2*TElektronColl/constProps(parcelP.
                    typeId()).mass();
1249
1250        }
1251        else if (IonIonColl)
1252        {
1253                if (diffCollIons > 0)
1254                {
1255                        deltaT = deltaTIonIon;
1256                }
1257                //Expected relative Velocity squared
1258                expRelVel = 3*physicoChemical::k.value()
                    *2*TIonColl/constProps(parcelP.typeId
                    ()).mass();
1259
1260        }
1261        else    //ElecIonCollision
1262        {
1263                //Expected relative Velocity squared
1264                expRelVel = 3*physicoChemical::k.value()*
                    TElektronColl/constProps(0).mass()+3*
                    physicoChemical::k.value()*TIonColl/
                    constProps(2).mass()
1265                + ((UElektronColl - UIonColl) & (
                    UElektronColl - UIonColl));
1266        }
1267
1268        //Reduced Mass
1269        scalar  muR = constProps(parcelP.typeId()).mass()
            *constProps(parcelQ.typeId()).mass()/(
            constProps(parcelP.typeId()).mass()+constProps
            (parcelQ.typeId()).mass());
1270        //Debye length as a function of e- quantities (
            Telek >> TIon)
```

```
1271                                lambdaD = sqrt((electromagnetic::epsilon0.value()
                                        *physicoChemical::k.value()*TElektronColl)
1272                                                                        /(
                                                                            rhoNElektronColl
                                                                            *constProps
                                                                            (0).charge()
                                                                            *constProps
                                                                            (0).charge()
                                                                            ));
1273                                scalar  b0 = mag(chargeP*chargeQ)/(4*constant::
                                        mathematical::pi*electromagnetic::epsilon0.
                                        value()*muR*expRelVel);
1274
1275                                if (b0 <0 || lambdaD <= 0)
1276                                {
1277                                     scalar errror = 1;
1278                                }
1279
1280                                scalar  CLog=log(lambdaD/b0);
1281
1282                                if (CLog < 0)
1283                                {
1284                                        ClogWarnings ++;
1285                                }
1286
1287                                scalar  AABeta = rhoNTotalPlasma*sqr(chargeP*
                                        chargeQ/(electromagnetic::epsilon0.value()*muR
                                        ))*CLog/(4*constant::mathematical::pi);
1288                                scalar  taoCollC = AABeta*deltaT;
1289
1290                                vector g = parcelP.U() - parcelQ.U();
1291                                scalar magg = mag(g);
1292                                scalar taoColl = taoCollC/pow(magg,3);
1293
1294                binaryCollision().collide
1295                (
1296                parcelP,
1297                                parcelQ,
1298                                A,              // A parameter
1299                                tao,            // tau parameter
1300                                taoColl
1301                );
1302
1303                //Update of Leapfrog velocity for t-Deltat/2
1304                vector EExt = EExt_[cellI];
1305                vector EInt = EInt_[cellI];
1306                vector ETotal = EExt + EInt;
1307                vector Fp = constProps(parcelP.typeId()).charge()*ETotal;
1308                vector Fq = constProps(parcelQ.typeId()).charge()*ETotal;
1309                parcelP.ULeapfrog() = parcelP.U() - 0.5*deltaT*Fp/constProps(
                    parcelP.typeId()).mass();
1310                parcelQ.ULeapfrog() = parcelQ.U() - 0.5*deltaT*Fq/constProps(
                    parcelQ.typeId()).mass();
1311
1312                                if (taoColl < taoCollMin)
1313                                {
1314                                        taoCollMin = taoColl;
```

```
1315                                     taoCollMin_[cellI] = taoCollMin;
1316                             }
1317                             if (taoColl > taoCollMax)
1318                             {
1319                                     taoCollMax = taoColl;
1320                                     taoCollMax_[cellI] = taoCollMax;
1321                             }
1322 
1323                             if (taoColl < taoCollMinIt)
1324                             {
1325                                     taoCollMinIt = taoColl;
1326                             }
1327                             if (taoColl > taoCollMaxIt)
1328                             {
1329                                     taoCollMaxIt = taoColl;
1330                             }
1331 
1332                             collisions++;
1333 
1334                             ///////////////////////////// End Coulomb
                                     Collisions //////////////////////
1335                     }
1336 
1337     }
```

The post-collision velocities for the charge carriers after a Coulomb interaction are calculated based on the collisional parameter τ (s in Section 5.3.4.1) in the sub-function *binaryCollision( ).collide* defined in the file CoulombCollision.C.

Listing B.4: File CoulombCollision.C - Coulomb collision.

```
117  template<class CloudType>
118  void Foam::CoulombCollision<CloudType>::collide
119  (
120      typename CloudType::parcelType& pP,
121      typename CloudType::parcelType& pQ,
122          scalarField A,           //      A parameter
123          scalarField tao,         //      tau parameter
124          scalar  taoColl
125  )
126  {
127      CloudType& cloud(this->owner());
128 
129      label typeIdP = pP.typeId();
130      label typeIdQ = pQ.typeId();
131      vector& UP = pP.U();
132      vector& UQ = pQ.U();
133      vector g = UP - UQ;
134      scalar magg = mag(g);
135 
136      Random& rndGen(cloud.rndGen());
137 
138      scalar mP = cloud.constProps(typeIdP).mass();
139      scalar mQ = cloud.constProps(typeIdQ).mass();
140 
```

```
141
142     //
            ///////////////////////////////////////---------------------------------
143     // Coulomb Collisions [Nanbu2000]
144
145     scalar AColl=-1;
146     scalar X = VSMALL;
147
148     if (taoColl <= 0)
149     {
150         X = acos(1+0.00001*log(rndGen.scalar01()));
151     }
152
153     else if (0 < taoColl && taoColl < 0.01)
154     {
155         X = acos(1+taoColl*log(rndGen.scalar01()));
156     }
157
158     else if (0.01 <= taoColl && taoColl <= 3)
159     {
160         scalar  ACollI(Foam::interpolateXY(taoColl, tao, A ));
161         AColl = ACollI;
162         X = acos(log(exp(-AColl)+2*rndGen.scalar01()*sinh(AColl))/AColl);
163     }
164     else if (3 < taoColl && taoColl <= 20)
165         {
166         AColl = 3*exp(-taoColl);
167         X = acos(log(exp(-AColl)+2*rndGen.scalar01()*sinh(AColl))/AColl);
168     }
169     else if (taoColl > 20)
170         {
171         X = acos(2*rndGen.scalar01()-1);
172     }
173
174     scalar epsilon = 2*constant::mathematical::pi*rndGen.scalar01();
175     scalar gQ = sqrt(sqr(g.y())+sqr(g.z()));
176
177     if (gQ == 0)
178     {
179         gQ = VSMALL;
180     }
181
182     scalar hx = gQ*cos(epsilon);
183     scalar hy = -(g.y()*g.x()*cos(epsilon)+magg*g.z()*sin(epsilon))/gQ;
184     scalar hz = -(g.z()*g.x()*cos(epsilon)-magg*g.y()*sin(epsilon))/gQ;
185     vector h = vector
186                 (
187                         hx,
188                         hy,
189                         hz
190                 );
191
192     UP = UP - mQ/(mP + mQ)*(g*(1-cos(X))+sin(X)*h);
193
194     UQ = UQ + mP/(mP + mQ)*(g*(1-cos(X))+sin(X)*h);
195
```

```
196 }
```

B.4 Electron-Neutral collisions

The electron-neutral collisional approach is implemented in *dsmcPlasmaFoam* in the *collisions* function of the *evolve* method, both located in the file DsmcCloud.C. Hence, electron-neutral collisions and Coulomb collisions are handled in parallel. The number of collision candidates is defined in lines 1346:1386 based on the NTC implementation for *dsmcPlasmaFoam* described in Section 5.3.5. The collisional electron-neutral pair is determined in lines 1392:1445 and the collisional cross sections for elastic, excitation and ionisation interactions are calculated based on the electron's kinetic energy in the next steps.

The collision outcome is defined in lines 1491:1517 based on the random number U. If an elastic or excitation collision takes place, the sub-function *binaryCollision().NEcollide* is called in lines 1523:1528, where the post-collision velocities of the electron and neutral are calculated. On the other hand, if ionisation takes place, the virtual pre-collision energy of the progeny electron is estimated in line 1564 and the energy of the parent electron is updated accordingly prior to its collision with the neutral. Next, the sub-function *binaryCollision().Ionizationcollide* is called, where the post-collision velocities of the parent electron and neutral are calculated. The post-collision velocity of the progeny electron is computed in lines 1587:1608 assuming an additional collision with the neutral particle. A new electron and new ion are created in lines 1611:1673, while the numerical weight of the neutral is updated in line 1632 to take into account the fraction of the neutral test particle becoming an ion. Note that an update of the Leapfrog velocity of the parent electron is performed based on the obtained post-collision velocity in lines 1579:1584.

Listing B.5: File: DsmcCloud.C - Collisions function - Electron-Neutral collision.

```
1342    ////////////////////////////////////////////////////////////////////////
1343    ///////////// Ionization Collisions (Neutral + Electron) //////////////
1344    ////////////////////////////////////////////////////////////////////////
1345
1346    if (NElec >= 1 && NNeutforIon >= 1 && ElecNeutralCollisions == true)
1347    {
1348
1349        //List of electrons in each subcell
1350        List<label> listElectronCandidates(NElec,-1);
1351        //List of neutrals in each subcell
1352        List<label> listNeutralCandidates(NNeutforIon,-1);
1353        //Inverse addressing for listElectronCandidates
```

```
1354            List<label> InverseNeutralCandidates(nC,-1);
1355
1356            label j=0;
1357            label stf = 0;
1358
1359            forAll(cellParcels, i)
1360            {
1361                const ParcelType& p = *cellParcels[i];
1362
1363                if (p.typeId()==0 || p.typeId()==3)
1364                {
1365                    listElectronCandidates[j] = i;
1366                    j+=1;
1367                }
1368                else if ((p.typeId()==1) & (p.Wt() >= plasmaWt))
1369                {
1370                    listNeutralCandidates[stf] = i;
1371                    InverseNeutralCandidates[i] = stf;
1372                    stf += 1;
1373                }
1374            }
1375
1376            scalar deltaT = mesh().time().deltaTValue();
1377            scalar sigmaTcRMax = sigmaTcRMax_[cellI];
1378
1379            // For mixture to only include Neutral-Electron Collisions (Eq. 11.5 Bird
                    , modified as in Boyd2007)
1380            //WtMax used as suggested in Sakiyama2001; Normal NNeut used
1381            scalar selectedPairs = CSR_[cellI]
1382                                            + 0.5*WtMax*NElec*NNeut*sigmaTcRMax*deltaT/
                                                    mesh().cellVolumes()[cellI];
1383
1384            label nCandidatesDsmc(selectedPairs);
1385            CSR_[cellI] = selectedPairs - nCandidatesDsmc;
1386            collisionCandidatesDsmc += nCandidatesDsmc;
1387
1388            for (label c = 0; c < nCandidatesDsmc; c++)
1389            {
1390
1391                ////////// Electron Selection /////////////
1392                label indexcandidateP = -1;
1393                label candidateP = -1;
1394
1395                do
1396                {
1397                    indexcandidateP = rndGen_.integer(0, NElec - 1);
1398                    candidateP = listElectronCandidates[indexcandidateP];
1399
1400                    if (listElectronCandidates[indexcandidateP]==-1)
1401                    {
1402                        candidateP=-1;
1403                    }
1404                }   while (candidateP == -1);
1405
1406                ////////// Neutral Selection /////////////
1407                label indexcandidateQ = -1;
1408                label candidateQ = -1;
```

```
1409
1410        //Neutrals in same subcell as electron P
1411        List<label> subCellNeutralP = subCellsAllNeutrals[whichSubCell[
                candidateP]];
1412        label subCellNeutralSize(subCellNeutralP.size());
1413        label NNeutinSC = varsubCellsizeDsmcNeut[whichSubCell[candidateP]];
1414
1415
1416        if (NNeutinSC > 0)  //Neutral selection in same subcell
1417        {
1418            do
1419            {
1420                candidateQ = subCellNeutralP[rndGen_.integer(0,
                        subCellNeutralSize - 1)];
1421                indexcandidateQ = InverseNeutralCandidates[candidateQ];
1422
1423                if (listNeutralCandidates[indexcandidateQ]==-1)
1424                {
1425                    candidateQ=-1;
1426                }
1427
1428            } while (candidateQ == -1);
1429        }
1430        else        //Neutral selection in whole cell
1431        {
1432            do
1433            {
1434                indexcandidateQ = rndGen_.integer(0, NNeutforIon - 1);
1435                candidateQ = listNeutralCandidates[indexcandidateQ];
1436
1437                if (listNeutralCandidates[indexcandidateQ]==-1)
1438                {
1439                    candidateQ=-1;
1440                }
1441            }       while (candidateQ == -1);
1442        }
1443
1444        ParcelType& parcelP = *cellParcels[candidateP];
1445        ParcelType& parcelQ = *cellParcels[candidateQ];
1446
1447        if (parcelQ.Wt() < plasmaWt )
1448        {
1449            FatalErrorIn("Foam::DsmcCloud<ParcelType>::Neutral-Electron
                    collisions")
1450                    << "Neutral weight < charge carrier weight - Not
                            supported" << nl
1451                    << abort(FatalError);
1452        }
1453
1454        vector UP = parcelP.U();
1455        vector UQ = parcelQ.U();
1456        scalar mQ = constProps(parcelQ.typeId()).mass();    //Neutral mass
1457        scalar mP = constProps(parcelP.typeId()).mass();    //Electron mass
1458
1459        scalar cR = mag(UP - UQ);
1460
1461        if (cR < VSMALL)
```

```
1462            {
1463                cR = 0;
1464            }
1465            scalar eEnergy = 0.5*mP*(UP&UP)/1.602177e-19;                    //
                    electronVolt
1466
1467            // calculating cross section from tabulated data:
1468            //Elastic and Ionization from Hayashi [http://jila.colorado.edu/~avp/
                    collision_data/electronneutral/hayashi.txt] (Ref 20 from Nanbu2000
                    - IEE Trans Plasma Science 28,3)
1469            //Composite excitation from Surendra 1990 Physical review A 41,2  /
                    Eggarter (32 from Surendra)
1470
1471            scalar sigmaEl = VSMALL;
1472            scalar sigmaEx = VSMALL;
1473            scalar sigmaIon = VSMALL;
1474            //scalar sigmaTot = VSMALL;
1475
1476            if (eEnergy <= 1000)                  //Tabulated data up to 1000ev =
                    1.16e7 K
1477            {
1478                scalar sigmaElInt(Foam::interpolateXY(eEnergy, electronEnergy,
                        sigmaElastic ));
1479                scalar sigmaExInt(Foam::interpolateXY(eEnergy, electronEnergy,
                        sigmaExcitation ));
1480                scalar sigmaIonInt(Foam::interpolateXY(eEnergy, electronEnergy,
                        sigmaIonization ));
1481                sigmaEl = sigmaElInt;
1482                sigmaEx = sigmaExInt;
1483                sigmaIon = sigmaIonInt;
1484            }
1485
1486            scalar sigmaElcR = sigmaEl*cR*1e-20;         //Sigma * relative
                    Velocity and conversion to SI Units
1487            scalar sigmaExcR = sigmaEx*cR*1e-20;         //conversion to SI Units
1488            scalar sigmaIoncR = sigmaIon*cR*1e-20;       //conversion to SI Units
1489            scalar sigmaTcR = sigmaElcR + sigmaExcR + sigmaIoncR;
1490
1491            double K=3; //Number of possible collisional events
1492            scalar U = rndGen_.scalar01();
1493
1494            if (sigmaTcR > sigmaTcRMax_[cellI])
1495            {
1496                sigmaTcRMax_[cellI] = sigmaTcR;
1497            }
1498
1499            label event =0;
1500            bool NECollision = false;
1501            bool IonColl = false;
1502
1503            if (U < sigmaElcR/sigmaTcRMax)
1504            {
1505                event = 1;
1506                NECollision = true;
1507            }
1508            else if (U < (sigmaElcR + sigmaExcR)/sigmaTcRMax)
1509            {
```

```
1510                event = 2;
1511                NECollision = true;
1512            }
1513            else if (U < (sigmaElcR + sigmaExcR + sigmaIoncR)/sigmaTcRMax)
1514            {
1515                event = 3;
1516                IonColl = true;
1517            }
1518
1519            if (NECollision)
1520            {
1521                scalar preE = 0.5*mP*parcelP.Wt()*(UP&UP) + 0.5*mQ*parcelQ.Wt()*(
                        UQ&UQ);
1522
1523                binaryCollision().NEcollide
1524                        (
1525                                parcelP,
1526                                parcelQ,
1527                                event
1528                        );
1529
1530                //Update of Leapfrog velocity for t-Deltat/2
1531                vector EExt = EExt_[cellI];
1532                vector EInt = EInt_[cellI];
1533                vector ETotal = EExt + EInt;
1534                vector F = constProps(parcelP.typeId()).charge()*ETotal;
1535                parcelP.ULeapfrog() = parcelP.U() - 0.5*deltaT*F/mP;
1536
1537                if (event == 1)
1538                {
1539                    elasticCollisions++;
1540                    scalar postE = 0.5*mP*parcelP.Wt()*(parcelP.U()&parcelP.U())
                            + 0.5*mQ*parcelQ.Wt()*(parcelQ.U()&parcelQ.U());
1541                    deltaEnergy_[cellI] += preE - postE;
1542                }
1543                else if (event == 2)
1544                {
1545                    excCollisions++;
1546                    scalar thresholdE = constProps(parcelQ.typeId()).thresholdEEx
                            ();
1547                    scalar postE = 0.5*mP*parcelP.Wt()*(parcelP.U()&parcelP.U())
                            + 0.5*mQ*parcelQ.Wt()*(parcelQ.U()&parcelQ.U());
1548                    deltaEnergy_[cellI] += preE - (postE + thresholdE*parcelP.Wt
                            ()*1.602177E-019);
1549                }
1550
1551            }
1552
1553            else if (event == 3 && IonColl == true)            //Ionization
1554            {
1555                scalar thresholdIon = constProps(parcelQ.typeId()).thresholdEIon
                        ();
1556                //electronVolt  -  Valid for Argon
1557                scalar constEps0 = 2-100/(eEnergy+10);
1558                 //electronVolt  -  Valid for Argon
1559                scalar constEps1 = 0.5*(eEnergy-thresholdIon)-constEps0;
1560                 //               -  Valid for Argon
```

```
1561                scalar constA = 10.3;
1562                scalar U = rndGen_.scalar01();
1563                //Energy of progeny electron
1564                scalar eProgeny = constEps0+constA*tan(U*(atan(constEps1/constA)+
                        atan(constEps0/constA))-atan(constEps0/constA));
1565                //Imaginary Precollision velocity of progeny electron
1566                vector UProg = UP/mag(UP)*sqrt(2*eProgeny*1.602177e-19/mP);
1567                //E reduction for parent electron
1568                scalar Echange = sqrt(1-(thresholdIon+eProgeny)/(0.5*mP*(UP&UP)
                        /1.602177e-19));
1569                //Save original neutral velocity
1570                vector UQOrg = parcelQ.U();
1571
1572                binaryCollision().Ionizationcollide
1573                        (
1574                                parcelP,
1575                                parcelQ,
1576                                Echange
1577                        );
1578
1579                //Update of Leapfrog velocity for t-Deltat/2
1580                vector EExt = EExt_[cellI];
1581                vector EInt = EInt_[cellI];
1582                vector ETotal = EExt + EInt;
1583                vector F = constProps(parcelP.typeId()).charge()*ETotal;
1584                parcelP.ULeapfrog() = parcelP.U() - 0.5*deltaT*F/mP;
1585
1586                // For progeny Electron
1587                vector UQ = parcelQ.U();    //Update velocity of Ion particle
1588                vector gProg = UProg - UQ;
1589                scalar cRProg = mag(UProg - UQ);
1590
1591                //Azimuthal Angle
1592                scalar epsilonProg = 2*constant::mathematical::pi*rndGen_.
                        scalar01();
1593                scalar gQProg = sqrt(sqr(gProg.y())+sqr(gProg.z()));
1594                scalar hxProg = gQProg*cos(epsilonProg);
1595                scalar hyProg = -(gProg.y()*gProg.x()*cos(epsilonProg)+cRProg*
                        gProg.z()*sin(epsilonProg))/gQProg;
1596                scalar hzProg = -(gProg.z()*gProg.x()*cos(epsilonProg)-cRProg*
                        gProg.y()*sin(epsilonProg))/gQProg;
1597                vector hProg = vector
1598                        (
1599                                hxProg,
1600                                hyProg,
1601                                hzProg
1602                        );
1603                U = rndGen_.scalar01();
1604                scalar XProg = acos((eProgeny+2-2*(pow(1+eProgeny,U)))/eProgeny);
1605                 //post collision velocity progeny electron
1606                UProg = UProg - mQ/(mP + mQ)*(gProg*(1-cos(XProg))+sin(XProg)*
                        hProg);
1607                //2nd post collision velocity Ion
1608                UQ = UQ + mP/(mP + mQ)*(gProg*(1-cos(XProg))+sin(XProg)*hProg);
1609
1610
1611                vector positionQ = parcelQ.position();
```

```
1612
1613            label tetFaceI=1;
1614            label tetPtI=1;
1615            mesh_.findCellFacePt(positionQ, cellI, tetFaceI, tetPtI);
1616            scalar Ei = 0;
1617            scalar Wt = plasmaWt;     //Same Wt for Progeny Electron and Ion
1618
1619            label newTypeId = 3;     //Progeny Electron
1620            label IonTypeId = 2;     //Created Ion
1621            label labelAlreadyCounted = 0;  //Switch used to mark electrons
                    on Anode (particleTemplates.C)
1622            label labelIonEvent = 0;        //Counter ionization events
1623            label parFirstLeapFrog = 1;                //Switch used to
                    identify new particles in move method and push their vel. back
                     for Leapfrog-
1624            vector ULeapfrog = vector (0, 0, 0);               //Leapfrog
                    Velocity
1625            scalar creationPosition = positionQ.x();          //for Ionization
                    validation
1626
1627            //Neutral handling //
1628            //Normal case
1629            if (parcelQ.Wt() > Wt)
1630            {
1631                //Update Neutral particles in cloud not involved in
                        ionization
1632                parcelQ.Wt() -= Wt;
1633                parcelQ.U() = UQOrg;
1634            }
1635
1636            else if (parcelQ.Wt() == Wt)
1637            {
1638                deleteParcel(parcelQ); // Delete Neutral
1639            }
1640
1641            //Create new clouds
1642            addNewParcel     //Electron
1643            (
1644                    positionQ,
1645                    UProg,
1646                    Ei,
1647                    Wt,
1648                    cellI,
1649                    tetFaceI,
1650                    tetPtI,
1651                    newTypeId,
1652                    labelAlreadyCounted,
1653                    labelIonEvent,
1654                    parFirstLeapFrog,
1655                    ULeapfrog,
1656                    creationPosition
1657            );
1658            addNewParcel     //Ion
1659            (
1660                    positionQ,
1661                    UQ,
1662                    Ei,
```

```
1663                        Wt,
1664                        cellI,
1665                        tetFaceI,
1666                        tetPtI,
1667                        IonTypeId,
1668                        labelAlreadyCounted,
1669                        labelIonEvent,
1670                        parFirstLeapFrog,
1671                        ULeapfrog,
1672                        creationPosition
1673                );
1674
1675                IonizCollisions++;
1676                //counter for validation Ionization
1677                parcelP.labelIonEvent()+=1;
```

The post-collision velocities for elastic and excitation collisions are calculated in the function *binaryCollision().NEcollide* defined in the file CoulombCollision.C. For the excitation case, the threshold energy of excitation is subtracted from the pre-collision energy of the electron (lines 215:221).

Listing B.6: File: CoulombCollision.C - Elastic and excitation collision.

```
198 template<class CloudType>
199 void Foam::CoulombCollision<CloudType>::NEcollide
200 (
201     typename CloudType::parcelType& pP,
202     typename CloudType::parcelType& pQ,
203         label event
204 )
205 {
206     CloudType& cloud(this->owner());
207
208     label typeIdP = pP.typeId();          //P = Electron
209     label typeIdQ = pQ.typeId();          //Q = Neutral
210     vector& UP = pP.U();
211     vector& UQ = pQ.U();
212     scalar mP = cloud.constProps(typeIdP).mass();
213     scalar mQ = cloud.constProps(typeIdQ).mass();
214
215     if (event == 2) //Electronic Excitacion
216     {
217         //threshold Energy of Excitation
218         scalar thresholdE = cloud.constProps(typeIdQ).thresholdEEx();
219         //EKin of electron reduced in the amount of thresholdEEx
220         UP = UP*sqrt(1-thresholdE/(0.5*mP*(UP&UP)/1.602177e-19));
221     }
222
223     //  Neutral spliting for unequal particle weights //
224     vector UQnC = UQ;   //Original velocity = Vel. of non-colliding particle
225     scalar WtFactor = pP.Wt()/pQ.Wt();
226
227     vector g = UP - UQ;
228     scalar cR = mag(UP - UQ);
```

```
229
230     Random& rndGen(cloud.rndGen());
231
232     scalar epsilon = 2*constant::mathematical::pi*rndGen.scalar01(); //
            Azimuthal Angle
233     scalar gQ = sqrt(sqr(g.y())+sqr(g.z()));
234     scalar hx = gQ*cos(epsilon);
235     scalar hy = -(g.y()*g.x()*cos(epsilon)+cR*g.z()*sin(epsilon))/gQ;
236     scalar hz = -(g.z()*g.x()*cos(epsilon)-cR*g.y()*sin(epsilon))/gQ;
237     vector h = vector
238             (
239                     hx,
240                     hy,
241                     hz
242             );
243     scalar ElecEnergy = 0.5*mP*(UP&UP)/1.602177e-19;         //electronVolt
244     scalar X = acos((ElecEnergy+2-2*(pow(1+ElecEnergy,rndGen.scalar01())))/
            ElecEnergy);
245
246     UP = UP - mQ/(mP + mQ)*(g*(1-cos(X))+sin(X)*h);
247
248 }
```

The post-collision velocities for an ionisation collision are calculated in the function *binaryCollision().Ionizationcollide* defined in the file CoulombCollision.C. The threshold energy of ionisation is subtracted from the pre-collision energy of the electron in line 269.

Listing B.7: File: CoulombCollision.C - Ionisation parent electron collision.

```
250 template<class CloudType>
251 void Foam::CoulombCollision<CloudType>::Ionizationcollide
252 (
253     typename CloudType::parcelType& pP,
254     typename CloudType::parcelType& pQ,
255     scalar Echange
256     //vector UQCorr
257 )
258 {
259     CloudType& cloud(this->owner());
260
261     Random& rndGen(cloud.rndGen());
262
263     label typeIdP = pP.typeId();          //P = Electron
264     label typeIdQ = pQ.typeId();          //Q = Neutral
265
266     vector& UP = pP.U();
267     vector& UQ = pQ.U();
268
269     UP = UP*Echange;        //EKin of parent electron reduced in the amount of
            thresholdEIon and E of progeny
270
271     scalar mP = cloud.constProps(typeIdP).mass();
272     scalar mQ = cloud.constProps(typeIdQ).mass();
```

```
273
274         vector g = UP - UQ;
275         scalar cR = mag(UP - UQ);
276         scalar ElecEnergy = 0.5*mP*(UP&UP)/1.602177e-19;                  //
                Energy of parent electron after Ionization // electronVolt
277
278         // For parent Electron
279         scalar epsilon = 2*constant::mathematical::pi*rndGen.scalar01();     //
                Azimuthal Angle
280         scalar gQ = sqrt(sqr(g.y())+sqr(g.z()));
281         scalar hx = gQ*cos(epsilon);
282         scalar hy = -(g.y()*g.x()*cos(epsilon)+cR*g.z()*sin(epsilon))/gQ;
283         scalar hz = -(g.z()*g.x()*cos(epsilon)-cR*g.y()*sin(epsilon))/gQ;
284         vector h = vector
285                 (
286                         hx,
287                         hy,
288                         hz
289                 );
290
291         scalar X = acos((ElecEnergy+2-2*(pow(1+ElecEnergy,rndGen.scalar01())))/
                ElecEnergy);
292
293         //post collision velocity parent electron
294         UP = UP - mQ/(mP + mQ)*(g*(1-cos(X))+sin(X)*h);
295         //post collision velocity Ion after collision with parent electron
296         UQ = UQ + mP/(mP + mQ)*(g*(1-cos(X))+sin(X)*h);
297     }
298
299
300 // ************************************************************************* //
```

B.5 Dynamic particle weighting

The particle reduction and weight update scheme is initialised during the *evolve* method if the total number of electrons exceeds the maximum *nElectronMax* defined in the *plasmaProperties* case file. To this end, the function *reduceParticles*, responsible for the particles deletion and weight update, is called in line 3877.

Listing B.8: File: DsmcCloud.C - Weight update initialisation.

```
3853 template<class ParcelType>
3854 void Foam::DsmcCloud<ParcelType>::evolve()
3855 {
3856     typename ParcelType::trackingData td(*this);
3857
3858     label nDsmcElectrons = nDsmcElectronsinSystem();
3859     label nDsmcElectronsProc = nDsmcElectrons;
3860     reduce(nDsmcElectrons, sumOp<label>());
3861
3862     label nDsmcIonsProc = nDsmcIonsinSystem();
```

```
3863
3864        // Reset the data collection fields
3865        resetFields();
3866
3867        if (debug)
3868        {
3869            this->dumpParticlePositions();
3870        }
3871
3872        //Reduce number of dsmcParticles if limit is reached
3873        if (nDsmcElectrons >= nElectronMax)
3874        {
3875            label nDsmcElectronsOld = nDsmcElectrons;
3876
3877            Cloud<ParcelType>::reduceParticles(nDsmcElectronsProc, nDsmcIonsProc,
                    rndGen_, nElectronMax, reductionFactor);
3878
3879            nDsmcElectrons = nDsmcElectronsinSystem();
3880            reduce(nDsmcElectrons, sumOp<label>());
3881
3882            label nDsmcIons = nDsmcIonsinSystem();
3883            reduce(nDsmcIons, sumOp<label>());
3884
3885            scalar WtNew = plasmaWt*(nDsmcElectronsOld/(reductionFactor*
                    nElectronMax));
3886            plasmaWt = WtNew;
3887
3888            Info<< "    Number of electron dsmc particles after reduction        =
                    "
3889                    << nDsmcElectrons << nl
3890                    << "    Number of ion dsmc particles after reduction        =
                            "
3891                    << nDsmcIons << nl
3892                    << endl;
3893
3894            //For correction of writing bug after weight adjustment
3895            buildCellOccupancy();
3896            buildCellOccupancywCorrection();
3897        }
```

The function *reduceParticles* is responsible for the random selection of particles and their subsequent deletion from computational memory. The amount of particles to be removed is calculated based on the *reductionFactor* defined in the *plasmaProperties* case file. Note that the numerical weights of electrons and ions are always equal in *dsmcPlasmaFoam*. Hence, the weight update step applies to both electrons and ions. Since *dsmcPlasmaFoam* supports parallel execution, the particles to be deleted must be defined on a *per computational core* basis. The amount of electrons to be removed per core is defined in lines 221:345 based on a probability based consideration linked to the ratio of particles in the computational core to the total amount of particles in the numerical domain. Deletion candidates for each core are randomly defined as label lists in lines 352:370 for electrons and lines 377:394 for ions. Once the deletion lists are created, the particles cloud is loaded in line 434. Next, the solver loops over all particles comparing the particles location in the stack with the randomly generated deletion list and deleting a particle if its position is contained in the list.

Listing B.9: File: Cloud.C - Weight update and random particle deletion.

```
181 template<class ParticleType>
182 void Foam::Cloud<ParticleType>::reduceParticles(label nDsmcElectronsProc, label
        nDsmcIonsProc, Random rndGen_, scalar nElectronMax, scalar reductionFactor)
183 {
184
185     int myrank = Pstream::myProcNo();
186     char buffer[10];                                                  //For Log
            Files
187     sprintf( buffer, "%d", myrank );                                  //For Log
            Files
188
189     label nDsmcElectronsProcc = nDsmcElectronsProc;                   //
            Electrons in Core
190     reduce(nDsmcElectronsProc, sumOp<label>());                       //Total
191     label nDsmcElectrons = nDsmcElectronsProc;                        //Total
192
193     label nDsmcIonsProcc = nDsmcIonsProc;                             // Ions in
            Core
194     reduce(nDsmcIonsProc, sumOp<label>());                            //Total
195     label nDsmcIons = nDsmcIonsProc;                                  //Total
196
197     scalar WtOld;
198
199     //Read current Weight of electrons/ions
200     forAllIter(typename Cloud<ParticleType>, *this, pIter)
201     {
202         ParticleType& p = pIter();                  //P is particle; p p.U_ =
                Velocity
203
204         if (p.typeId() == 0 || p.typeId() == 3) //If particle is electron
205         {
206             WtOld = p.Wt();
207             break;
208         }
209     }
210
211     Info<< "    Info: Particles reduced and weight adjusted        "<< nl
212             << "    Number of electron dsmc particles before reduction       =
                "
213             << nDsmcElectrons << nl
214             << "    Number of ion dsmc particles before reduction            =
                "
215             << nDsmcIons << nl
216             << "    Old Weight                                               =
                "
217             << WtOld << nl
218             << endl;
219
220     scalar WtNew = WtOld*(nDsmcElectrons/(reductionFactor*nElectronMax));
```

```
221        label toRemove = nDsmcElectrons - reductionFactor*nElectronMax;      //To
               remove Electrons
222
223        //Ions to be removed in order to keep same weight as electrons
224        label toRemoveIons = nDsmcIons - WtOld*nDsmcIons/WtNew;
225
226        // List for electrons in core; For checking whether electron marked twice
227        List<label> listDsmcElectrons(nDsmcElectronsProcc,1);
228        // List for ions in core; For checking whether ion marked twice
229        List<label> listDsmcIons(nDsmcIonsProcc,1);
230
231        //  Determination of number of particles to be deleted per core //
232
233        scalar toRemoveCoreSC = nDsmcElectronsProcc*double(toRemove)/nDsmcElectrons
               ;
234        label toRemoveCore(toRemoveCoreSC); //Electrons to be removed per Core
235        label toRemoveTotal = toRemoveCore  ;        //Total Electrons to be removed
236        reduce(toRemoveTotal, sumOp<label>());       //sum over cores
237        label diff = toRemove - toRemoveTotal;       //since toRemoveCore=integer
               --> some electrons missing
238
239        scalar toRemoveCoreIonSC = nDsmcIonsProcc*double(toRemoveIons)/nDsmcIons;
240        label toRemoveCoreIon(toRemoveCoreIonSC);   //Ions to be removed per Core
241        label toRemoveTotalIon = toRemoveCoreIon;   //Total Ions to be removed
242        reduce(toRemoveTotalIon, sumOp<label>());   //sum over cores
243        label diffIon = toRemoveIons - toRemoveTotalIon;     //since toRemoveCoreIon
               =integer --> some Ions missing
244
245        // Increase number of electrons/ions to be removed in given core based on
               probability so that diff = 0
246        label n=Pstream::nProcs();
247        //Probability list with values for each core - electrons
248        List<scalar> coreProbList(n);
249        //Probability list with values for each core - ions
250        List<scalar> coreProbListIon(n);
251
252        forAll(coreProbList, i)
253        {
254            coreProbList[i] = 0;
255        }
256
257        forAll(coreProbListIon, i)
258        {
259            coreProbListIon[i] = 0;
260        }
261
262        //Probability for each core - electron
263        scalar coreProb = double(nDsmcElectronsProcc)/nDsmcElectrons;
264        //probabilities of individual processors
265        coreProbList[Pstream::myProcNo()] = coreProb;
266
267        //Probability for each core - ion
268        scalar coreProbIon = double(nDsmcIonsProcc)/nDsmcIons;
269        //probabilities of individual processors
270        coreProbListIon[Pstream::myProcNo()] = coreProbIon;
271
272        //Gather probabilities of individual processors on master proc
```

```
273     Pstream::gatherList(coreProbList);
274     //Gather probabilities of individual processors on master proc
275     Pstream::gatherList(coreProbListIon);
276 
277     if (Pstream::parRun() & Pstream::master())
278     {
279         for (label a = 1; a < n; a++)
280         {
281             //Cumulative probabilities calculated on master core
282             coreProbList[a] = coreProbList[a] + coreProbList[a-1];
283             coreProbListIon[a] = coreProbListIon[a] + coreProbListIon[a-1];
284         }
285     }
286 
287     //Scatter list of probabilities to all processors
288     Pstream::scatter(coreProbList);
289     Pstream::scatter(coreProbListIon);
290 
291     if (diff > 0)
292     {
293         do {
294             scalar U = rndGen_.scalar01();      //Random number for selection
                    of core
295             Pstream::scatter(U);
296 
297             for (label stf = 0; stf < n; stf++) //iterator over cores
298 
299             {
300                 if (U <= coreProbList[stf])     //Random number falls in core
                        stf
301                 {
302 
303                     if (myrank == stf)  //Additional electron marked as to be
                            removed in core stf
304                     {
305                         toRemoveCore ++;
306                     }
307 
308                     break;
309                 }
310             }
311 
312             diff -= 1;
313         } while (diff > 0);     //Repeat until diff = 0
314 
315     }
316 
317     if (diffIon > 0)
318     {
319         do {
320             scalar U = rndGen_.scalar01();      //Random number for selection
                    of core - ions
321             Pstream::scatter(U);
322 
323             for (label stf = 0; stf < n; stf++) //iterator over cores
324 
325             {
```

```
326                 if (U <= coreProbListIon[stf])  //Random number falls in core
                        stf
327                 {
328
329                     if (myrank == stf)  //Additional ion marked as to be
                            removed in core stf
330                     {
331                         toRemoveCoreIon ++;
332                     }
333
334                     break;
335                 }
336             }
337
338             diffIon -= 1;
339         } while (diffIon > 0);  //Repeat until diff = 0
340     }
341
342     //Sortable lists (one per core) created; Size = number of e- to be removed
            in core
343     SortableList<label> ElecToRemove(toRemoveCore);
344     //Sortable lists (one per core) created; Size = number of ions to be
            removed in core
345     SortableList<label> IonToRemove(toRemoveCoreIon);
346
347     //   Selection of electrons to be removed //
348     bool check;          //boolean for checking whether particle already marked
349     label toRemoveElec = -1;     // Declaration of candidate for deleting
350     label countermarkedElec = 0;          // Counter for selected candidates for
            log file
351
352     for (label c = 0; c < toRemoveCore; c++)
353     {
354         do {
355
356             check = true;
357             //Candidate for deleting
358             toRemoveElec = rndGen_.integer(0, nDsmcElectronsProcc-1);
359
360             if (listDsmcElectrons[toRemoveElec] == 0)   //already marked
361             {
362                 check = false;
363             }
364
365         } while (!check);
366
367         listDsmcElectrons[toRemoveElec] = 0;     //marked as already selected
368         ElecToRemove[c] = toRemoveElec; //Added to List of electrons to be
                removed
369         countermarkedElec ++;    //Sum of already selected electrons for log
                file
370     }
371
372     //   Selection of ions to be removed //
373     bool checkIon;       //boolean for checking whether particle already marked
374     label toRemoveIon = -1;      // Declaration of candidate for deleting
375     label countermarkedIons =0; // Counter for selected candidates for log file
```

```
376
377     for (label c = 0; c < toRemoveCoreIon; c++)
378     {
379         do {
380
381             checkIon = true;
382             toRemoveIon = rndGen_.integer(0, nDsmcIonsProcc-1); //Candidate for
                    deleting
383
384             if (listDsmcIons[toRemoveIon] == 0) //already marked
385             {
386                 checkIon = false;
387             }
388
389         } while (!checkIon);
390
391         listDsmcIons[toRemoveIon] = 0;  //marked as already selected
392         IonToRemove[c] = toRemoveIon;   //Added to List of ions to be removed
393         countermarkedIons ++;   //Sum of already selected ions for log file
394     }
395
396     ElecToRemove.sort();        //List sorted
397     IonToRemove.sort();         //List sorted
398
399     reduce(countermarkedElec, sumOp<label>());
400     reduce(countermarkedIons, sumOp<label>());
401     Info    << "    Number of electrons to be deleted                   =
         "
402             << countermarkedElec << nl
403             << "    Number of ions to be deleted                        =
                 "
404             << countermarkedIons << nl
405             << endl;
406
407     /////////#########################    LOGS
            #########################//////////
408
409 //    std::string filename1 = "deletedElectrons_" + std::string(buffer) + ".txt
    ";
410 //    ofstream printDeletedElec(filename1.c_str());
411 //
412 //    std::string filename2 = "toDeleteElectrons_" + std::string(buffer) + ".
    txt";
413 //    ofstream printToDeleteElec(filename2.c_str());
414
415 //    std::string filename3 = "deletedIons_" + std::string(buffer) + ".txt";
416 //    ofstream printDeletedIon(filename3.c_str());
417 //
418 //    std::string filename4 = "toDeleteIons_" + std::string(buffer) + ".txt";
419 //    ofstream printToDeleteIon(filename4.c_str());
420
421     /////////#########################    LOGS
            #########################//////////
422
423     // Loop over particles
424
425     label CounterElec = 0;      //For Loop electrons
```

```
426      label CounterIon = 0;          //For Loop ions
427
428      label i = 0;                   //For Loop electrons
429      label j = 0;                   //For Loop ions
430
431      label deleteElec = 0;          //For log File: How many electrons were deleted
432      label deleteIon = 0;           //For log File: How many ions were deleted
433
434      forAllIter(typename Cloud<ParticleType>, *this, pIter)
435      {
436          ParticleType& p = pIter(); //P ist particle; p p.U_ = Velocity
437
438          //If particle is electron and still particles to be removed
439          if ((p.typeId() == 0 || p.typeId() == 3) & (i < toRemoveCore))
440          {
441              //If electron number = electron marked as to be deleted
442              if (CounterElec == ElecToRemove[i])
443              {
444                  deleteParticle(p);
445                  deleteElec ++;  //Counter for number of electrons already
                         deleted (Log file)
446                  i ++;           //Move to next position in list of electrons to
                          be removed
447              }
448              CounterElec ++;     //Counter for electron in proc
449          }
450
451          //If particle is ion and still particles to be removed
452          if ((p.typeId() == 2) & (j < toRemoveCoreIon))
453          {
454              //If ion number = ion marked as to be deleted
455              if (CounterIon == IonToRemove[j])
456              {
457                  deleteParticle(p);
458                  deleteIon ++;   //Counter for number of ions already deleted (
                         Log file)
459                  j ++;           //Move to next position in list of ions to be
                         removed
460              }
461              CounterIon ++;      //Counter for ion in proc
462          }
463
464
465      }
466
467      reduce(deleteElec, sumOp<label>()); //Sum over procs of deleted electrons
468      reduce(deleteIon, sumOp<label>());  //Sum over procs of deleted ions
469
470      //Weight correction
471      forAllIter(typename Cloud<ParticleType>, *this, pIter)
472      {
473          ParticleType& p = pIter(); //P ist particle; p p.U_ = Velocity
474
475          if (p.typeId() != 1)         //If particle is electron or ion
476          {
477              p.Wt() = WtNew;
478
```

```
479         }
480     }
481
482     Info    << "    deleted Electrons                                                    =
          "
483             << deleteElec << nl
484             << "    deleted Ions                                                         =
                  "
485             << deleteIon << nl
486             << "    New Weight                                                           =
                  "
487             << WtNew << nl
488             << endl;
489 }
```

Bibliography

[ACG+02] A. Alexeenko, R. Collins, S. Gimelshein, D. Levin, and B. Reed. Numerical modeling of axisymmetric and three-dimensional flows in microelectromechanical systems nozzles. *AIAA J.*, 40(5):897–907, 2002.

[ADH+93] F. Assous, P. Degond, E. Heintze, P.-A. Raviart, and J. Segré. On a finite-element method for solving the three-dimensional maxwell equations. *Journal of Computational Physics*, 109(2):222–237, 1993.

[AFG+06] A. Alexeenko, D. Fedosov, S. Gimelshein, D. Levin, and R. Collins. Transient heat transfer and gas flow in a MEMS-based thruster. *J. Microelectromech. Syst.*, 15(1):181–194, 2006.

[AGLC00] A. Alexeenko, S. Gimelshein, D. Levin, and R. Collins. Numerical modeling of axisymmetric and three-dimensional flows in MEMS nozzles. In *Proceedings of the 36th AIAA/ASME/SAE/ASEE Joint Propulsion Conference & Exhibit- Huntsville. July 16-19 (2000)*, 2000.

[AH07] J. Allen and T. Hauser. foamDSMC: An object oriented parallel DSMC solver for rarefied flow applications. In *45th AIAA Aerospace Sciences Meeting & Exhibit*, page 1106, 2007.

[AI91] Y. Arakawa and K. Ishihara. A numerical code for cusped ion thrusters. *IEPC Paper*, pages 91–118, 1991.

[AJ91] J. D. Anderson Jr. *Fundamentals of aerodynamics*. MacGraw-Hill, New York, 1991.

[AKGH+98] M. Auweter-Kurtz, T. Golz, H. Habiger, F. Hammer, H. Kurtz, M. Riehle, and C. Sleziona. High-power hydrogen arcjet thrusters. *Journal of Propulsion and Power*, 14(5):764–773, 1998.

[ALFG05] A. Alexeenko, D. Levin, D. Fedosov, and G. Gimelshein. Performance analysis of microthrusters based on coupled thermal-fluid modeling and simulation. *J. Propuls. Power*, 21(1):95–101, 2005.

[ALG$^+$02] A. Alexeenko, D. Levin, S. Gimelshein, R. Collins, and G. Markelov. Numerical simulation of high-temperature gas flows in a millimeter-scale thruster. *J. Thermophys. Heat Transfer*, 16(1):10–16, 2002.

[BAB97] R. Bayt, A. Ayon, and K. Breuer. A performance evaluation of MEMS-based micronozzles. In *Proceedings of the 33rd AIAA/ASME/SAE/ASEE Joint Propulsion Conference & Exhibit- Seattle. July 7-9 (1997)*, 1997.

[Bal04] R. Balakrishnan. An approach to entropy consistency in second-order hydrodynamic equations. *J. Fluid Mech.*, 503:201–245, 2004.

[Bat75] D. R. Bates. Recombination. *Case Studies in Atomic Collision Physics IV*, 1975.

[BHG$^+$04] J. Bareilles, G. Hagelaar, L. Garrigues, C. Boniface, J. Boeuf, and N. Gascon. Critical assessment of a two-dimensional hybrid hall thruster model: Comparisons with experiments. *Physics of Plasmas (1994-present)*, 11(6):3035–3046, 2004.

[BHG$^+$05] C. Boniface, G. Hagelaar, L. Garrigues, J. Boeuf, and M. Prioul. Modeling of double stage hall effect thruster. *Plasma Science, IEEE Transactions on*, 33(2):522–523, 2005.

[Bio82] M. Biondi. Electron - ion recombination in gas lasers. In E. W. McDaniel and N. W. L., editors, *Applied Atomic Collision Physics Vol. 3: Gas Laser.* Academic Press, New York, 1982.

[Bio17] M. A. Biondi. Electron - ion recombination in gas lasers. *Gas Lasers: Applied Atomic Collision Physics*, 3:173, 2017.

[Bir81] G. Bird. Monte-carlo simulation in an engineering context. *Progress in Astronautics and Aeronautics*, 74:239–255, 1981.

[Bir89] G. Bird. Perception of numerical methods in rarefied gas dynamics. *Rarefied gas Dynamics: Theoretical and Computational Techniques, Eds. EP Muntz, and DP Weaver and DH Capbell*, 118:374–395, 1989.

[Bir91] C. K. Birdsall. Particle-in-cell charged-particle simulations, plus monte carlo collisions with neutral atoms, PIC-MCC. *Plasma Science, IEEE Transactions on*, 19(2):65–85, 1991.

[Bir94] G. Bird. *Molecular Gas Dynamics and the Direct Simulation of Gas Flows*. Oxford University Press, New York, 1994.

[Bir13] G. Bird. *The DSMC method.* CreateSpace Independent Publishing Platform, 2013.

[Bit04] J. A. Bittencourt. *Fundamentals of plasma physics*. Springer Science & Business Media, 2004.

[BK92] G. Butler and D. King. Single and two fluid simulations of arcjet performance. In *Joint Propulsion Conference & Exhibit*, 1992.

[BL04] C. K. Birdsall and A. B. Langdon. *Plasma physics via computer simulation*. CRC Press, 2004.

[Bog09] A. Bogaerts. Hybrid monte carlo—fluid model for studying the effects of nitrogen addition to argon glow discharges. *Spectrochimica Acta Part B: Atomic Spectroscopy*, 64(2):126–140, 2009.

[Boy97a] I. D. Boyd. Extensive validation of a monte carlo model for hydrogen arcjet flowfields. *Journal of propulsion and power*, 13(6):775–782, 1997.

[Boy97b] I. D. Boyd. Monte carlo simulation of nonequilibrium flow in a low-power hydrogen arcjet. *Physics of Fluids (1994-present)*, 9(10):3086–3095, 1997.

[Boy05] I. D. Boyd. Numerical modeling of spacecraft electric propulsion thrusters. *Progress in Aerospace Sciences*, 41(8):669–687, 2005.

[Bre05] H. Brenner. Navier–Stokes revisited. *Physica A*, 349(1-2):60–132, 2005.

[Bre06] H. Brenner. Fluid mechanics revisited. *Physica A*, 370(2):190–224, 2006.

[CC70] S. Chapman and T. Cowling. *The Mathematical Theory of Non-uniform Gases*. Cambridge University Press, 1970.

[CG14] P. Coche and L. Garrigues. A two-dimensional (azimuthal-axial) particle-in-cell model of a hall thruster. *Physics of Plasmas (1994-present)*, 21(2):023503, 2014.

[CS68] A. Clarke and E. Smith. Low–temperature viscosities of argon, krypton, and xenon. *J. Chem. Phys.*, 48(9):3988–3991, 1968.

[CS69] A. Clarke and E. Smith. Low–temperature viscosities and intermolecular forces of simple gases. *J. Chem. Phys.*, 51(9):4156–4161, 1969.

[Dei64] R. Deissler. An analysis of second-order slip flow and temperature-jump boundary conditions for rarefied gases. *Int. J. Heat Mass Transfer*, 7(6):681–694, 1964.

[DGS06] F. Durst, J. Gomes, and R. Sambasivam. Thermofluiddynamics: Do we solve the right kind of equations? In *Proceeding of the international symposium on turbulence, heat and mass transfer - Dubrovnik. September 25-29 (2006)*, pages 3–18, september 25-29 2006.

[DSD09] N. Dongari, A. Sharma, and F. Durst. Pressure-driven diffusive gas flows in micro-channels: from the knudsen to the continuum regimes. *Microfluid. Nanofluid.*, 6(5):679–692, 2009.

[DZR11] N. Dongari, Y. Zhang, and J. Reese. Modeling of knudsen layer effects in micro/nanoscale gas flows. *J Fluids Eng*, 133(7):071101, 2011.

[EBWL95] D. J. Economou, T. J. Bartel, R. S. Wise, and D. P. Lymberopoulos. Two-dimensional direct simulation monte carlo (DSMC) of reactive neutral and ion flow in a high density plasma reactor. *IEEE Transactions on plasma science*, 23(4):581–590, 1995.

[Egg75] E. Eggarter. Comprehensive optical and collision data for radiation action. II. Ar. *The Journal of Chemical Physics*, 62(3):833–847, 1975.

[Fas11] F. Fastabend. *Numerische Modellierung der Strömung in einer Kaltgasdüse unter Feinvakuum mittels der DSMC-Methode. Master's thesis*. Universität Bremen, 2011.

[Fef06] C. L. Fefferman. Existence and smoothness of the Navier-Stokes equation. *The millennium prize problems*, pages 57–67, 2006.

[FMS95] J. M. Fife and M. Martinez-Sanchez. Two-dimensional hybrid particle-in-cell (PIC) modeling of hall thrusters. In *24th International Electric Propulsion Conference,(Moscow, Russia)*, pages 1213–1224, 1995.

[FMSS+97] J. Fife, M. Martinez-Sanchez, J. Szabo, J. Fife, M. Martinez-Sanchez, and J. Szabo. A numerical study of low-frequency discharge oscillations in hall thrusters. In *33rd Joint Propulsion Conference and Exhibit*, page 3052, 1997.

[FP02] J. H. Ferziger and M. Peric. Computational methods for fluid dynamics, 2002.

[GA15] A. Giovannini and R. Abhari. Rarefied flow expansion in linear aerospikes. *Phys. Fluids*, 27(6):062003, 2015.

[GdA13] J. Gonzalez del Amo. European space agency activities in electric propulsion. In *The 33rd International Electric Propulsion Conference, Washington D.C., USA*, 2013.

[GdA15] J. Gonzalez del Amo. European space agency (ESA) electric propulsion activities. In *The 34th International Electric Propulsion Conference and 6th Nano-satellite Symposium, Hyogo-Kobe, Japan*, 2015.

[GHB+03] L. Garrigues, G. Hagelaar, J. Bareilles, C. Boniface, and J. Boeuf. Model study of the influence of the magnetic field configuration on the performance and lifetime of a hall thruster. *Physics of Plasmas (1994-present)*, 10(12):4886–4892, 2003.

[GK08] D. M. Goebel and I. Katz. *Fundamentals of electric propulsion: ion and Hall thrusters*, volume 1. John Wiley & Sons, 2008.

[Gre15] C. J. Greenshields. *OpenFOAM User Guide*. OpenFOAM Foundation Ltd., 2.4.0 edition, 2015.

[Hän04] D. Hänel. *Molekulare Gasdynamik*. Springer-Verlag, Berlin Heidelberg, 2004.

[Hay] M. Hayashi. A set of electron-ar cross sections with 25 excited states. Website, Retrieved 2016-10-04. `http://jila.colorado.edu/~avp/collision_data/electronneutral/hayashi.txt`.

[HBGB02] G. Hagelaar, J. Bareilles, L. Garrigues, and J.-P. Boeuf. Two-dimensional model of a stationary plasma thruster. *Journal of Applied Physics*, 91(9):5592–5598, 2002.

[HLO08] F. Hermeline, S. Layouni, and P. Omnes. A finite volume method for the approximation of maxwell equations in two space dimensions on arbitrary meshes. *Journal of Computational Physics*, 227(22):9365–9388, 2008.

[HT59] E. Harrison and W. Thompson. The low pressure plane symmetric discharge. *Proceedings of the Physical Society*, 74(2):145, 1959.

[Ins] C. M. Institute. Millennium problems. Website, Retrieved 2017-02-14. `http://www.claymath.org/millennium-problems`.

[Iss86] R. I. Issa. Solution of the implicitly discretised fluid flow equations by operator-splitting. *Journal of computational physics*, 62(1):40–65, 1986.

[JH01] M. Jugroot and J. Harvey. Numerical modeling of neutral and charged particles within a gridded ion thruster. In *27th International electric propulsion conference. Paper*, pages 01–100, 2001.

[JS01] S. Jin and M. Slemrod. Regularization of the burnett equations via relaxation. *J. Stat. Phys.*, 103(5-6):1009–1033, 2001.

[KB04] J. W. Koo and I. D. Boyd. Computational model of a hall thruster. *Computer physics communications*, 164(1):442–447, 2004.

[KH92] K. Kosaki and M. Hayashi. Electron-argon atom collision cross sections. In *Preprint Nat. Mtg. Inst. Elect. Eng.*, pages 2–9, 1992.

[KKM+84] J. Kestin, K. Knierim, E. Mason, B. Najafi, S. Ro, and M. Waldman.

Equilibrium and transport properties of the noble gases and their mixtures at low density. *J. Phys. Chem. Ref. Data*, 13(1):229–303, 1984.

[Kre10] G. M. Kremer. *An introduction to the Boltzmann equation and transport processes in gases*. Springer Science & Business Media, 2010.

[KSBY00] V. Kapila, A. G. Sparks, J. M. Buffington, and Q. Yan. Spacecraft formation flying: dynamics and control. *Journal of Guidance, Control, and Dynamics*, 23(3):561–564, 2000.

[KT00] A. Kurganov and E. Tadmor. New high-resolution central schemes for nonlinear conservation laws and convection–diffusion equations. *J. Comput. Phys.*, 160(1):241–282, 2000.

[LH10] W. Louisos and D. Hitt. Transient simulations of 3-D supersonic micronozzle flow. In *Proceedings of the FEDSM-ICNMM2010 - Montreal. August 1-5 (2010)*, pages 491–501, 2010.

[Li08] S. Li. High order central scheme on overlapping cells for magneto-hydrodynamic flows with and without constrained transport method. *Journal of Computational Physics*, 227(15):7368–7393, 2008.

[LL05] M. A. Lieberman and A. J. Lichtenberg. *Principles of plasma discharges and materials processing*. John Wiley & Sons, 2005.

[LR08] D. Lockerby and J. Reese. On the modelling of isothermal gas flows at the microscale. *J. Fluid Mech.*, 604:235–261, 2008.

[LREB04] D. Lockerby, J. Reese, D. Emerson, and R. Barber. Velocity boundary condition at solid walls in rarefied gas calculations. *Phys. Rev. E*, 70:017303, 2004.

[LWY+10] H. Liu, B. Wu, D. Yu, Y. Cao, and P. Duan. Particle-in-cell simulation of a hall thruster. *Journal of Physics D: Applied Physics*, 43(16):165202, 2010.

[MAKF+14] C.-D. Munz, M. Auweter-Kurtz, S. Fasoulas, A. Mirza, P. Ortwein, M. Pfeiffer, and T. Stindl. Coupled particle-in-cell and direct simulation monte carlo method for simulating reactive plasma flows. *Comptes Rendus Mécanique*, 342(10):662–670, 2014.

[Max79] J. Maxwell. On stress in rarefied gases arising from inequalities of temperature. *Philos. Trans. R. Soc. London*, 170:231–256, 1879.

[MCL+10] S. Mahalingam, Y. Choi, J. Loverich, P. Stoltz, M. Jonell, and J. Menart. Dynamic electric field calculations using a fully kinetic ion thruster discharge chamber model. In *46th AIAA/ASME/SAE/ASEE Joint Propulsion Conference & Exhibit*, page 6944, 2010.

[MCL+11] S. Mahalingam, Y. Choi, J. Loverich, P. Stoltz, B. Bias, and J. Menart. Fully coupled electric field/PIC-MCC simulation results of the plasma in the discharge chamber of an ion engine. In *Paper Number AIAA-2011-6071, 47th AIAA/ASME/SAE/ASEE Joint Propulsion Conference & Exhibit, San Diego, CA*, 2011.

[MEK+72] A. Morozov, Y. V. Esipchuk, A. Kapulkin, V. Nevrovskii, and V. Smirnov. Effect of the magnetic field on a closed-electron-drift accelerator. *Sov. Phys.-Tech. Phys.(Engl. Transl.) 17: No. 3, 482-7 (Sep 1972).*, 1972.

[MKB96] T. W. Megli, H. Krier, and R. L. Burton. Plasmadynamics model for nonequilibrium processes in N2/H2 arcjets. *Journal of thermophysics and heat transfer*, 10(4):554–562, 1996.

[MKH11] I. Mikellides, I. Katz, and R. Hofer. Design of a laboratory hall thruster with magnetically shielded channel walls, phase I: Numerical simulations. In *47th AIAA/ASME/SAE/ASEE Joint Propulsion Conference & Exhibit*, page 5809, 2011.

[MM05] S. Mahalingam and J. Menart. Computational model tracking primary electrons, secondary electrons and ions in the discharge chamber of an ion engine. In *41st AIAA/ASME/SAE/ASEE Joint Propulsion Conference & Exhibit*, page 4253, 2005.

[MM06] S. Mahalingam and J. A. Menart. Ion engine discharge chamber plasma modeling using a 2-D PIC simulation. In *42nd AIAA/ASME/SAE/ASEE Joint Propulsion Conference & Exhibit*, page 4488, 2006.

[MMS96] S. Miller and M. Martinez-Sanchez. Two-fluid nonequilibrium simulation of hydrogen arcjet thrusters. *Journal of Propulsion and Power*, 12(1):112–119, 1996.

[MnHQ08] J. Moriñigo and J. Hermida-Quesada. Analysis of viscous heating in a micro-rocket flow and performance. *J. Therm. Sci.*, 17(2):116–124, 2008.

[MTJW03] J. Maurer, P. Tabeling, P. Joseph, and H. Willaime. Second-order slip laws in microchannels for helium and nitrogen. *Phys. Fluids*, 15(9):2613–2621, 2003.

[Nan94] K. Nanbu. Simple method to determine collisional event in monte carlo simulation of electron-molecule collision. *Japanese journal of applied physics*, 33(8R):4752, 1994.

[Nan97] K. Nanbu. Theory of cumulative small-angle collisions in plasmas. *American Physical Society*, 55(4):4642–4652, 1997.

[Nan00] K. Nanbu. Probability theory of electron-molecule, ion-molecule, molecule-molecule, and coulomb collisions for particle modeling of materials processing plasmas and cases. *IEEE Transactions on plasma science*, 28(3):971–990, 2000.

[NK97] K. Nanbu and S. Kondo. Analysis of three-dimensional dc magnetron discharge by the particle-in-cell/monte carlo method. *Japanese journal of applied physics*, 36(7S):4808, 1997.

[Nol93] B. Noll. *Numerische Strömungsmechanik: Grundlagen*. Springer-Verlag, 1993.

[OLRE07] L. O'Hare, D. Lockerby, J. Reese, and D. Emerson. Near-wall effects in rarefied gas micro-flows: some modern hydrodynamic approaches. *Int. J. Heat Fluid Flow*, 28:37–43, 2007.

[PAFMS06] F. Parra, E. Ahedo, J. Fife, and M. Martinez-Sanchez. A two-dimensional hybrid model of the hall thruster discharge. *Journal of Applied Physics*, 100(2):023304, 2006.

[Phe] A. Phelps. Swarm parameters. Website, Retrieved 2016-10-13. `http://jila.colorado.edu/~avp/collision_data/electronneutral/eletrans.txt`.

[Pop00] S. Pope. *Turbulent flows*. Cambridge University Press, Cambridge, 2000.

[PP99] A. Phelps and Z. L. Petrovic. Cold-cathode discharges and breakdown in argon: surface and gas phase production of secondary electrons. *Plasma Sources Science and Technology*, 8(3):R21, 1999.

[PS72] S. V. Patankar and D. B. Spalding. A calculation procedure for heat, mass and momentum transfer in three-dimensional parabolic flows. *International Journal of Heat and Mass Transfer*, 15(10):1787–1806, 1972.

[Rae71] W. Rae. Some numerical results on viscous low density nozzle flows in the slender-channel approximation. *AIAA J.*, 9(5):811–820, 1971.

[SB10] G. P. Sutton and O. Biblarz. *Rocket Propulsion Elements*. John Wiley & Sons, 2010.

[SBM01] C. Sabol, R. Burns, and C. A. McLaughlin. Satellite formation flying design and evolution. *Journal of Spacecraft and Rockets*, 38(2):270–278, 2001.

[SGJ90] M. Surendra, D. Graves, and G. Jellum. Self-consistent model of a direct-current glow discharge: Treatment of fast electrons. *Physical Review A*, 41(2):1112, 1990.

[SJ01] J. J. Szabo Jr. *Fully kinetic numerical modeling of a plasma thruster*. PhD thesis, Massachusetts Institute of Technology, 2001.

[Son02] Y. Sone. *Kinetic Theory and Fluid Dynamics*. Birkhäuser, Boston, 2002.

[SS13] F. Sharipov and J. Strapasson. Benchmark problems for mixtures of rarefied gases. I. Couette flow. *Phys. Fluids*, 25(2):027101, 2013.

[ST03] H. Struchtrup and M. Torrilhon. Regularization of Grad's 13 moment equations: derivation and linear analysis. *Phys. Fluids*, 15(9):2668–2680, 2003.

[Sti15] T. Stindl. *Entwicklung und Untersuchung eines Partikelverfahrens zur Simulation elektromagnetischer Wechselwirkungen in verdünnten Plasmaströmungen*. PhD thesis, Stuttgart, Universität Stuttgart, Diss., 2015, 2015.

[STM01] Y. Sakiyama, S. Takagi, and Y. Matsumoto. Full simulation of silicon chemical vapor deposition process. In *AIP Conference Proceedings*, volume 585, pages 206–213. AIP, 2001.

[Sto70] D. Stops. The mean free path of gas molecules in the transition regime. *J. Phys. D: Appl. Phys.*, 3(5):685–696, 1970.

[Sto13] A. Stock. *A High-Order Particle-in-Cell Method for Low Density Plasma Flow and the Simulation of Gyrotron Resonator Devices*. PhD thesis, Stuttgart, Universität Stuttgart, Diss., 2013, 2013.

[Stu04] T. Stueber. Discharge chamber primary electron modeling activities in 3-dimension. In *40th AIAA/ASME/SAE/ASEE Joint Propulsion Conference & Exhibit*, page 4105, 2004.

[Stu05] T. Stueber. Ion thruster discharge chamber simulation in three dimension. In *41st AIAA/ASME/SAE/ASEE Joint Propulsion Conference & Exhibit*, page 3688, 2005.

[Sut93] W. Sutherland. The viscosity of gases and molecular force. *Philos. Mag.*, 36:507–531, 1893.

[TLCS05] F. Taccogna, S. Longo, M. Capitelli, and R. Schneider. Plasma flow in a hall thruster. *Physics of Plasmas (1994-present)*, 12(4):043502, 2005.

[Tsk08] D. Tskhakaya. The particle-in-cell method. In *Computational Many-*

Particle Physics, pages 161–189. Springer, 2008.

[VDI13] VDI. *VDI-Wärmeatlas*. Springer Vieweg, Berlin Heidelberg, 2013.

[Ver05] J. P. Verboncoeur. Particle simulation of plasmas: review and advances. *Plasma Physics and Controlled Fusion*, 47(5A):A231, 2005.

[VM95] H. K. Versteeg and W. Malalasekera. *An introduction to computational fluid dynamics: the finite volume method.* Longman Scientific & Technical, New York, 1995.

[Whi06] F. M. White. *Viscous fluid flow*, volume 3. McGraw-Hill New York, 2006.

[Win08] H. Windisch. *Thermodynamik: Ein Lehrbuch für Ingenieure*. Oldenbourg Verlag, 2008.

[WK03] R. Wirz and I. Katz. A preliminary 2-D computational model of an ion thruster discharge chamber. In *Presented at the 39th AIAA/ASME/-SAE/ASEE Joint Propulsion Conference AIAA*, volume 2003, page 5163, 2003.

[WK04] R. Wirz and I. Katz. 2-D discharge chamber model for ion thrusters. In *40th AIAA/ASME/SAE/ASEE Joint Propulsion Conference and Exhibit*, page 4107, 2004.

[ZMC93] X. Zhong, R. MacCormack, and D. Chapman. Stabilization of the burnett equations and application to hypersonic flows. *Aiaa J.*, 31(6):1036–1043, 1993.

www.ingramcontent.com/pod-product-compliance
Ingram Content Group UK Ltd.
Pitfield, Milton Keynes, MK11 3LW, UK
UKHW022001190726
13853UKWH00004B/1673